METHODS IN BONE BIOLOGY

METHODS IN BONE BIOLOGY

Edited by

TIMOTHY R. ARNETT
Department of Anatomy and Developmental Biology
University College London
London
UK

and

BRIAN HENDERSON
Eastman Dental Institute
University College London
London
UK

CHAPMAN & HALL

London · Weinheim · New York · Tokyo · Melbourne · Madras

Published by
Chapman & Hall, an imprint of Thomson Science, 2–6 Boundary Row, London
SE1 8HN, UK

Thomson Science, 2–6 Boundary Row, London SE1 8HN, UK

Thomson Science, 115 Fifth Avenue, New York, NY 10003, USA

Thomson Science, Suite 750, 400 Market Street, Philadelphia, PA 19106, USA

Thomson Science, Pappelallee 3, 69469 Weinheim, Germany

First edition 1998

© 1998 Chapman & Hall Ltd

Thomson Science is a division of International Thomson Publishing **I**Ⓣ**P**

Typeset in 10/12pt Palatino by Florencetype Limited, Devon

Printed by TJ International Ltd, Padstow, Cornwall

ISBN 0 412 75770 2

A catalogue record for this book is available from the British Library

♾ Printed on acid-free text paper, manufactured in accordance with ANSI/NISO
Z39.48-1992 (Permanence of Paper).

Contents

Foreword

The field of bone biology has expanded dramatically in the last decade. In part, this is due to the interest in osteoporosis as one of the most common disease states of the late 20th century, and has coincided with the biotechnology revolution and application of the techniques in molecular biology to the bone field. This has resulted in striking insights into osteoclast and osteoblast biology, and the function of a number of genes and proteins critical to normal bone remodelling. This has been a boon to academic investigators who have long struggled with the complexities and difficulties in studying bone as a tissue. In view of the obvious enormity of the markets involved, the pharmaceutical industry and smaller biotechnology companies have recently devoted large amounts of resources to the identification of new therapeutic agents and diagnostic tests in diseases of bone, and this has also fuelled interest in the existing techniques in bone biology, as well as the development and modification of these techniques and the development of newer techniques to answer these needs. This book therefore is very timely, and should serve a useful purpose for all those pharmaceutical and academic investigators interested in this important field.

The topics have been well chosen and the editors Dr Arnett and Professor Henderson have assembled experts in the topics assigned to them. It will provide a useful reference not only for newcomers to the field, but also experts wishing to keep abreast of the newer techniques.

Professor G.R. Mundy
San Antonio, November 1997

Contributors

Timothy R. Arnett
Senior Lecturer
University College London
London
UK

Jane E. Aubin
Professor and Chair
Department of Anatomy and Cell Biology
University of Toronto
Toronto
Canada

Brendan F. Boyce
Professor of Pathology
Department of Pathology
University of Texas Health Science Center
San Antonio, TX
USA

J. Chris Buckland-Wright
Reader in Radiological Anatomy and Head,
Unit of Radiological Anatomy
Division of Anatomy and Cell Biology
United Medical and Dental Schools of
Guy's and St Thomas's Hospitals
London
UK

Juliet Compston
Senior Research Associate and Honorary Consultant Physician
University of Cambridge School of Clinical Medicine
Cambridge
UK

Janice R. Connor
Department of Cellular Biochemistry
SmithKline Beecham Pharmaceuticals
King of Prussia, PA
USA

David W. Dempster
Professor of Clinical Pathology
College of Physicians and Surgeons
Columbia University
New York
  and
Director, Regional Bone Center
Helen Hayes Hospital
West Haverstraw, NY
USA

Robert A. Dodds
Department of Cellular Biochemistry
SmithKline Beecham Pharmaceuticals
King of Prussia, PA
USA

Colin R. Dunstan
Research Scientist
Pathology Department
AMGEN
Thousand Oaks, CA
USA

Adrienne M. Flanagan
Senior Lecturer
Department of Histopathology and Experimental Pathology
Imperial College School of Medicine at St Mary's
London
UK

Malcolm Harris
Professor of Oral and Maxillofacial Surgery
Eastman Dental Institute
University College London
London
UK

Brian Henderson
Maxillofacial Surgery Research Unit
Eastman Dental Institute
University College London
London
UK

Peter A. Hill
Lecturer in Orthodontics
Department of Orthodontics and Pediatric Dentistry
United Medical and Dental Schools of Guy's and St Thomas's Hospitals
London
UK

David E. Hughes
Lecturer in Pathology
Department of Pathology
University of Sheffield
Sheffield
UK

Ian E. James
Department of Cellular Biochemistry
SmithKline Beecham Pharmaceuticals
King of Prussia, PA
USA

John A. Lynch
Research Fellow
Unit of Radiological Anatomy
Division of Anatomy and Cell Biology
United Medical and Dental Schools of
Guy's and St Thomas's Hospitals
London
UK

Sajeda Meghji
Senior Lecturer
Maxillofacial Surgery Research Unit
Eastman Dental Institute
University College London
London
UK

Bruce H. Mitlak
Clinical Research Physician
Lilly Research Laboratory
Eli Lilly and Co.
Indianapolis, IN

and

Assistant Professor of Medicine
Indiana School of Medicine
Indianapolis, IN
USA

Gregory R. Mundy
Professor and Head of Division of Endocrinology
Department of Medicine
University of Texas Health Science Center
San Antonio, TX
USA

Richard J. Murrills
Senior Research Scientist
Department of Bone Metabolism and Osteoporosis Research
Wyeth-Ayerst Research
Philadelphia, PA
USA

Simon P. Robins
Head of Skeletal Research Unit
Rowett Research Institute
Aberdeen
UK

Usha Sarma
Department of Histopathology and Experimental Pathology
Imperial College of Medicine at St Mary's
London
UK

Masahiko Sato
Research Scientist
Department of Endocrine Research
Lilly Research Laboratories
Eli Lilly and Co.
Indianapolis, IN
USA

Timothy M. Skerry
Professor of Biology
Department of Biology
University of York
York
UK

Kenneth R. Wright
Assistant Professor of Anatomy
Division of Human Anatomy
Loma Linda University
Loma Linda, CA
USA

Introduction

The skeleton, out of sight and often out of mind, is a formidable mass of tissue occupying about 9% of the body by bulk and no less than 17% by weight. The stability and immutability of dry bones and their persistence for centuries, and even millions of years after the soft tissues have turned to dust, give us a false idea of bone during life. Its fixity after death is in sharp contrast to its ceaseless activity during life.

*A.M. Cooke (1955) Osteoporosis. Lancet **I** 878–882, 929–937*

Forty years on, the living skeleton is still apt to be thought of, even by otherwise intelligent people, as a dull, inanimate, rock-like material and bone biology perceived as a subject lacking the appeal of, for example, the neurosciences. However, research into the cellular and molecular biology of bone has increased enormously in the last two decades, driven largely by the need to comprehend the mechanisms underlying the pathogenesis of bone diseases. With the avalanche of papers, conferences and abstracts, public and media awareness of bone and its problems have begun to improve. We approach the millennium keenly aware of the enormous social and medical costs of common bone disorders such as osteoporosis, the periodontal diseases, osteoarthritis, rheumatoid arthritis, less common but serious conditions including Paget's disease, bone malignancies, hypercalcaemia and hyperparathyroidism as well as a spectrum of orthopaedic and implant problems. Many of these conditions are associated with the ageing process, and as life expectancy continues to rise in developed countries their prevalence will increase. In consequence, the pharmaceutical industry now has a keen interest in skeletal pathobiology and most of the world's major drug companies have developed active bone research programmes.

The techniques needed for the study of the skeleton, with its mineralized matrix and inaccessible or sparse cell populations, are necessarily specialized. Many of these techniques seem to have acquired a daunting mystique, sometimes because methods and their underlying rationales are not always described fully in research papers. In common with many researchers in the bone field, the editors are frequently asked for advice in setting up or troubleshooting research techniques. We therefore felt that there was a pressing need for a single source that would cover the main experimental techniques used in skeletal research, both *in vitro* and *in vivo*. This volume, *Methods in Bone Biology*, consists of 12 chapters written by internationally recognized scientists from academia and industry. Contributors were asked to provide an intellectual framework in which the methodologies they were describing fitted.

The first six chapters cover the basic cellular and organ culture techniques used to study bone. Chapter 1 provides a detailed description of the methods used to culture cells of the osteoblast lineage, with a review of the current molecular and cellular biology of this lineage and of the matrix components they produce.

Chapters 2 and 3 review the methodologies used to measure the recruitment and activity of osteoclasts from avian, rodent and human sources. Chapter 4 describes the history and provides a detailed methodology which is increasingly used in assessing bone-modulating drugs. Chapter 5 deals with the assessment of apoptosis and the application of such measurements for the study of bone remodelling and bone pathology. The final chapter in this section details the cellular, explant and *in vivo* methods used to determine how bone responds to mechanical stress.

The second section deals with the skeletal system and the techniques used to measure its response to disease. Chapter 7 describes bone histomorphometry, a technique widely used to quantify *in situ* bone remodelling. Chapter 8 provides a detailed methodological treatment of the use of immunocytochemistry and *in situ* hybridization in the study of skeletal function and the use of these techniques as research tools. Chapter 9 details the assay of biochemical markers of bone turnover, including those involved in bone synthesis (e.g. osteocalcin) and those signifying bone breakdown (e.g. deoxypyridinolinium crosslinks). Chapters 10 and 11 review radiographic techniques (microfocal radiography and DXA) used to assess skeletal status in small animals and in humans. The final chapter gives a critical overview of *in vivo* techniques for assessing agents capable of modulating skeletal turnover.

This book will be of interest to bone and connective tissue researchers at all levels in academia and industry and to all those who wish to enter this fascinating realm of bone biology, but were put off by the complexity of the methodologies involved.

CHAPTER ONE

Culture of cells of the osteoblast lineage

Francis J. Hughes and Jane E. Aubin

1.1 INTRODUCTION

Osteoblastic cell cultures have been utilized to study a wide range of issues in bone biology, including the regulation of the differentiation and metabolic activity of bone cells. These studies have included, for example: the investigation of the action of hormones, cytokines and other signalling molecules on bone cells; the molecular mechanisms of action of such factors; studies of the osteoblast cell lineage and differentiation-dependent changes in phenotype; the synthesis of matrix proteins and other secreted molecules; and studies of the interaction of bone cells with biomaterials. Some obvious advantages of such methods over experiments conducted *in vivo* include the ability to control more fully the cells being assessed and their environment, the relative ease of sampling and analysing changes in the parameters under investigation, and the ability to avoid or reduce animal experimentation.

The literature documents the use of a large number of different cell culture models for the investigation of osteoblast activity and function, and these different systems often produce complex, apparently contradictory findings. Thus a thorough understanding of the model system employed is required in order to gain the maximum information from laboratory experiments. In the case of osteoblastic cell cultures, such an understanding is somewhat hampered by the relative lack of information of the specific stages of osteoblast differentiation and the lack of unambiguous markers to identify these putative stages. Further complications may arise when considering differences between primary vs. established cell cultures, interspecies differences and the possibility of phenotypic heterogeneity of osteoblastic cells – for example, from animals of different ages or bone from different anatomical sites.

This chapter starts with an overview of the current understanding of the osteoblast lineage and osteoblast differentiation, including an overview of the biochemical markers that may be used for characterization of these cells and how these markers may change during differentiation. Although these topics have been extensively discussed previously, we believe that their understanding is an essential foundation on which to consider subsequently specific aspects of osteoblastic

Methods in Bone Biology. Edited by Timothy R. Arnett and Brian Henderson.
Published in 1997 by Chapman & Hall, London. ISBN 0 412 75770 2.

cell culture. (For a full discussion of these important issues see, for example, [1–3].). Some general issues on selecting appropriate cell culture models are considered, followed by a discussion of culture conditions that support growth and differentiation of cells. Techniques for isolating primary cultures are then described, followed by a description of the specific properties of some of the commonly used established cell lines. Finally we describe some of the basic assays that are routinely used in the field. Throughout this chapter an understanding of the basic principles of anchorage-dependent cell culture is assumed; the reader is referred to one of the excellent general texts on animal cell culture (e.g. [4]) for further information.

1.2 THE OSTEOBLAST LINEAGE AND OSTEOBLASTIC DIFFERENTIATION

The **osteoblast** is the mature differentiated cell responsible for the formation and mineralization of bone matrix. Histologically, osteoblasts are recognized as a single layer of cuboidal, basophilic, polarized cells with abundant rough endoplasmic reticulum on the periosteal or endosteal surfaces at sites of active bone deposition. These cells express a variety of characteristic features, such as high alkaline phosphatase activity and synthesis of type I collagen and the non-collagenous proteins of bone matrix, which are considered in more detail below. Immediately subjacent to the osteoblasts, an additional layer (or layers) of cuboidal cells, the **preosteoblasts**, which express some but not all of these features, can be recognized. In addition two other cell types of the osteoblast lineage are readily recognized histologically: osteocytes and bone lining cells. The **osteocytes** are embedded in the bone matrix and communicate with each other by cytoplasmic processes within interconnecting channels (canaliculi) in the matrix. Although their function is not unequivocally known they may have roles in the maintenance of bone mass, in regulating osteoclastic bone resorption, and in modulating responses to mechanical stress. **Bone lining cells**, the flattened cells found on bone surfaces during periods when bone deposition is not taking place, are considered to be inactive osteoblasts that may be reactivated to become osteoblasts during periods of new bone formation.

1.2.1 The osteoblast lineage

Studies of the kinetics of bone cells *in situ* with ^{3}H-thymidine labelling and subsequent autoradiography of tissue sections suggest that preosteoblasts retain some proliferative potential, but osteoblasts, lining cells and osteocytes do not normally undergo mitosis [5]. Use of 'pulse-chase' ^{3}H-thymidine labelling techniques, where the fate of labelled cells is followed over an extended time period, suggests that mature osteoblasts have a defined life span, since only a small proportion (~20%) of osteoblasts are estimated to become osteocytes. The rest are thought to be eliminated, presumably by apoptosis [5–7]. (Chapter 5 describes methods used to study bone cell apoptosis.) The osteoblast pool is replaced by the further differentiation of preosteoblasts. In addition these experiments suggest that new preosteoblasts are recruited from a mesenchymal cell pool in the subjacent connective tissue and bone marrow stroma. These cells do not initially express any of the known markers of the osteoblastic phenotype and appear to represent an earlier stage osteoprogenitor cell and/or stem cell which subsequently undergoes osteoblastic differentiation.

By definition, a **stem cell** is an undifferentiated cell which exhibits unlimited self-renewal capacity, and the potential to give rise to precursor cells capable of undergoing differentiation to cells of a mature phenotype. Thus a stem cell may undergo an asymmetric mitosis resulting in two distinct daughter cells: one retaining the stem cell phenotype and the other being a cell which is committed to differentiate to express a mature phenotype (for discussion, see [8]). Many stem cells may be multipotential, such that the daughter cell may give rise to one of a number of different phenotypes. For example, in haemopoiesis a stem cell can give rise to all the different mature cell phenotypes of the haemopoietic system [9]. General principles of stem cell biology suggest the probability of the existence of a stem cell which would give rise to committed cells that will ultimately differentiate into osteoblasts and other mesenchymal cell phenotypes.

There is now substantial evidence to demonstrate the existence of multipotential mesenchymal cells which may give rise to cells of both the osteoblast and other connective tissue cell lineages. Dispersed bone marrow stromal cells give rise to colonies of fibroblastic cells (colony-forming unit fibroblast, CFU-F; or colony-forming cell fibroblast, CFC-F) which, when placed in diffusion chambers and implanted into rats, can give rise to a range of differentiated cell phenotypes, including osteoblasts, chondroblasts, adipocytes and fibroblasts [7, 10, 11]. Subsequent experiments have also demonstrated the formation of a range of differentiated cell phenotypes *in vitro* in marrow stromal populations which may be regulated by steroid hormones such as dexamethasone [7, 12–14]. Further evidence for the existence of multipotential mesenchymal cells has been obtained by analysis of the differentiation outcomes of clonally derived cell lines *in vitro*, such as the mouse embryonic fibroblast line C3H 10T1/2, the rat calvaria-derived cell lines ROB-C26 and RCJ 3.1, and the mesodermally derived C1 line [15–19]. Whilst the results of experiments with these lines differ in detail, they demonstrate the capacity of a clonal population to give rise to multiple differentiated cell phenotypes including osteoblasts, chondroblasts, myoblasts and adipocytes. Further analysis of subclones of RCJ 3.1 and C3H 10T1/2 cells also suggests the existence of a lineage hierarchy in which the multipotential cell gives rise to more restricted bi- or tri-potential cells, and these ultimately give rise to monopotential progenitors [20, 21]. The use of these models in investigating the issues of osteoblast lineage and regulation of cell commitment is considered further in section 1.6.

Cells showing a restricted potential (commitment) to the osteoblast lineage are defined as **osteoprogenitor cells**. Evidence from bone marrow stromal cell cultures and rat cavaria cell bone nodule assays suggests the existence of at least two distinct populations of osteoprogenitors. One population appears capable of constitutive differentiation *in vitro*, i.e. in appropriate culture conditions the cells will undergo a series of steps leading to a mature phenotype; while the other, apparently less differentiated population may show osteoblastic differentiation only following the addition of specific inductive stimuli [22]. Thus the addition of dexamethasone or other factors such as bone morphogenetic proteins (BMPs) increases the number of bone nodules or bone colonies in rat calvaria cell populations and increases the number of osteoblastic colonies in bone marrow stromal cell cultures, suggesting the presence of 'inducible' osteoprogenitor cell populations [12, 14, 23, 24]. As mentioned above, some experiments in the rat calvaria system have suggested that at least some of these latter cells may be a more primitive progenitor stage than the former.

The issue of whether any of these cells exhibit genuine stem cell kinetics (i.e. show unlimited self-renewal capacity) has not been rigorously demonstrated. On the other hand, at least some evidence has been presented that a majority of the osteoprogenitors present in both rat calvaria cell and bone marrow stromal cell cultures and assayed by the bone nodule assay have only a limited self-renewal capacity in culture [25–27].

Taken together, the evidence suggests the existence of distinct stages along the osteoblastic lineage as shown in Table 1.1. However, the recognition of these stages is currently hampered by the inability to identify directly any of the early stage cells in this lineage, and thus their existence is deduced from studies that show their ability to give rise to mature differentiated cell phenotypes in long-term cultures. There may also be a number of distinct intermediate osteoprogenitor cell differentiation stages, which are only beginning to be identified by the detailed analysis of proliferative capacity and marker expression in individual cells with time as bone nodules form.

1.2.2 The osteoblast phenotype and markers of differentiation

Clearly, under many or most culture conditions, osteoblastic cells may not express their normal histological features making this tool of little value for identification purposes *in vitro*. However, there has been extensive characterization of other features of the osteoblast phenotype, e.g. their synthesis of bone matrix and other proteins and responsiveness to specific hormones and cytokines, and a number of these are of considerable value for use as markers of the mature osteoblast phenotype *in vitro*.

(a) Bone matrix proteins

The production of the organic matrix of bone is largely the result of synthetic activity of the osteoblasts. The matrix consists of approximately 90% type I collagen, with the remaining 10% comprising a number of non-collagenous proteins. These matrix proteins include osteocalcin, matrix Gla protein, bone sialoprotein, osteopontin, osteonectin, fibronectin, thrombospondin, tenascin and certain proteoglycans (e.g. decorin and biglycan). Of these proteins, only osteocalcin and bone sialoprotein (BSP) are relatively restricted to osteoblasts, but a number of them are highly expressed by osteoblasts during osteogenesis and have proved to be useful markers of the osteoblast phenotype.

The function of these matrix proteins remains largely unclear, though most exhibit calcium-binding activities and some (e.g. BSP, osteopontin) contain the integrin-binding cell attachment motif RGD (Arg-Gly-Asp). Thus it has been proposed that these proteins may have roles in regulating mineralization and in mediating cell attachment to bone matrix. The recent production of 'knock-out' mice for both osteocalcin and BSP may provide new insights into the function of these molecules [28, 29]. (For a recent review of the nature of the non-collagenous proteins of bone matrix see [30].)

Osteocalcin (bone Gla protein, BGP) is currently the most 'bone-specific' of the non-collagenous proteins, its expression being restricted to mineralized tissue cells including osteoblasts, the odontoblasts and cementoblasts of teeth and those in hypertrophic chondrocytes. It makes up over 10% of the non-collagenous protein in mature bone but is present in much lower concentrations in embryonic bone. Given

Table 1.1 Changes in phenotypic markers during osteoblast differentiation

Marker	Multi-potent progenitor	Committed osteoprogenitor	Preosteoblast	Osteoblast – early bone formation	Osteoblast – later bone formation	Lining cell	Osteocyte
Collagen type I	++	++	++++	++++	+++	+	+
Collagen types III, V	++	+	-	-	-	-	-
Fibronectin	++	++	++	+	+	?	?
Tenascin, thrombospondin	-	-	++	++	+	?	?
Osteocalcin	–	–	–	++	++++	-	-
Osteonectin	-	-	+	+++	++	?	+
Osteopontin	-	++	+	++++	++	-	?
BSP	-	-	++	++++	++	-	?
Decorin, biglycan	-	-	-	++	+++	?	?
Alkaline phosphatase	-	-	+++	++++	++	-	-
PG responsive	+++	+++	++	+	+	?	?
PTH responsive	-	-	++	++++	+++	?	++
Ly-6a	?	++	++	?	?	?	?
CD44	?	?	+	++	++	?	++++
E-11 antigen	-	-	-	-	+++	-	++++

This is an approximate schema of the changes seen in various markers during osteoblast differentiation. There may be significant variations in these details between different species or between different cell lines, and possibly a degree of plasticity even between different cells in the same culture.

the relatively specific expression of osteocalcin by osteoblasts, it has been widely used as a phenotypic marker for these cells and is used clinically where it can be detected in the serum by immunoassay, as a diagnostic marker of bone formation.

Osteopontin and bone sialoprotein II are glycoproteins that both contain the integrin-binding cell attachment motif RGD (Arg-Gly-Asp). Osteopontin is known by an array of pseudonyms which include bone sialoprotein I (BSPI), secreted phos-phoprotein (SPP), 2ar and pp69. Bone sialoprotein is also referred to as BSP and as bone sialoprotein II (BSPII). Osteopontin and BSP are both expressed by hyper-trophic chondrocytes and the odontoblasts and cementoblasts of teeth; in some species osteoclasts also express osteopontin. In addition osteopontin is expressed in a wide range of other tissues, including renal tubular cells, neurons, T-lympho-cytes and platelets. BSP has not been found to be so widely distributed in non-mineralized tissues, though it has been reported in platelets and trophoblasts. Despite the fact that they are not uniquely expressed by osteoblasts, their high expression has made them useful markers of the phenotype, especially when used in conjunction with other markers.

Osteonectin (Secreted Protein Acid Rich in Cysteine - SPARC, BM-40) is a 32 kDa protein originally isolated from bone matrix but which has subsequently been shown to be expressed in a wide range of both embryonic and adult tissues and cultured cells, including connective tissue fibroblasts, cultured endothelial cells, lung, ovaries and testis. In view of its wide tissue distribution, on its own it is of limited value as an osteoblast marker despite its high abundance in bone matrix.

A number of other proteins are expressed by osteoblasts but are less useful as markers either because of their lack of tissue specificity or because little is known about their expression patterns during growth and differentiation. These include the putative cell attachment molecules such as fibronectin, thrombospondin and tenascin. The proteoglycans of bone – decorin and biglycan – are small proteo-glycans whose core protein is expressed in a wide range of tissues with dermatan sulphate glycosaminoglycan side chains, whereas in bone they have chondroitin sulphate side chains [31]. There is relatively little known about their stage-specific expression during osteoblast differentiation.

(b) Alkaline phosphatase (EC 3.1.3.1)
Alkaline phosphatase (AlP) is widely used as a marker of the osteoblast phenotype, not least because of the simplicity with which it can be assayed (section 1.7). Osteoblasts express the tissue-non-specific form of the enzyme which is expressed by many cell types, but the levels are particularly high in bone, liver and kidney and hence it is usually referred to as BLK. The BLK AlP enzyme is one of at least four isoforms, with the other types showing a more restricted tissue expression pattern. High levels of AlP are seen in both preosteoblasts and osteoblasts *in vivo* and in dif-ferentiating osteoblasts *in vitro*. AlP is localized to the cell membrane on osteoblastic cells and is covalently bound to phosphatidyl inositol (PI) phospholipid complexes. It can thus be released from cells by the PI-specific phospholipase C enzyme [22, 32].

(c) Hormone and cytokine responsiveness
Osteoblasts, like other cells, respond to a wide array of hormones and cytokines and some of these interactions are potentially useful as markers of their pheno-type. In particular, the binding of parathyroid hormone (PTH) to PTH/PTHrP receptors and subsequent activation of adenylate cyclase (resulting in cAMP

production) has been widely used as a marker of the osteoblast phenotype (e.g. [33–39]). In cell cultures an increase in cAMP following PTH stimulation is generally associated with differentiated osteoblasts, and this response is often increased by factors such as dexamethasone which upregulate expression of other markers of the mature osteoblast phenotype (e.g. [39]). However, as discussed in section 1.2.3, there are also contradictory data concerning the distribution of PTH receptors in earlier stage osteoblastic cells.

Many cells, including those of the osteoblast lineage, both synthesize and respond to prostaglandins through their interaction with a family of specific cell surface receptors. Prostaglandin E_2 (PGE$_2$) is the major PG synthesized by osteoblasts, and may mediate a number of the effects of diverse stimuli such as cytokines, hormones and mechanical deformation that may act on bone cells [40–44]. Although a wide range of osteoblast populations may respond to PGE2 stimulation *in vitro*, recent evidence suggests that PGE$_2$ responsiveness may be particularly associated with less differentiated osteoblastic cells [41, 45–47].

Other paracrine factors that may act on osteoblastic cells include interleukin 1, 6 and 11, TNFα, interferon gamma (IFNγ) and growth factors including epidermal (EGF), platelet-derived (PDGF), insulin-like (IGFs), transforming beta (TGFβ), fibroblast-derived (FGFs) and BMPs. Many of the data from different studies with at least some of these factors appear contradictory. As an example, IL-1 has been shown both to stimulate and inhibit osteoblastic proliferation, to inhibit or stimulate expression of markers of osteoblast differentiation including AlP, osteocalcin, and osteopontin, and to inhibit or stimulate bone formation *in vitro* [41, 42, 48–56]. Furthermore, there is evidence of complex biphasic effects of some factors according to either the timing or concentration of the stimulating factor [41, 57]. These conflicting results may reflect differences between different osteoblast culture systems and conditions of culture, but growing evidence suggests that there is a marked heterogeneity of responses amongst osteoblastic cells according to their differentiation stage.

(d) Cell surface markers

Given the successful use of cell membrane-associated markers in identifying cells of, for example, the haematopoietic lineages, recent studies have described the investigation of the expression of various adhesion molecules and other surface markers in osteoblastic cells [58–64]. It is clear that further work may be required to determine the value of such molecules as osteoblastic markers and to ascertain their possible functions in regulating bone cell function. Studies of the expression of integrin family members both *in situ* and *in vitro* suggest the possibility of qualitatively different expression patterns in different cells of the osteoblast lineage but more work is needed to clarify these data, which to date do not give a clear consensus of such patterns [58–60]. Expression of CD44, an adhesion molecule which binds to hyaluronan, type I collagen and fibronectin, has been reported in osteoprogenitor cells, in differentiated osteoblasts and in osteocytes both *in situ* and *in vitro* [61, 63–65]. However, there is little evidence of this pattern being regulated by factors that may alter differentiation stage *in vitro* [63]. Finally, a recent report describes the enrichment of osteoblastic cells from mouse bone marrow cell populations on the basis of their expression of the early leukocyte antigen Sca-1 and their ability to bind wheat germ agglutinin [66], and from human marrow cells using the STRO-1 antibody [67, 68].

(e) Monoclonal antibodies
Several examples have now been reported of monoclonal antibodies recognizing subpopulations of osteoblastic cells in chicks, rats and humans, with diverse patterns of reactivity, including cell surface, cytoplasmic and extracellular matrix-related staining patterns. Many label the most mature cells (osteoblasts and osteocytes) in the lineage most intensely, sometimes with weaker labelling in less mature osteoblastic cells, including preosteoblasts in the periosteum. Some of these have been shown to label AlP [22, 69–72] whilst others appear to have novel distribution patterns amongst cells of the lineage.

Many of the antibodies reported to date have been reviewed recently [73], but it may be useful to discuss a few representative examples here. For instance, of monoclonal antibodies that recognize later-stage markers of the osteoblast lineage, two label the surface of chick osteocytes specifically but not other earlier stages of the lineage (the OB 7.3 antibody [73, 74]; the SB-5 antibody [75]). Although the molecules recognized by these antibodies have not been identified, they have been useful in gaining new insights into the osteocyte. For example, they have been used to isolate osteocytes from chick mixed bone cell populations [76, 77] and, in combination with other monoclonal antibodies (SB-1, SB-2, SB-3), to help to discriminate stages in the transition from mature osteoblasts through to the terminally differentiated osteocyte [70]. Transitional steps from osteoblast to osteocyte are evident with another monoclonal antibody, designated E11 [78], which is raised against rat osteoblastic cells. The E11 antigen has recently been shown to be homologous to OTS-8, which is highly expressed in lung and is expressed in phorbol ester-treated MC3T3 osteoblasts [79, 80]. The role of E11/OTS-8 in osteoblast function is unknown, though a recent report described evidence for its role in the shape changes associated with transition to the stellate shape of an osteocyte [81].

Antibodies have also been isolated that label the cell cytoplasm rather than, or in addition to, the cell surface of mature cells in the lineage. For example, RCC 455.4 labels the cell cytoplasm in a vesicular pattern in mature osteoblasts, osteocytes and chondrocytes *in vivo* and *in vitro* [71]. Recently, expression cloning was used to identify the macromolecule recognized by RCC 455.4 as galectin 3, an S-type, β-galactoside-binding lectin with binding domains for matrix proteins including laminin and collagen [73].

To date, few antibodies have been raised that recognize cells earlier than the preosteoblast. It is unclear whether this is technical artefact resulting from the nature of the populations injected or the detection schema used, or implies a relative lack of detectable antigenic determinants on these cells. Exceptions to this include the isolation of three antibodies (SH-2, SH-3, SH-4) that react on a subset of human marrow stromal cells and a variety of tissues *in vivo* but do not label osteoblasts or osteocytes [82]. Another antibody that appears to label a subset of stromal cells, including early mesenchymal precursor cells and osteoprogenitor cells, amongst others, is the anti-human monoclonal antibody STRO-1 [67, 68].

Few (if any) of the antibodies raised to date appear to recognize antigenic determinants completely restricted to osteoblastic cells, although many cross-react with only a limited number of other tissue and cell types. Clear exceptions are the antibodies giving osteocyte-restricted staining as summarized above. (For further discussion, see [73].)

(f) Bone formation and induction of mineralization

The defining characteristic of the mature osteoblast is its ability to produce a mineralized bone matrix. Early evidence for the presence of osteoblastic cells in some cell cultures came from their ability to produce a mineralized bone matrix when implanted into suitable recipient animals in diffusion chambers. Such a model has been widely employed to investigate the phenotypes of bone marrow stromal cells by determining the differentiated cell phenotypes expressed following implantation [7, 10]. The factors that permit bone formation in implanted diffusion chambers are not faithfully reproduced in cell culture systems, as true bone formation is not seen in many osteoblastic culture systems *in vitro*. However, many such cultures have been shown to be able to induce matrix mineralization *in vitro*, particularly when cells are cultured in the presence of organic phosphate supplements. The role of AlP in initiating mineralization has been the subject of a number of studies which suggest that this enzyme is responsible for a local increase in inorganic phosphate ion concentration by its action on the organic phosphate supplement, but is not required for mineralization to proceed once initiated [83, 84]. The initiating action of AlP on organic phosphate substrates can be replaced by supplementing the medium with high concentrations of inorganic phosphate [84, 85].

Although the induction of matrix mineralization is often considered to be a specific property of the osteoblastic matrix, there is conflicting evidence as to whether high AlP levels in the presence of organic phosphate or high inorganic phosphate levels may result in mineralization of collagen matrices from similar fibroblastic cell cultures [83, 85–87]. Against this, the presence of a mature osteoblastic matrix may be sufficient to support mineralization, even in the absence of viable cells, when exposed to appropriate concentrations of inorganic phosphate [85]. In addition to true matrix mineralization, it has been shown that high AlP activity may result in 'ectopic mineralization' where calcium phosphate crystals are deposited in cultures but without the intimate relationship of crystals to matrix described above. It is thus important to distinguish between these two phenomena if using mineralization as a marker of osteoblast [85, 87].

A few osteoblastic culture systems can produce discrete, three-dimensionally organized mineralized matrices which are recognisably bone like [14, 88–91]. These bone nodules consist of a woven bone matrix covered by cuboidal osteoblastic cells and containing osteocyte-like cells embedded in the matrix. Characterization of bone nodules has demonstrated that the processes of nodule formation, matrix deposition and subsequent mineralization follow a well ordered, temporally defined pattern which appears analogous to bone formation and mineralization *in vivo*. Bone nodule assays are considered in more detail in section 1.5.3.

1.2.3 Changes in expression of phenotypic markers during osteoblast differentiation

In an attempt to describe the stages occurring as cells progress through the osteoblast lineage, a number of studies have described changes in phenotypic expression during osteoblast differentiation and bone formation both *in vivo* and *in vitro*. Although particular details of some of these studies vary, there is some consensus on the sequence of expression of markers at least for the more differentiated osteoblastic cells.

(a) Early stages of osteoblast differentiation
As already discussed, the absence of definitive markers of mesenchymal stem cells and early osteoprogenitors means that their identification *in vitro* is usually based on their ability to express subsequently a more mature and identifiably osteoblastic phenotype. In practical terms, this means that their phenotype is characterized experimentally by the absence of bone markers (e.g. AlP, PTH response), combined with later evidence of osteoblast differentiation and, in the case of mesenchymal stem cells, evidence of multiple potential by the ability to give rise to other mature phenotypes as well [7, 16, 21]. In terms of matrix synthesis the immature cells may express types III and V collagen in addition to type I, together with fibronectin, but do not express proteins characteristic of bone, such as osteocalcin and BSP. As discussed earlier, relatively undifferentiated cells have also been fractionated along with other cells from bone marrow as a non-adherent low density cell population and by expression of leukocyte markers such as Ly-6a and SCA1/WGA binding [66, 92].

(b) Later stages of osteoblast differentiation
A number of studies have described changes in expression of recognizably osteoblastic markers during the later stages of osteoblast differentiation by examination of bone formation *in situ* or by examination of osteoblastic cells *in vitro* over a time course during which differentiation occurs. In general, osteoblastic cells express AlP and type I collagen relatively early in the maturation sequence; some authors have reported considerable expression of osteopontin at these early stages, consistent with the expression of osteopontin in other cycling populations. Their expression is followed by that of other bone matrix proteins such as BSP and a second peak of osteopontin; the expression of osteocalcin is very late in osteoblast development and often appears with the onset of matrix mineralization. However, there are noticeable variations in the temporal pattern of expression for a variety of markers reported from different studies using different cell systems, culture conditions and methodologies. For example, many studies demonstrate the expression of PTH receptor mRNA and protein *in situ* largely confined to the osteoblast and preosteoblast layers [93–95] , whilst other studies show maximum PTH binding to relatively undifferentiated cells [96, 97]. Further studies show the acquisition of a cAMP response to PTH in cultured cells with the onset of expression of osteoblastic markers such as AlP [98]. A recent series of labelling and *in situ* hybridization studies strengthened the idea that PTH receptor may be present at low levels relatively early, but be markedly up-regulated late in the differentiation sequence.

In culture, most osteoblastic markers such as AlP are down-regulated during cell proliferation and show increased expression after confluence. Studies of the relationship between cell growth, proliferation and differentiation in rat calvaria cell cultures have described stage-specific expression of a number of osteoblastic marker genes associated with proliferation, matrix deposition and mineralization. During cell growth, proliferation-associated markers such as H4 histone are expressed, whilst osteoblastic markers such as AlP are low. During matrix deposition proliferation markers such as H4 are down-regulated and matrix deposition is associated with up-regulation of AlP and type I collagen. Finally, mineralization is associated with expression of matrix proteins such as osteocalcin and the bone sialoproteins [99–102].

The precise pattern of changes seen in this system differs in some respects from those described in other systems, including studies of bone formation *in situ*. For example, in many studies type I collagen expression is not markedly down-regulated during later stages of differentiation, and studies *in situ* suggest that BSP and osteopontin are expressed by newly mature osteoblasts during early bone matrix deposition, with osteocalcin being expressed at a later stage by osteoblasts during matrix mineralization. In addition, there is contradictory evidence as to the direct regulatory role of matrix mineralization on the osteoblast [93]. Some of the apparent contradictions seen with these studies may result from the altered proliferative status of cells in culture compared with cells *in situ*, and may also be the result of the heterogeneity of cells present in cultures. In an attempt to address this latter problem directly, analysis of gene expression of single cells by reverse transcription–polymerase chain reaction (RT-PCR) and immunocytochemistry has been used on colonies of primary rat calvaria cells [103]. These data suggest a general pattern of expression in which AlP, type I collagen and osteopontin are expressed early during osteoblastic differentiation, with BSP and osteocalcin expression being largely associated with the mature osteoblast phenotype. The results also show a degree of heterogeneity in gene expression patterns between individual cells which suggests the possibility of considerable plasticity in the precise sequence of gene expression during osteoblast differentiation and maturation.

Overall, the data provide a general overview of the changing gene expression patterns of osteoblastic markers with time, as shown in Table 1.1. These studies do not allow the construction of a precise lineage map of identifiable phenotypic stages during osteoblast differentiation at the present time, particularly in the absence of well characterized markers that are characteristic of less differentiated cells of the lineage. However, there is good evidence to suggest that the changes in gene expression are closely regulated by factors which include proliferation status and presence and deposition of extracellular matrix and that multiple stages of this sequence can be regulated by a variety of hormones such as dexamethasone, 1,25 dihydroxyvitamin D_3 and growth factors such as EGF, TGFβ and BMPs. In some cases, agents have biphasic effects – for example, stimulating progenitor proliferation while inhibiting matrix production or mineralization (e.g. [57, 104]).

1.3 SELECTION OF CELL MODELS FOR STUDIES OF OSTEOBLASTIC REGULATION

A large number of osteoblastic culture systems have been used as models for studies in bone metabolism and, as is apparent from the discussion above, many of these models vary considerably. Intrinsically, a cell culture system is an artificial model of the situation *in vivo*; how closely its behaviour mimics the *in vivo* situation will be influenced by how well it reflects or controls many factors, including the cell's normal micro-environment, cell matrix and proliferation status (Table 1.2). In general, the available cell systems fall into two main categories: primary cell cultures isolated directly from tissues; and established, immortal cell lines which are generally cells derived clonally from normal or malignant cell cultures. The specific properties of some of the more commonly used cell systems are considered in more detail in section 1.6.

Table 1.2   Examples of how the culture environment may influence cell behaviour

Culture parameter	Mechanisms	Affected behaviour
Cell density	Proliferation status Changes in cell–cell contacts	Differentiation stage: hormone and cytokine responsiveness
Time in culture	Proliferation status Matrix synthesis Matrix mineralization	Differentiation stage: hormone and cytokine responsiveness
Culture medium/ supplements	Composition of medium/serum Requirement for ascorbic acid Requirement for organic phosphate (β-glycerophosphate)	Differentiation stage: ability to differentiate Matrix synthesis Matrix mineralization
Extracellular matrix	Cell attachment to plastic vs. biological ECM	Alterations in cell proliferation and differentiation
Isolated culture of cells	Loss of normal regulatory signals from other adjacent cell types present *in vivo*	Unknown, but could alter cellular responses and differentiation status

1.3.1 Primary cell cultures

In principle, primary cell cultures should most closely reflect the osteoblastic cell populations present *in vivo*; they should retain the normal (inverse) relationship between cell growth and differentiation and, if appropriate cell culture conditions are found, should express patterns of gene expression consistent with those found *in vivo*. In view of these issues, the use of primary cell cultures is often particularly appropriate for investigation of osteoblastic differentiation or where differentiation status may be an essential determinant of cell behaviour. Against this, primary cell cultures are, by their nature, a heterogeneous mixture of cells which may include other cell lineages such as fibroblasts and there is a risk that cells isolated on different occasions may give inconsistent results. Because they tend to be phenotypically unstable with time in culture, either due to overgrowth of one population by another or because of loss of differentiated cell function with increasing passages, it is usually necessary to use such cultures for experiments directly after initial isolation or at best after very few cell passages.

1.3.2 Established cell lines

Osteoblastic cell lines have been established from a wide variety of sources and in general are derived from clones of an original cell line. In contrast to human cultures, rodent cells may become spontaneously transformed with continuous passage, and this has been exploited for the establishment of, for example, the well characterized mouse line MC3T3-E1. Many other cell lines are derived from cells cloned from osteosarcomas and others have been immortalized by use of viral vectors. Many cell lines have the merits of ease of culture and increased pheno-typic stability with serial passages, which should theoretically result in increased reproducibility of results in independently conducted experiments, though it should be recognized that such clonal lines may drift considerably in phenotype with time and generate considerable subclone diversity. Whilst established cell lines tend to approximate a particular stage of osteoblast differentiation during culture, many such lines show an increased expression of a differentiated cell phenotype following, for example, treatment with hormones and growth factors, or in post-confluent cultures which cease mitosis. However, given the altered growth control in established cell lines, it may be expected – and has been documented – that the normally closely regulated association between growth and differentiation would be altered in such cells. For example, normal regulation of osteocalcin gene expression is altered in the rat osteosarcoma cell line ROS 17/2.8, resulting in a disassociation of growth and differentiation controls in these cells [105, 106]. Thus, it may not be possible to list those functions that can be considered 'normal' in any transformed cell line, and those that may be aberrant as a result of the transformation process.

1.3.3 Other variables to be considered in selecting osteoblastic culture models

(a) Species of cell sources
There are marked variations in the reported behaviour of cells derived from different sources, and it is tempting to ascribe many of these differences to the various species from which the cells are derived. Although many important functional domains remain highly or relatively conserved, some variation may be expected to arise from variations in sequence of genes and proteins in different species; there are also reported differences in the relative amounts of proteins expressed by apparently comparable cells of different species. However, there are a number of other variables which may profoundly affect the suitability of a particular cell system for any given type of experiment. In fact, until recently there was little decisive evidence to demonstrate that osteoblastic cells derived from different animal species were the same or were fundamentally different from each other; accumulating evidence suggests there may be quite considerable species differences. Secondly, some of the reported differences seen between cell cultures may also be the result of differences in factors such as the anatomical source of the cells, which may influence both the nature of the osteoblasts (i.e. anatomical site-specific phenotype and/or the differentiation status of the majority of cells obtained) and the presence and frequency of 'contaminating' cell populations. Other considerations include the age of the donor tissue and the culture conditions employed. It is clear that one reasonably chooses specific cell models to study in depth and does not routinely try to replicate all experiments with different model systems with which one may have less knowledge and experience, but ultimately the comparisons will have to be done.

One very practical issue that may determine the species of donor cells chosen for a particular study is the availability of reagents that will work in cells of that species. Many hormones, cytokines and growth factors show cross-reactivity against cells of many species, though this observation does not address possible variations in, for example, receptor binding affinity between species. One notable exception to this is interferon-gamma, which shows marked species specificity in its action. Some cDNA probes may be of little use in detection of genes in different species when used with hybridization conditions of high stringency. Sequence data for design of primers for PCR reactions may be limited, though they are often available for mouse species and humans. While workable solutions for all of these problems are available, some, such as re-cloning specific homologues in other species, can be time-consuming and the availability of a particular cDNA probe or sequence information may be a significant influence when selecting a particular species from which to use cells.

Antisera used for immunohistochemistry, immunocyotochemistry or ELISA or other binding assays may or may not display cross-reactivity with similar proteins from other species. This may be particularly (but not exclusively) the case with monoclonal antibodies. For example, commercially available radioimmunoassay kits for detection of osteocalcin are generally specific for the human form of osteocalcin. Suppliers often have details of cross-reactivity of their antisera and the range of assay types in which they are known to work. Problems associated with the use of *in situ* techniques are described in Chapter 8.

(b) Age of donor tissue
The age of the donor tissue may have significant effects on the properties of primary cultures that are isolated from them. For example, and perhaps not surprisingly, cells derived from fetal sources often tend to show a greater ability to give rise to a range of differentiated phenotypes than those from older donors, suggesting an increased number of undifferentiated progenitor-type cells in these cultures. Studies also show a reduced number of total colony-forming cells and bone-forming cells in bone marrow stromal cell cultures from older donor sources than from younger donors [107]. On this basis, it is likely that some of the variation seen between primary culture systems is because of variations in age of donor tissue, resulting in differences in the relative proportions of cells at specific differentiation stages, though in most cases this has not been rigorously investigated.

(c) Anatomical sources of cells
Osteoblastic cells have been isolated from a range of anatomical sources including calvariae, femoral heads, endosteal and periosteal surfaces of femurs, vertebrae, marrow stroma, mandibles, etc. In addition to the likely variations in the proportions of cells at distinct differentiation stages, there is the possibility of true phenotypic variations between cells at these sites. The variation in gene expression patterns seen between individual cells derived from the same source lends further support to the possibility of the existence of phenotypically distinct osteoblastic subpopulations whose proportions might vary according to their anatomical source. Given the paucity of information in this area one can only speculate as to whether these cells remain phenotypically stable in culture, when grown in isolation from their normal anatomical relationships with the surrounding extracellular matrix and adjacent tissues.

1.3.4 Summary

When taken together, it is clear that the selection of an appropriate model for osteoblastic cell culture studies can only be made using as much information as possible on that system, accepting that variations between systems may result from a range of factors including species, age and anatomical donor site of cells for primary culture, and the possibility of departures from normal phenotypic expression in established cell lines. For example, osteoblastic cell cultures derived from human trabecular bone (section 1.5) appear to have low numbers of undifferentiated osteoprogenitor cells and as such would be a poor model to investigate factors that may regulate early differentiation events. Similarly, established human cell lines, which are either derived from osteosarcomas or are virally transformed cells, are likely to be poor models for the study of physiological growth regulation or programmed cell death. The uncertainty of these factors illustrates the care required in extrapolating data obtained from one system to general application to all systems and the situation *in vivo*.

1.4 GENERAL CONSIDERATION OF CELL CULTURE CONDITIONS

1.4.1 Growth medium and serum supplements

For successful culture of osteoblastic cells, the growth medium used (with appropriate supplements if required) should be able to support both the proliferation and differentiation of the cells. A wide range of basic growth media have been used successfully and the requirements for an adequate medium may not be too critical, particularly with established cell lines. We have obtained consistently good results using alpha-modified MEM (minimal essential medium) with nucleotides. This is a relatively complex (and expensive) medium, but we have found in some of the more fastidious assays, such as the rat calvaria cell bone nodule assays (section 1.5.3), that it produces better results than with some of the more commonly used media such as Dulbecco's MEM (FH, unpublished observations). Both established and primary cell cultures are routinely grown in medium supplemented with 10–15% fetal bovine serum (FBS), though lower concentrations down to 5% are also often successful. In terms of specific assays, there is notable variation between batches of serum which can only be established by testing, and in view of this it is wise practice to purchase large stocks of a particular batch of serum once its efficacy is established. Indeed in bone nodule assays some batches of FBS are completely unable to support nodule formation, whilst others support abundant nodule formation (see Fig. 1.3). Most suppliers will provide small samples of serum for batch testing whilst holding a stock in reserve, awaiting the outcome of such tests. Testing should include the comparison of serum batches in supporting growth of primary cultures together with an assay of osteoblast differentiation. Typically, this can be carried out by plating replicate cell cultures at low density and measuring population doubling times for cells exposed to medium with different serum batches, while concomitantly assessing bone nodule formation under standard conditions.

One obvious problem with all culture media containing serum supplements is that there is no way of knowing all the constituents to which the cells are exposed. Some companies (e.g. Hyclone) do offer serum batches with defined levels of many known hormones and growth factors. It is also possible to maintain cells under serum-free conditions for short periods (~ 1–2 days) and this is commonly used to investigate the effects of a specific factor in isolation on parameters such as proliferation (particularly as estimated by DNA synthesis) or AlP activity. In addition, chemically defined media (without serum supplements) and media designed for use with low serum concentrations, or defined serum substitutes, can support the proliferation of osteoblastic cell cultures. However, there is as yet little evidence of the successful use of chemically defined media supporting osteoblast differentiation; and we have had little success with chemically defined media, commercially available serum substitutes or enriched media using low serum concentrations in supporting differentiation *in vitro*, for example, in bone nodule assays. Against this, a recent study has described the culture of rat bone marrow stromal cells in a fully defined medium containing supplements which included PDGF and bFGF [108]. Although these cells were not shown to differentiate *in vitro*, they retained their ability to express osteoblast and chondroblast phenotypes subsequently when implanted in calcium phosphate-derived carriers *in vivo*. The observation that the cells retained the ability to express mature phenotypes following culture in a defined medium is potentially of value, for example, for studies investigating factors that may regulate stem cell commitment and initiation of differentiation pathways.

1.4.2 Antimicrobial supplements

The use of antimicrobial supplements cannot be used as cover for inadequate aseptic technique when feeding or manipulating cells. Strictly speaking, such supplements should not be required in culture media, particularly for propagation of established cell lines, though it is common practice to use such supplements routinely. A stronger case can be made for the use of antimicrobial supplements for primary cultures, where it is sometimes difficult or impossible to be assured of total asepsis during the initial collection of tissue from the donor. Medium may be routinely supplemented with a combination of antibiotics such as 50 units of penicillin G (benzyl penicillin)/ml and 50 µg streptomycin/ml, and with an antifungal agent such as 0.3 µg amphotericin B (Fungizone)/ml. These reagents are readily available from all the major suppliers of tissue culture media, and are not significantly toxic to the cells when used at these concentrations.

To address the potential problem of contamination of material during collection for establishment of primary cultures, it is valuable to use a specific 'transport medium' containing high concentrations of antimicrobials for transport and 'disinfection' of tissue. A suitable transport medium would consist of Dulbecco's MEM with 25 mM HEPES buffer (bicarbonate buffer is not suitable for transport media as CO_2 concentration will not be regulated in these circumstances) supplemented with 500 units of penicillin G/ml, 500 µg streptomycin/ml and 3 µg amphotericin B/ml, the equivalent of 10 times the normal working concentrations of these antimicrobials. Transport medium does not require addition of serum or other supplements. Donor tissue can be safely left in this medium for periods of up to a few hours, preferably refrigerated, although such high concentrations of antimicrobials can be toxic, particularly to isolated cells, when left for longer periods of time.

1.4.3 Ascorbic acid

Ascorbic acid (AA, vitamin C) is frequently used as a routine supplement in osteoblast cultures and has been shown to have important effects on cell behaviour *in vitro*. AA promotes collagen maturation, and hence extracellular matrix (ECM) deposition, but does not appear to affect collagen gene expression. Studies with a range of osteoblastic cultures, including fetal rat calvaria-derived primary osteoblastic cells, porcine osteoblastic cells and the MC3T3-E1 cell line, suggest that AA-regulated ECM synthesis may have important regulatory effects on the cells. Treatment of cells with AA during logarithmic growth increases proliferation, whereas AA increases expression of differentiation markers such as AlP and osteocalcin in post-confluent cells and is necessary for production of nodules and mineralized matrix in long-term cultures [88, 91, 102, 109–114]. Furthermore, these effects can be blocked using inhibitors of collagen maturation, suggesting that the action of AA on osteoblastic cells is mediated by its direct action on matrix deposition [110, 114]. This is further supported by evidence demonstrating that the effects of AA can be blocked with an RGD-containing peptide which inhibits interactions by the cellular integrins with the collagenous ECM [111]. There is an interesting series of papers documenting ascorbate uptake and effects on ion transport in osteoblastic cells that also suggest other activities or modes of action of AA that will need to be addressed further [115, 116].

AA is generally used at a working concentration of 50 µg/ml in isolated cell culture systems, at which concentration it is not toxic nor does it lead to ectopic mineralization. There has been discussion that this concentration may be high under certain other culture conditions, such as chondrocyte culture [117] (see also below). AA is readily made up using phosphate buffered saline (PBS) or stock medium as the diluent. It is rather unstable and readily oxidizes, and so sterile 100× stock solutions should be dispensed into single-use aliquots, immediately stored at –20°C (where they are stable for many months) and thawed immediately prior to addition to cultures at least three times per week. Alternatively, long-life analogues of AA such as Asc-2-P have now been described which may be a valuable alternative to AA for use in osteoblast cultures [118], though they have not been evaluated extensively for this purpose.

1.4.4 Organic phosphates

A number of long-term osteoblast culture systems are able to produce a mineralized matrix, but in general this occurs only after supplementing alpha-MEM with the organic phosphate sodium β-glycerophosphate (βGP), which acts as a substrate for AlP, resulting in an increased concentration of inorganic phosphate and subsequent mineral deposition [83–85, 118]. It has been questioned whether mineralization occurring only in the presence of high concentrations of exogenous organic phosphate can be considered 'physiological' [87, 118]. The nature of this mineral deposition was discussed in section 1.2. βGP is usually used to a final concentration of 10 mM and, as with AA, a 100× stock solution can conveniently be made up, with PBS as diluent, and stored in single-use aliquots at –20°C.

1.4.5 Dexamethasone

Glucocorticoids such as dexamethasone have complex effects on osteoblast metabolism which may include increased expression of a differentiated phenotype, together with suppression of production of a number of extracellular signalling molecules such as interleukin 1, prostaglandins and nitric oxide. The synthetic glucocorticoid dexamethasone, at concentrations comparable to the physiological concentrations of natural glucocorticoids (10^{-8} M), promotes bone formation in a number of primary culture systems including chick, some mouse cells and rat cells and results in the up-regulation of osteoblastic features such as AlP in various established cell lines, such as ROS 17/2.8, MC3T3-E1 and UMR 106.06 [23, 119–122]. In a number of cell systems, dexamethasone has been shown to be an important medium supplement to achieve expression of a recognizably osteoblastic phenotype. For example, the addition of dexamethasone to bone marrow stromal cell cultures from rat, rabbit and human sources results in formation of colonies of cells expressing both osteoblastic and adipocytic phenotypes, whereas in the absence of dexamethasone these colonies remain predominantly fibroblastic *in vitro*. In rat calvaria cell bone nodule assays, the addition of 10^{-8} M dexamethasone increases proliferation and self-renewal of osteoprogenitor cells, and results in an increase in both the number and size of bone nodules. Given recent reports that 10^{-8} M dexamethasone is toxic to mouse bone marrow stromal cells, which have been said to differentiate without added glucocorticoids [123], and that mouse calvaria cells also differ in their glucocorticoid requirements compared with rat calvaria cells, it may be useful to assess the need, appropriate concentration and appropriate glucocorticoid in any system being established *de novo*. (For a fuller discussion of the effects of dexamethasone on osteoprogenitor cell populations, see [3].)

As a result of these findings, dexamethasone can be a useful supplement of media in a number of different situations, including as a positive control treatment in experiments to determine the effect of specific factors on osteoblast differentiation. As dexamethasone is a steroid-based molecule, it is not water soluble and thus it is necessary to dissolve it first in absolute (Analar grade) ethanol. This alcoholic solution is subsequently added to serum-free culture medium to make a stock solution of 10^{-6} M ($\approx$ 100× strength of working concentration) and stored at –20°C in single-use aliquots, or it can be stored as high concentration alcohol stocks at –20°C and added fresh to medium at the appropriate concentration at each medium change.

1.4.6 Propagation and storage of cells

In general, osteoblastic cultures can be passaged (subcultured) by methods that are common to all anchorage-dependent cell lines; for example, by use of 5 minutes digestion of trypsin or trypsin/EDTA solutions, which can be obtained commercially from many suppliers or prepared as stock solutions from commercially available enzyme powders. On occasion, it can be difficult to obtain single-cell suspensions by trypsin treatment, owing to the extensive extracellular matrix that is produced in many osteoblastic cell cultures. This is particularly the case in cultures which have been maintained at confluence for prolonged periods, and in these cases it is not uncommon to obtain an intact sheet of cells embedded in matrix which separates from the culture dish, or the persistent adherence of some cells to the culture dish. In such cases more prolonged trypsin treatment may be useful, together with mechanical disaggregation of cells by repeated aspiration through a pipette nozzle. Alternatively these cells may be released by digestion with a collagenase solution (section 1.5.3) or a combined collagenase–trypsin mixture.

As noted earlier, the major limitation on subculture of osteoblastic cells is the question of their phenotypic stability. By their nature, established cell lines tend to be phenotypically stable over many cell doublings, though their absolute stability cannot be guaranteed *ad infinitum* and published reports have documented considerable subclone variation in several well characterized osteoblastic lines, e.g. the ROS17/2 line. However, primary cell cultures exhibit marked phenotypic instability, resulting in the loss of osteoblastic characteristics in a small number of passages.

Suspended cells from both primary and established cell lines can also be stored by freezing, using conventional methods. Primary cell cultures which have been carefully frozen according to this protocol retain their osteoblastic characteristics [37] but again the usefulness of freezing such cultures may be limited principally by their phenotypic instability with increasing passage number.

1.5 ESTABLISHMENT AND PROPAGATION OF PRIMARY CELL CULTURES

A variety of methods have been described for the establishment of primary osteoblastic cell cultures from a range of donor sources, both from different species and different anatomical sites. In general these techniques either use methods where cells grow out of the explanted tissue or involve release of cells from the donor tissue by enzymatic digestion or mechanical disaggregation. As already noted, these cultures may show significant differences in their properties which may be the result of the factors discussed in section 1.3 such as species, age and anatomical site of donor tissue, in addition to the method of cell isolation used. The issue of how closely the make-up of the cells that grow *in vitro* reflects the cells present *in vivo* has not been fully addressed but it is likely that only a subset of cells present *in vivo* is present in culture and it is certain that the proportion of cells at various stages of differentiation changes significantly with time in culture. The fact that the mature osteoblast *in vivo* does not appear to undergo mitosis suggests that these cells either will not be heavily represented in proliferating cultures or will be present in an altered, perhaps less differentiated state.

In this section, some specific methods for the establishment of primary cell cultures are considered, including methods that have been successfully used for the isolation of bone marrow stromal cells and for isolation of osteoblastic cells by enzymatic digestion of tissues or outgrowth of cells from explanted tissue.

1.5.1 Bone marrow stromal cells

Bone marrow consists of cells of the haemopoietic system together with the fibroblast-like cells of the marrow stroma. Stromal cells comprise a heterogeneous mixture of cell lineages and consist of cells at a variety of differentiation stages, including multi-potential and bi-potential mesenchymal precursor cells together with cells committed to restricted lineages, such as osteoprogenitor cells and pre-adipocytes. At its simplest, establishment of marrow stromal cell cultures depends on the mechanical disaggregation of the marrow and placement into culture to allow cells to adhere to the culture dish and to allow growth of the fibroblastic cells to occur. In addition, other adherent cells such as those of the macrophage/monocyte series are present in the primary cultures; while the proportion of these cells may be lower after passage, under at least some culture conditions, they remain a significant proportion of the cells present [124]. As the process of adherence may take a few days, cultures are left undisturbed for 2–3 days after initiation of culture before moving dishes and changing culture medium. Once growth has occurred, cells can be subcultured by trypsin and plated into dishes at known cell densities for subsequent analyses.

Although these cultures have been used as a source of cells for the study of osteoblast function, bone marrow stromal cell cultures (particularly from rabbits but also from other species such as rats, mice and humans) have been of particular value for the study of questions related to the osteoblast lineage and that of connective tissue stem cells. When initially isolated, the cultures contain few, if any, identifiably osteoblastic cells. Differentiation outcome of marrow stromal cells has been analysed either by placing cells in diffusion chambers, which are implanted into animals, or by using colony-forming assays [7, 10–13]. For these assays *in vitro,* cells are plated at low density and kept in culture for periods of 2–4 weeks, over which time discrete colonies form whose phenotype can be identified by staining with suitable markers, such as AlP and osteocalcin for osteoblasts, alcian blue or toluidine blue and type II collagen and other cartilage markers for chondroblasts, and oil red O or other adipocyte markers for adipocytes. Alternatively, bone marrow stromal cells (at least from rats) can be used for bone nodule assays, which can be considered a form of colony-forming assay of osteoblast differentiation [14]. In these circumstances it is also necessary to ensure that the culture conditions will support differentiation. Osteoblastic differentiation of bone marrow stromal cells generally requires the use of dexamethasone as a supplement in the culture medium, and for bone-forming assays will additionally require the use of ascorbic acid and β-glycerophosphate. Other factors, such as the presence of 1,25 dihydroxyvitamin D_3, may also influence the ability of osteoblasts to differentiate *in vitro.*

A full protocol for the isolation of rat bone marrow stromal cells (Fig. 1.1) is given in Box 1.1. This protocol is also suitable for the isolation of marrow stromal cells from other small mammals, such as mice or rabbits. Although studies demonstrate

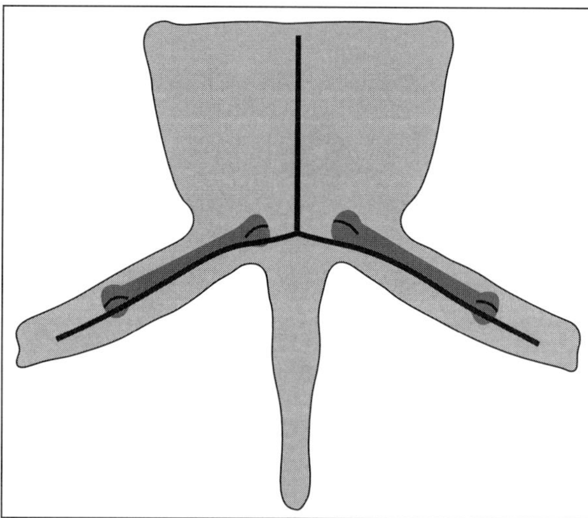

Figure 1.1   Dissection to expose rat femurs for isolation of bone marrow stromal cells. Skin incisions are made medially and extending over the legs, and the skin is reflected to expose femurs.

Box 1.1 Isolation and culture of rat bone marrow stromal cells [14]

Materials

Transport medium   Dulbecco's MEM with 25mM HEPES buffer; 500 units penicillin G/ml, 500 µg streptomycin/ml and 3 µg amphotericin B/ml.

Culture medium   Alpha-modified MEM with nucleotides; 15% fetal bovine serum; 50 units penicillin G/ml, 50 µg streptomycin/ml and 0.3 µg amphotericin B/ml; 50 µg ascorbic acid/ml, 10 mM Na β-glycerophosphate, 10–8M dexamethasone.

General   Simple dissecting kit including scalpels, tweezers, dissecting scissors; bone cutters (optional); culture plasticware; 10 ml syringes with 20 gauge hypodermic needles.

Animals   Young adult Wistar rats (100–120 g, approximately 4–6 weeks old).

Methods
 1. Rats killed by cervical dislocation or asphyxiation with CO_2, immobilized, abdomen facing upwards, on a dissecting board.
 2. Soak skin in 70% alcohol and then expose femurs by making incisions through skin over hindlegs and centrally up the abdomen (Fig. 1.1). Reflect skin by blunt dissection.
 3. Dissect major muscles away from femur and remove femurs intact. Place femurs in transport medium and keep refrigerated for not more than 1 hour prior to use.
 4. In a tissue culture hood, remove epiphyses of femurs with bone cutters (or with a scalpel) to expose medullary cavity of shafts of the bones.
 5. Expel bone marrow by ejecting a stream of medium into the medullary cavity of the femoral shaft using a 10 ml syringe fitted with a 20 gauge hypodermic needle. Collect bone marrow in a sterile tube and repeat as necessary until the bone shaft loses its red colour and takes on a translucent whitish appearance.
 6. Disperse bone marrow by repeated aspiration and expulsion through a 10 ml pipette and finally with a syringe and 20 gauge needle.
 7. Adjust concentration of dispersed marrow with culture medium and plate into 75 cm² flasks at a final concentration of the contents of 1–2 femurs/flask.
 8. Culture undisturbed for 3 days at 37°C and then change medium, removing unattached cells in the process. Continue to change medium every 2 days.
 9. Culture for 7 days from time of isolation and then release cells with trypsin/EDTA. (Cells often take 10 minutes or more to release from the flask into suspension.)
 10. For bone nodule assays, plate cells at density between 10^4 and 5×10^4 cells/dish and plate in 2 ml of medium into 30 mm culture dishes. Feed every two days and maintain cultures for 21 days.

evidence of at least limited self-renewal of the osteoprogenitor cells within these cultures, particularly in the presence of dexamethasone [26, 119], repeated subculture results in the loss of osteoblastic potential and thus experiments with these cells are normally done on passage 1 cells.

(a) Human bone marrow stromal cell cultures and selection of
osteoblastic cells from bone marrow
Most of the work to date on osteoblast cell lineage using marrow stromal cells has been carried out on cells derived from small mammals, but in recent years there have been a number of reports describing the successful establishment of marrow stromal cell cultures from human sources that can exhibit osteoblastic properties [12, 108]. These cultures have been established from human marrow aspirate biopsies using protocols similar to those described for rat cells, though these cells do not usually form true bone nodules. As with other species, analysis of differentiation by colony-forming assays shows greatly increased osteoblastic differentiation following the addition of dexamethasone supplements, and cells are also responsive to regulation with the osteotropic hormone 1,25 dihydroxyvitamin D_3 [12].

Further studies have described enrichment for cells of the osteoblast lineage from bone marrow on the basis of cell density and surface antigen characteristics. Cells able to express osteoblast markers have been enriched from human bone marrow by density gradient separation, absence of cell adherence and the absence of My10 (a surface antigen expressed by most haemopoietic progenitor cells) [92]. The relationship of these low-density/non-adherent cells and the adherent fibroblastic cells seen in other marrow stromal cell cultures is not at all clear at present, but despite their non-adherent properties they appear to be at a later stage of differentiation than the typical CFU-F seen in marrow stromal cell cultures. In another study, murine osteoblastic cells have been sorted from mouse bone marrow cells by flow cytometry using their expression of the Sca-1 surface antigen and their ability to bind wheat germ agglutinin. These manipulations do not result in pure populations, but amongst the populations isolated are osteoblastic cells that exhibit characteristics of later stage osteoblasts relative to classic CFU-F cells and this may be interpreted as demonstrating the presence of low numbers of committed osteoblastic cells within the bone marrow [66]. Such a hypothesis is also supported by studies that describe the isolation of later stage osteoblastic cell lines from bone marrow stromal cells [45, 125].

1.5.2 Explant cultures from mature bone

Osteoblastic cultures have been isolated by allowing outgrowth of cells from many donor sources, including human trabecular bone, rat femurs and various bovine and porcine bone sources [34, 109, 126, 127]. Technically, this method of cell isolation is relatively simple, though once again the cells will consist of a heterogeneous mix of osteoblastic cells at various stages of differentiation together with non-osteoblastic cells such as fibroblasts. It is not known what the effect of cell migration from the donor tissue is in relation to selecting cells of specific phenotype, but it is likely that the isolated cells will not mirror the overall proportions of different cell types initially present in the donor tissue.

Successful culture of cells with this technique generally depends firstly on cutting or mincing the original tissue into small fragments, which are placed in culture dishes in suitable medium; the tissue fragments are then left undisturbed for a sufficient period for cells to grow out and attach to the culture dish. This technique is now widely used for the establishment of human osteoblast cultures from trabecular bone samples removed during surgical procedures, such as samples from femoral heads removed during total joint replacement or mandibular bone removed during dento-alveolar surgery. A protocol for the establishment of human osteoblastic cell cultures from femoral bone explants is given in Box 1.2. A similar protocol has also been described with the addition of a collagenase digestion stage of the bone fragments, though we have not found this additional step to be of great significance. Alternatively, some authors have placed small explant pieces in

---

Box 1.2 Isolation of human osteoblastic cells by explant culture

Materials

Transport medium   Dulbecco's MEM with 25 mM HEPES buffer; 500 units penicillin G/ml, 500 µg streptomycin/ml and 3 µg amphotericin B/ml.

Culture medium   Alpha-modified MEM with nucleotides; 15% fetal bovine serum; 50 units penicillin G/ml, 50 µg streptomycin/ml and 0.3 µg amphotericin B/ml.

General   Tissue forceps and tweezers; bone cutters (rongeurs).

Methods
1. Obtain fresh bone sample in transport medium such as femoral head removed for total hip replacement surgery. (Femoral heads are in many ways ideal if there is a sympathetic orthopaedic surgeon, as they are large samples which are removed electively and do not usually require histopathological analysis.)
2. Store refrigerated until ready for preparation of cultures. (In principle it is best to process tissue as soon as possible, but we have successfully established cultures from tissue even left overnight before processing on occasions.)
3. Use bone cutters to remove fragments of trabecular bone from the tissue. (It may be necessary to cut some of the outer cortical bone to improve access to the inner trabecular bone.)
4. Mince bone fragments to pieces of approximately 3 mm or less.
5. Place in 75 cm^2 flasks and add 10 ml culture medium. (We put the tissue from one femoral head into a total of 4 flasks.)
6. Place in incubator at 37°C and leave undisturbed for at least 3 days.
7. Remove medium and replace twice weekly. During the first 2 weeks of culture try not to disturb bony fragments which settle on bottom of flask whilst moving and feeding cells.
8. Outgrowths of spindle-shaped cells become visible after 1 week or more. Cells grow relatively slowly and it may take 4 weeks or so to approach confluence, after which cells can be passaged by trypsin treatment.

fibrin clots to facilitate the adherence of pieces to the culture dishes; in these cases, cells can often be seen to migrate from the explant pieces along fibrin fibrils.

   Cells isolated in this way exhibit typical features of osteoblastic cells, including expression of AlP and type I collagen (without type III collagen in the case of human); they show a cAMP response to PTH stimulation and responsiveness to 1,25 dihydroxyvitamin D_3, and express bone matrix proteins such as osteocalcin and osteonectin [34, 38, 126]. Indeed, unlike many similar cultures, we have observed that even in mid log-phase cultures the cells stain positively for AlP histochemically, suggesting that these cells are preosteoblastic in phenotype with few less differentiated osteoprogenitor cells present. Using the protocol described, the cells retain their osteoblastic characteristics at least up to two passages when split at a ratio of 1 : 3, but by passage 3 they start to lose osteoblastic characteristics such as AlP activity and osteocalcin expression (unpublished observations).

1.5.3 Isolation of osteoblastic cells by enzymatic digestion of tissue

Enzymatic digestion of donor tissue in order to release cells for primary culture is widely used for cells of many different cell types. In the case of bone cells, the pioneering work of Wong and Cohn, who showed that it was possible to obtain cultures that were selectively enriched for cells of the osteoblastic lineage by sequential collagenase digestions of mouse calvariae, has been a major influence on methods for the establishment of primary osteoblastic cell cultures [128]. These methods may be particularly applicable to tissue derived from fetal sources exhibiting incomplete mineralization and have been widely used for obtaining cells from calvariae of neonatal or fetal rats, mice and chicks and to a lesser extent from older tissues from diverse species including rats, mice and human, porcine and bovine sources [33, 38, 77, 88, 90, 91, 109, 127].

(a) Fetal rat calvaria cells

As with other primary cultures, cells isolated by the enzymatic digestion of tissue consist of a heterogeneous mixture comprising osteoblastic cells at different stages of differentiation, together with cells of other cell lineages. By collecting cells released from the tissue at sequential time points it is possible to obtain populations which, though still a heterogeneous mix of cells, are selectively enriched or depleted for cells of the osteoblast lineage. In the case of rat calvarial cells, small variations are reported in the lengths of each sequential digestion that is used. However, using the protocol described in Box 1.3 and Fig. 1.2, cells released in the first and second 10-minute collagenase digestions (designated populations I and II) are relatively devoid of cells of the osteoblastic lineage, as judged by low levels of expression of AlP and other typical osteoblastic markers – even following stimulation with factors that might influence osteoblast commitment, such as dexamethasone or BMPs [23, 24, 88]. In long-term cultures, these cells often give rise to formation of myotubes and adipocytes in addition to the predominantly fibroblastic cells present. In contrast, cells released from the three subsequent 20-minute digestions (designated populations III, IV and V) are selectively enriched for cells of the osteoblastic lineage, as judged by their expression of high levels of AlP and bone matrix proteins such as osteocalcin and osteopontin, the adenylate cyclate responsiveness to PTH stimulation and the ability to produce mineralized

Box 1.3 Isolation of osteoblast-enriched cultures and establishment of bone nodule assays from fetal rat calvariae [88]

Materials

Transport medium Dulbecco's MEM with 25 mM HEPES buffer; 500 units penicillin G/ml, 500 μg streptomycin/ml and 3 μg amphotericin B/ml.

Culture medium Alpha-modified MEM with nucleotides; 15% fetal bovine serum; 50 units penicillin G/ml, 50 μg streptomycin/ml and 0.3 μg amphotericin B/ml; 50 μg ascorbic acid/ml, 10 mM Na β-glycerophosphate.

Collagenase solution

1.5 mg collagenase/ml (Sigma Type II)	100 mg
9.7 units DNAse/ml	0.25 mg
0.12 mM chondroitin sulphate	79 mg
100 mM sorbitol	1200 mg
111.2 mM KCl	500 mg
1.3 mM MgCl2	17.4 mg
13 mM glucose	170 mg
0.5 mM ZnCl2	4.5 mg
50 mM TRIS HCl buffer pH 7.4	66 ml

Filter-sterilize through a 0.45 μm pore filter. Place in 4 ml aliquots; stable for many months stored at –20°C.

Alizarin Red stain
1. Heat distilled water to 45°C.
2. Add 2 g Alizarin Red stain/100 ml.
3. Continuously stir whilst cooling. Allow to reach room temperature.
4. Adjust pH to 4.2 with NH_4OH.

Stable at room temperature.

General Tissue forceps; 2 pairs fine tweezers; 1 pair straight-pointed dissecting scissors (sharp); small magnetic flea; magnetic stirrer.

Animals Three to five 'date-mated' pregnant Wistar rats, killed by cervical dislocation to obtain 21-day-old fetuses (just pre-term).

Method
1. Collect fetus heads in transport medium, keep on ice.
2. Hold head with tweezers down side of face and make small nick with scissors over bridge of nose. Hold cut skin above nick with second pair of tweezers and reflect back to expose calvarium.
3. Make cut right through calvaria from external auditory meatus across behind parietal bones. Make second cuts forward to above eye sockets (Fig. 1.2).

4. Carefully remove calvaria and free from loosely adherent tissue. Store in PBS on ice.
5. When all calvaria are isolated, mince tissue with scissors. Rinse two or three times with PBS.
6. Remove PBS and add 4 ml collagenase solution preheated to 37°C (4 ml collagenase for approximately 30–40 calvariae). Incubate at 37°C for 10 minutes with continuous magnetic stirring.
7. Remove from incubator, let tissue fragments settle and remove supernatant. Add supernatant to 4 ml fetal bovine serum (FBS) on ice. (Designated as population I.)
8. Repeat with fresh collagenase for further 10 minutes (population II) and then three further periods of 20 minutes each (populations III–V).
9. Centrifuge supernatants at 1500 rpm for 5 minutes and resuspend cell pellets in 10 ml full medium. Plate each population into a T75 flask and incubate for 24 hours.
10. Wash thoroughly with PBS to remove dead cells and debris and trypsinize adherent cells. Pool populations III–V and resuspend in medium with freshly defrosted 50 μg vitamin C/ml and 10 mM Na β-glycerophosphate.
11. For bone nodule cultures, plate into 30 mm culture dishes at density approximately between 2×10^4 and 10^5 cells/dish in 2 ml medium.
12. Feed three times/week and culture for 21 days. Nodules should become macroscopically visible at around 10 days and will commence mineralization a few days after that.
13. After 21 days culture, wash cells with PBS and fix for a minimum of 5 minutes with 10% formal saline.
14. To demonstrate mineralization, stain with Alizarin Red (for calcium) or von Kossa (for phosphate) staining. For Alizarin Red, add stain for 5 minutes and wash with running tap water.
15. Quantification of nodule numbers can be made by direct counting of stained nodules, which has the merit of being able to distinguish between ectopic mineralization and true nodules. Where it is possible to ascertain that stained material represents true nodules, quantification can be carried out using a computer-assisted image analysis system, which has the advantage of ease and objectivity. For this, culture dishes can be transilluminated on a light-box and the image captured by video camera connected to the system.

bone nodules [88, 91, 100, 102, 103]. These cultures have been used extensively for studies of osteoblast lineage, changes in gene expression during differentiation and investigation of the action of many factors in regulating osteoblast metabolism. As an osteoblast culture model they are particularly valuable in that the cells undergo a clear and predictable pattern of differentiation that has been well characterized and in that their extensive use by researchers provides a large existing literature on their behaviour. Similar protocols have been employed to isolate cells from mouse and chick calvariae [76, 90, 129].

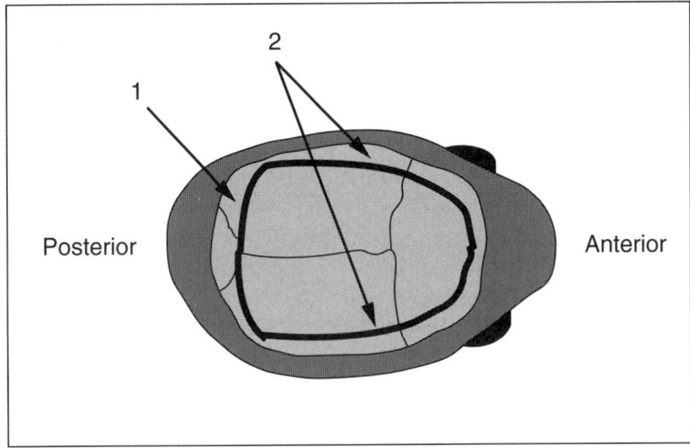

Figure 1.2 Isolation of rat calvaria. Following reflection of overlying skin, cuts are made (1) posteriorly and (2) laterally towards the midline. Calvaria can then be carefully removed.

(b) Bone nodule assays

As discussed briefly in section 1.2.1, the ability of cells to form mineralized bone *in vitro* has been described in a number of systems from various species including human, rat, mouse and chicken. Such systems provide a valuable model for the indirect assay of osteoblast differentiation, particularly in view of the difficulties in directly determining the presence of less differentiated osteoprogenitor cell populations, and the ease of demonstrating the presence of a mineralized matrix compared with many of the other biochemical assays of the osteoblast phenotype. When fetal rat cell populations isolated according to the protocol in Box 1.3 are maintained in medium supplemented with AA and βGP for up to 3 weeks, they produce discrete bone nodules in a predictable and reproducible manner (Fig. 1.3) [23, 88, 91, 102]. Following initiation of cultures, cells exhibit a normal logarithmic growth phase, and areas of cuboidal osteoblastic cells can be seen approximately 2–3 days after reaching confluence. Although the time scale for nodule formation may vary slightly according to initial plating density, at about 10 days after initiation of cultures multilayered nodules can be seen both microscopically and macroscopically as whitish condensations within the culture. After about 14 days the nodules become increasingly dense microscopically as mineralization occurs. Mineralization of the cultures is confined to the nodules and is not seen in the monolayer of cells surrounding the nodules.

The nature of these bone nodules has been extensively characterized to demonstrate that they resemble true osseous structures. The nodules are lined with plump cuboidal AlP-positive cells and osteocyte-like cells are embedded within the matrix. The cells associated with the nodules stain positively for collagen type I and the non-collagenous proteins including osteonectin, osteocalcin, bone sialoprotein and osteopontin; they also stain positively with the osteoblast/osteocyte-specific antibody E11 (FH, unpublished observations). The deposited matrix contains the same collagenous and non-collagenous proteins listed above, and the mineral deposited has been shown to consist of hydroxyapatite crystals [23, 88, 89, 103, 112].

The number of nodules formed in a culture shows a linear correlation with the total number of cells initially plated and more detailed limiting dilution analysis

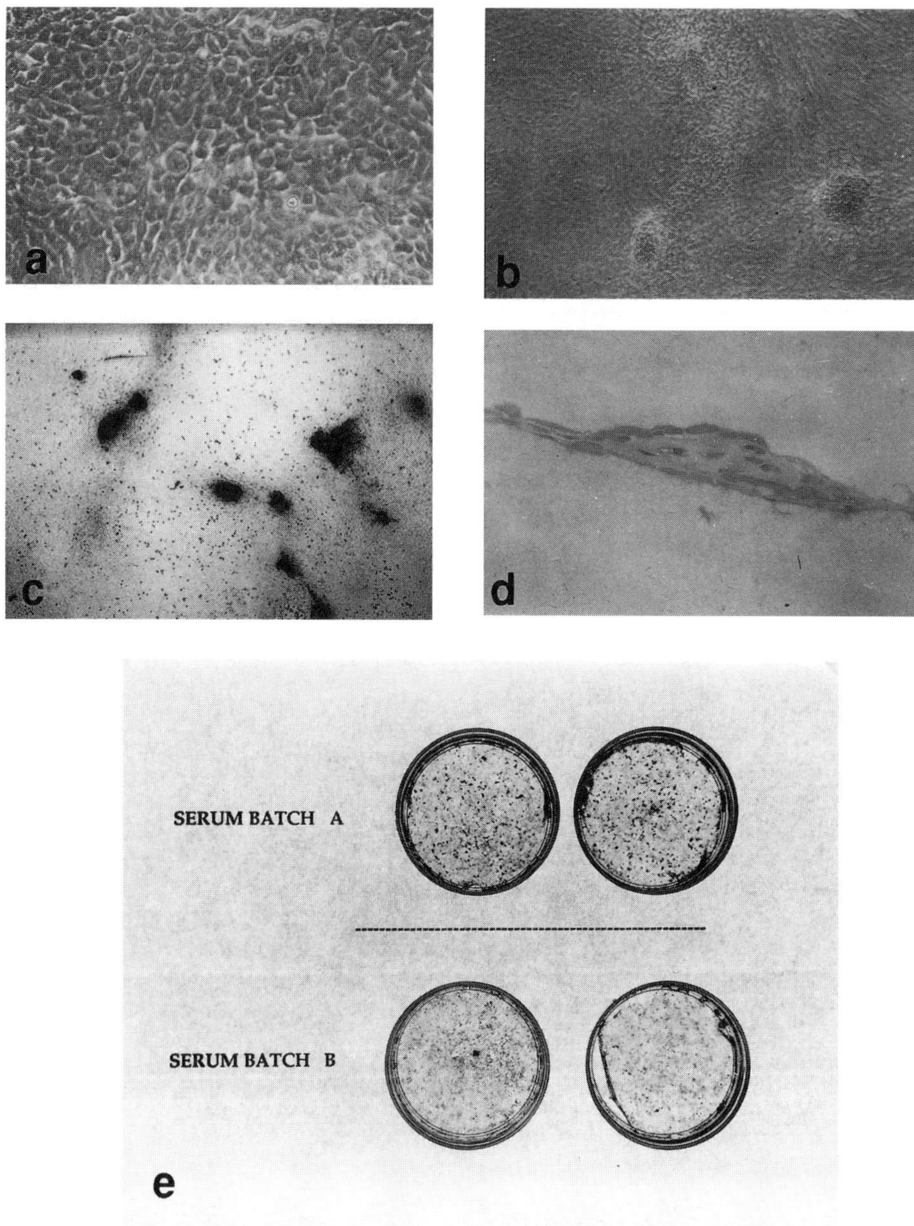

Figure 1.3   Bone nodule formation by osteoblast-enriched fetal rat calvaria cells. (a) Early condensation of cuboidal osteoblastic cells in confluent culture (approximately day 7); phase contrast. (b) Multi-layered bone nodules forming culture (approximately days 10–12); phase contrast. (c) Mineralized bone nodules in long-term cultures (day 21); bright field. (d) Histological section of bone nodule showing cuboidal osteoblasts on superficial surface of nodule – extensive bone matrix with osteocytes incorporated; haematoxylin and eosin. (e) Effect of different serum batches on bone nodule formation; mineralized bone nodules stained *in situ* with Alizarin Red. Original magnifications: (a) ×250; (b)–(d) ×100.

suggests that a single cell type is limiting for the formation of bone nodules within the cultures [27, 130]. Taken together, the findings suggest that each bone nodule is formed from the progeny of an osteoprogenitor cell type which is present in the original cultures at a frequency of approximately one in 200–300 cells. By measuring the average number of cells in each nodule it has been estimated that this osteoprogenitor cell undergoes approximately six to seven doublings during nodule formation under standard conditions [27]. These studies further suggest that nodule formation is essentially a colony-forming assay of osteoblast differentiation and can be used to assess the regulation of the differentiation of early stage osteoprogenitors *in vitro* simply by counting the number of bone nodules formed within a culture.

A wide range of hormones, including dexamethasone, 1,25 dihydroxyvitamin D_3, parathyroid hormone and cytokines, such as IL-1, IL-6, IL-11, TGFβ, EGF and BMPs, have been shown to regulate the number of bone nodules formed in rat calvaria cell cultures [23, 24, 41, 57, 104, 131–134]. Many of these factors have complex biphasic effects according to both the time point and total length of time that cultures are stimulated. These observations suggest that the responses depend on the cell differentiation stage from osteoprogenitor to mature osteoblast, and are consistent with the apparently diverse range of responses to external stimuli seen in studies using other methodologies. The role of dexamethasone in the regulation of bone nodule formation has attracted particular attention, as noted earlier. In general, increased bone nodule formation may be the result of either increased proliferation of additional progenitor cells or recruitment of less differentiated cells to proliferate and differentiate. In the case of dexamethasone, evidence suggests that the former does occur but that the increase in nodule number is at least in part due to the recruitment of earlier stage cells. Two distinct populations of cells were isolated by positive and negative immunoselection with an antibody to AlP; those cells expressing AlP constitutively formed bone nodules, whereas the AlP negative population required the presence of dexamethasone to form nodules [22]. In addition, although bone nodules form constitutively in rat calvaria cell cultures, dexamethasone is generally a requirement for nodule formation in rat bone marrow stromal cells [14, 25, 112, 135], but small numbers will form in its absence [124] or on stimulation with some other (ill-defined) osteoblast-derived factors [136]. Whether the action of other factors that stimulate nodule formation is due to recruitment of earlier stage cells to form nodules has not been rigorously demonstrated.

1.6 PROPERTIES OF ESTABLISHED CELL LINES

1.6.1 Multipotent cell lines

A number of permanent clonal cell lines have been isolated by limiting dilution methods which exhibit the capacity to give rise to a range of mature cell phenotypes, including osteoblasts, chondrocytes, adipocytes, fibroblasts and myoblasts, under appropriate culture conditions. These multipotent cell lines include the mouse embryonic fibroblast line C3H 10T1/2 clone 8, the rat calvaria cell lines RCJ 3.1 and ROB-C26, and the mouse mesodermally derived line C1. These lines may be of particular value in studying the events associated with cell commitment and in studying early stages of the osteoblast lineage. There are various general limitations with the use of such lines as models of multipotency and commitment from a putative multipotent mesenchymal stem cell which still need to be addressed. Firstly, demonstration of lineage ultimately depends on the cells expressing a sufficiently differentiated phenotype to express lineage-specific markers. However, it is possible that in any given set of culture conditions not all committed progenitor cells will show their potential to express a recognizable phenotype. In addition, the expression of a mature fibroblast cell phenotype cannot readily be demonstrated *in vitro* and at present might only be achieved using diffusion chamber experiments [20]. Secondly, it is difficult to demonstrate that the specific cells which give rise to one recognizable phenotype are the same as those which also give rise to other phenotypes. Thus, although the finding of a range of phenotypes within one clonally derived culture suggests their original derivation from a common multipotent cell, it does not directly demonstrate that factors which regulate the differentiation outcome of such cells were acting at the level of the multipotent cell rather than on cells of more restricted differentiation potential. Such issues might be addressed by more detailed analysis of subclones and to some extent by the demonstration of colonies (a type of subclone analysis) simultaneously expressing a range of mature phenotypes. Finally, the long-term growth of these rodent cells in culture results in their spontaneous immortalization and is associated with genomic alterations which may also have significant effects on their differentiation potentials.

The C3H 10T1/2 clone 8 line is apparently the first established mesenchymal cell line described to show multipotentiality [15]. Exposure of these murine embryonic fibroblast-like cells to 5-azacytidine (which blocks DNA methylation) results in the formation of differentiated tissues including myotubes, adipocytes and chondrocytes. Subclone analyses demonstrated the presence of populations of these cells which show tri-, bi- or monopotentiality, but only monopotential subclones remain stable in culture [21, 137, 138]. Although osteoblasts are not seen under these conditions, stimulation of 10T1/2 cells with BMP-2 or BMP-4 or with 10^{-6} M retinoic acid results in a marked increase in the expression of AlP and the formation of osteoblastic in addition to chondrocytic and adipocytic colonies [139–141]. In contrast to some other models of osteoblast commitment, these cells are not affected by dexamethasone stimulation [16, 139, 141]. 1,25 Dihydroxyvitamin D_3, TGFβ and IGF-I also failed to influence osteoblast commitment of these cells [141].

The RCJ 3.1 cell line is also capable of differentiating into osteoblast, chondrocyte, myoblast and adipocyte phenotypes but shows a number of differences from the C3H 10T1/2 cells [16, 20, 142]. RCJ 3.1 cells constitutively give rise to some differentiated myoblasts and osteoblastic cells, as judged by the ability to form bone nodules, and their expression of differentiated phenotypes is markedly upregulated by dexamethasone [16, 142]. Unlike C3H 10T1/2 cells, they are unresponsive to 5-azacytidine stimulation. Subclone analyses have again shown the presence of subpopulations of cells with tri-, bi- and monopotentials, but at least some of these subclones are stable in longer term culture [20].

The ROB-C26 line also expresses osteoblastic, chondrocytic and adipocytic phenotypes in the presence of dexamethasone [17]. Stimulation of C26 cells with both retinoic acid and BMP-2 increases osteoblastic differentiation, as judged by AlP expression, cAMP response to PTH and osteocalcin synthesis in the presence of 1,25 dihydroxyvitamin D_3 [143, 144]. In addition to its effects on osteoblast formation, BMP-2 simultaneously inhibits myotube formation in these cells. Subclone analyses with this cell line have not been reported.

The C1 SV40 transformed mesodermal line is able to make fat, cartilage and bone *in vitro*. When this line is grown under conditions in which cell aggregates form and then an inducer (e.g. dexamethasone) is added, either bone or cartilage or fat forms, rather than combinations of phenotypes. This has led to the suggestion of a different kind of lineage relationship in which the terminally differentiated cells are related to the multipotential percursor by a single commitment step influenced by cell density and regulatory factors [18, 19, 145, 146].

(a) Culture conditions for multipotent cell lines
These multipotential cell lines can be readily maintained in continuous culture with a simple medium such as Dulbecco's MEM or alpha-modified MEM with 10-15% fetal bovine serum. As discussed above, their analysis depends critically on their ability to express a more mature recognizable phenotype and this is profoundly influenced by the use of media supplements. For investigation of osteoblastic phenotype, culture medium can be additionally supplemented with ascorbic acid and β-glycerophosphate. As already described, commitment events may be influenced by dexamethasone (10^{-8} M), BMPs and retinoic acid (10^{-6} M) (retinoic acid is dissolved in dimethyl sulfoxide prior to addition to serum-free medium at a stock concentration of 10^{-4} M and stored at –20°C in single-use aliquots). In the case of BMPs there is evidence that concentration may differentially influence committed phenotype, low concentrations (10–50 ng/ml) selectively enhancing adipocytic phenotypes and high concentrations (100–500 ng/ml) enhancing osteoblast and chondrocyte phenotypes [139, 140, 147]. In addition, it is likely that the ability of committed cells to express their potential differentiated phenotype will also be influenced by factors that do not influence cell commitment *per se*, such as 1,25 dihydroxyvitamin D_3 and growth factors such as TGFβ and IGF-I [139, 144, 148]. It is also striking that some factors (e.g. RA) that stimulate commitment events in the multipotential progenitors may selectively inhibit the differentiation of lineage-restricted precursors, as shown for the chondrogenic C5.18 subclone derived from the multipotential RCJ 3.1 cell line.

1.6.2 Established osteoblastic cell lines

The general properties of established osteoblastic cell lines were discussed in section 1.3.2. The issue of how closely a cell line resembles the properties of its untransformed counterpart is of obvious importance in considering the characteristics of established lines. However, despite the general caveats given when considering these lines, much useful information has been derived from using such cultures. In this section some of the general properties of some of the more widely used cell lines are considered and some of the important properties of these cell lines are summarized in Table 1.3. In general, these cell lines are readily cultured in a medium consisting of Dulbecco's or alpha-modified MEM with 10% fetal bovine serum, though they may require further supplements such as AA and β-glycerophosphate to express their full repertoire of osteoblastic features.

(a) Spontaneously immortalized rodent cell cultures
The MC3T3-E1 cell line has become one of the most utilized of all osteoblastic cell lines. This line was originally obtained from the cloning of cells isolated by collagenase digestion of newborn mouse calvariae [149]. In many respects it appears to possess many of the properties of primary osteoblastic cultures. During logarithmic growth the cells express low levels of AlP and other differentiated osteoblastic

Table 1.3 Some phenotypic characteristics of commonly used established osteoblastic cell lines

Characteristic	MC3T3-E1	ROS17/2.8	UMR106.06	HOS/TE-85	MG-63	SaOS-2
Bone formation *in vivo*	ND	+	+	ND	ND	+
Mineralized matrix formation *in vitro*	+	-	-	-	-	+
Alkaline phosphatase activity	+	+	+	-	-*	+
cAMP response to PTH stimulation	+	+	+	+	+	+
1,25 dihydroxyvitamin D_3 response or receptor	+	+	+	+	+	+
Type I collagen synthesis	+	+	+	+	+	ND
Type III collagen synthesis	+	+	-	ND	+	ND
Osteocalcin synthesis	+	+	-	-	-*	-
Osteonectin	ND	+	ND	ND	ND	+
Osteopontin	+	+	ND	ND	ND	ND

ND = not done
* Both AlP activity and osteocalcin are absent or very low in MG-63 cells but can be induced by 1,25 dihydroxyvitamin D_3 stimulation.

markers. Stationary phase cultures show marked up-regulation of AlP, osteocalcin synthesis and a cAMP response to PTH, and can produce a mineralized bone-like matrix in long-term cultures [150]. Although the effects of dexamethasone have not been studied extensively with this line, it has been shown that dexamethasone acts synergistically with BMP-2 to increase AlP expression [122]. Taken together, these findings suggest a cell line which undergoes a predictable pattern of differentiation from osteoprogenitor to osteoblast to embedded osteocyte, while its immortal nature provides a continuous source of progenitors. The role of ascorbic acid and β-glycerophosphate supplements in supporting the growth and differentiation of MC3T3-E1 cells has been extensively described (section 1.4.3). Ascorbic acid increases proliferation of logarithmic phase cells and greatly increases AlP expression in post-confluent cultures by its action on matrix maturation [85, 110, 111, 113, 114]. Addition of βGP may act synergistically with ascorbic acid to increase AlP expression further and is required for mineralized tissue formation [113, 150].

The MBA-15 cell line (and its subclones) is an osteoblastic line derived from mouse bone marrow stroma which, although not as yet used extensively by many workers, possesses many interesting properties suggesting its potential value as an osteoblastic model [125]. The parent MBA-15 line expresses markers of the osteoblast phenotype including AlP, osteonectin, osteopontin, cAMP response to PTH and to PGE_2, responsive to dexamethasone and 1,25 dihydroxyvitamin D_3, and formation of bone *in vivo* and a mineralized matrix *in vitro* [125, 151, 152]. Of particular interest is the isolation of phenotypically stable clonal subpopulations from this line with distinct phenotypic expression patterns suggestive of both early (MBA-15.33, MBA-15.4) and later stage (MBA-15.6) cells of the osteoblast lineage [45]. The demonstration of differential effects of hormones and growth factors on these different cell lines suggests that they may be of particular value in addressing questions of the influence of differentiation stage on responses to regulatory stimuli [45, 151, 153].

(b) Lines derived from osteosarcomas
As mentioned briefly above, and as exemplified by the descriptions of properties outlined below, there are a number of osteoblastic lines originally derived from osteosarcomas. While extremely valuable for many studies, the cells do display aberrant proliferation and gene expression profiles, including, for example, expression of post-proliferation markers in proliferating cells and co-expression of genes not normally observed together. Several examples of these are given below.

Rat osteosarcoma cell lines   The rat osteosarcoma cell line ROS 17/2.8 is a widely used osteoblastic model that expresses features of a late stage osteoblastic cell. These include high AlP activity, cAMP response to PTH and synthesis of osteo-calcin, osteopontin, bone sialoprotein and osteonectin [120, 121, 154–156]. The cells are responsive to dexamethasone and 1,25 dihydroxyvitamin D_3 and show increased expression of AlP during time in culture, which suggests that the cells undergo partial differentiation [121, 156]. However, the cells do not appear to differentiate terminally *in vitro* with deposition of a mineralizable matrix and, as noted earlier, the normal relationship between proliferation and differentiation may be lost in these cells [105, 106].

The UMR 106.06 is one of four subclones isolated from the parent UMR 106 osteosarcoma cell line. These cells have high AlP activity and cAMP response to PTH and form osteogenic sarcomas *in vivo* [36]. They express bone sialoprotein and osteopontin, respond to dexamethasone and 1,25 dihydroxyvitamin D_3, but not do express osteocalcin [120, 154, 155, 157]. With continuous culture these cells have also been found to exhibit responsiveness to calcitonin [158–160]. These features together might be considered as combining both very early and very late stage osteoblastic markers [2].

Human osteosarcoma cell lines   A number of cell lines have been derived by cloning of cells from human osteosarcomas, including (amongst others) SaOS-2, HOS/TE-85 and MG-63. The SaOS-2 line has many features of a well differentiated osteoblastic cell, including constitutive expression of high levels of (BLK) AlP, cAMP response to PTH stimulation, and expression of osteonectin, bone sialoprotein and decorin [35, 39, 161–163]. It can form bone when implanted *in vivo* and a mineralized matrix in long-term cultures *in vitro* [39, 162]. However, despite the presence of functional receptors, 1,25 dihydroxyvitamin D_3 or glucocorticoid stimulation reportedly has little effect on AlP expression or cAMP response to PTH [35, 39]. In addition, these cells do not appear to express osteocalcin at any stage of their culture [164].

The MG-63 cell line has fewer characteristics of a mature osteoblast than SaOS-2 cells and was originally isolated from a human osteosarcoma because of its production of high quantities of interferon-γ [165]. Its mixture of features make it particularly difficult to assign a putative differentiation stage along the osteoblast lineage to it. Constitutively, the cells have very low AlP activity, do not express osteocalcin and do not form a mineralized matrix *in vitro*, but 1,25 dihydroxyvitamin D_3 stimulates AlP and osteocalcin synthesis [58, 156, 164, 166–168]. The cells show a cAMP response to PTH stimulation and treatment with dexamethasone has a small potentiating effect of 1,25 dihydroxyvitamin D_3 on both AlP and PTH responsiveness [168]. The osteoblastic nature of these cells has been questioned by the observations that relatively low levels of total collagen are produced and that up to 40% of that produced is type III collagen [169, 170]. Interestingly, a subpopulation of these cells has been isolated (MG-63.3A) by their resistance to detachment with RGD-containing peptides and which has a markedly more osteoblastic phenotype. MG-63.3A cells show increased production of type I collagen; they show AlP expression and form a mineralized matrix *in vitro* [167].

The HOS/TE-85 line is another widely used cell line which has been relatively poorly characterized with respect to its osteoblastic features. It has very low AlP activity which can be up-regulated by 1,25 dihydroxyvitamin D_3 and dexamethasone. The cells do not express osteocalcin [35, 58, 161, 171]. Recent evidence has suggested that some of the detectable AlP from this cell line may be of the placental AlP isoform, rather than the expected BLK isoform [171].

Comparative studies of these three cell lines and primary human osteoblasts have also revealed differences in expression of IGFI and II, IGF-binding proteins and integrin expression, which are cell-line specific [58, 172, 173] and which raise further questions about how closely any of these cell lines represent a specific stage of osteoblastic cell development rather than an altered phenotype as a result of their transformation.

(c) Virally immortalized cell lines
The use of viral transformation for establishment of permanent cell lines has been extensively employed in cell biology (e.g. [4] for discussion). Cells are permanently transfected with the (simian virus) SV40 large T-antigen and this may have the effect of permanently 'freezing' proliferating cells at a specific differentiation state, while allowing later differentiation and maturational events to be observed under appropriate culture conditions. This approach has been used to establish a number of osteoblastic cell lines such as the rat RCT-1 and RCT-3 lines and the human HOBIT line [2, 98, 174]. The HOBIT cells exhibit a wide range of osteoblastic markers including AlP, osteocalcin, osteopontin and osteonectin expression, cAMP response to PTH, and formation of a mineralized matrix *in vitro* [174]. However, in view of the abrogation of growth control that occurs during transformation and the unknown effects of the transformation process on phenotypic stability, it is not possible to place such immortalized cells definitively at a particular stage of differentiation or to be confident that a given characteristic is genuinely representative of a normal cell type.

In an attempt to address this issue and to provide a renewable source of osteoblastic cells, recent studies have described the conditional immortalization of cells where the SV40 large T-antigen can be inactivated to allow normal function to be resumed without the influence of the transforming agent. In these studies cells have been stably transfected with a temperature-sensitive mutant of SV40 large T-antigen which is active when cells are cultured at the permissive temperature (33°C), but when transferred to a restrictive temperature (39°C) the T-antigen is inactivated and cells may exhibit normal function and differentiation [175, 176]. Early reports of cell lines using this approach look promising and it will be interesting to see how valuable this approach will prove in the longer term.

1.7 ROUTINE ASSAYS OF OSTEOBLAST MARKERS

It will be apparent from the earlier discussion that it is neccessary to measure a range of osteoblastic markers when characterizing a culture. A full description of methods for the characterization of osteoblastic cells is beyond the scope of this chapter but some of the frequently used methods are listed in Table 1.4.

Some of these methods can be carried out using commercially available reagents or assay kits, which are usually supplied with detailed protocols. Most suppliers can also provide further technical information on the use of these reagents. Some additional specific information is provided here which may be useful for the investigator new to osteoblast biology.

1.7.1 Alkaline phosphatase

Demonstration of AlP activity is probably the most frequently used assay in characterizing osteoblast function. Histochemical demonstration of AlP staining and measurement of total AlP activity are simple techniques to perform which depend on the release of a coloured reaction product by the action of the AlP enzyme. A protocol for the histochemical demonstration of AlP by azo-dye coupling is given in Box 1.4.

Measurement of total AlP activity can be performed using the protocol described in Box 1.5. This method relies on the conversion of the (colourless) p-nitrophenyl phosphate substrate to p-nitrophenol (PNP – yellow) + inorganic phosphate. The amount of PNP released over a timed period is determined by comparison with a set of PNP standards of known concentration and is controlled by simultaneously determining the amount of total protein. Total AlP activity is usually expressed as PNP released (μM/minute) per milligram of total protein. Both of these assays measure total AlP activity and do not distinguish between the BLK isoform and other isoforms of the enzyme. Where AlP isoform-specific assays are required, this can be achieved by immunohistochemistry, electrophoretic separation of isoforms or measurement of mRNA expression.

Table 1.4 Some methods commonly used for determination of osteoblastic markers

Marker	Method	Notes
Alkaline phosphatase	Histochemistry	Described in Box 1.4
	Total enzyme activity	Described in Box 1.5
	Immunohistochemistry	BLK isoform specific
	Northern blot / RT PCR/ISH	Total/localization of mRNA
PTH receptor/	RIA/EIA for cAMP	Kit from Amersham International
responsiveness	Northern blot / RT PCR/ISH for PTH/PTHrP receptor mRNA	Total/localization of mRNA
Osteocalcin	Radioimmunoassay	Total osteocalcin; human specific
	Immunohistochemistry	Commercial Ab from biogenesis
	Northern blot / RT PCR/ISH	Total/localization of mRNA
Other matrix proteins	Immunohistochemistry	Localization of protein
	Western blot	Total protein
	Northern blot / RT RCR/ISH	Total/localization of mRNA
Mineralization	Von Kossa / Alzarin Red staining	Section 1.5.3
	^{45}Ca uptake	Total Ca deposition
	Electron probe (EDAX)	Determination of Ca : P ratios
	X-ray diffraction	Determination of Ca/P structure
Bone formation	Bone nodule assays	Section 1.5.3
	Diffusion chamber implantation	

Box 1.4 Histochemical determination of alkaline phosphatase staining [180]

Materials

Stock solution   0.2 M TRIS HCl buffer, pH 9.0.

Substrate solution   Dissolve 60 mg naphthol AS phosphate in 1 ml dimethyl formamide. Add slowly to 200 ml buffer. (Check pH at 9.0.) Store refrigerated; stable for up to 6 months.

Method
1. Fix cells in formalin for minimum of 10 minutes. Staining is best carried out immediately after fixing.
2. Wash cells with TRIS HCl buffer.
3. Add 20 mg Fast Blue BB to substrate solution, and filter.
4. Incubate cells with filtered solution for up to 1 hour.
5. Wash thoroughly with tap water.
6. If mounting slides with coverslip, do not use xylene-based mountants; use an aqueous mountant.

1.7.2 Hormone and cytokine responsiveness

Determination of responsiveness to hormones, cytokines and other signalling molecules can, in principle, be determined by measuring the specific response of a cell to stimulation or can be inferred by the demonstration of receptors for that molecule, either by binding studies or indirectly by showing the expression of mRNA for the receptor molecule (though the presence of the mRNA does not necessarily imply the presence of functional receptors).

As already discussed, an adenylate cyclase response of osteoblasts to PTH is widely used as a marker of the osteoblast phenotype by measurement of cAMP production. As the main fragment of the PTH molecule that reacts with the PTH/PTHrP receptor is at the N-terminal end, cells are stimulated with either intact PTH or 1-34 PTH fragment in serum-free medium. Following an appropriate incubation period (e.g. 15 minutes), cAMP is measured; there are now commercially available radioimmunoassay or (non-radioactive) enzyme immunoassay kits that have made this assay very routine. These kits come with a comprehensive protocol for collection and analysis of samples. A similar protocol can be used for measurement of cAMP responses to signalling molecules which react with other 7-transmembrane domain-containing cell surface receptors such as PGE_2 binding to the EP2 prostaglandin receptor [46].

1.7.3 Bone matrix proteins

Collagen synthesis can be measured by a number of relatively simple methods which include determination of uptake of ^{3}H-labelled proline and measurement of collagenase-digestible total protein (or a combination of both methods). Although these

---

Box 1.5 Determination of total alkaline phosphatase activity

Materials
0.05 M TRIS HCl buffer, pH 7.4
0.5 M NaOH
p-nitrophenol (PNP)
Substrate solution, available as kit from Sigma Chemical Company. (Alternatively, can be made from the following for each 96-well plate: 8.4 ml distilled water; 2.4 ml reaction buffer Sigma No. 221; 30 mg *p*-nitrophenylphosphate (Sigma No 104); 2.7 mg $MgCl_2$.)
Protein concentration measurement kit (e.g. from BioRad, Sigma)
Cell sonicator
96-Well spectrophotometric plate reader.

Method
1. Wash (unfixed) cells in TRIS buffer. Scrape cells into buffer with a rubber cell scraper.
2. Sonicate cells for 20 seconds on ice.
3. Centrifuge sonicates for 30 minutes at 2000 rpm and collect supernatant. Assay immediately or freeze at $-20°C$.
4. Prepare series of PNP standards by dilution in TRIS buffer (0.025–1 µM/ml) and add 150 µl to duplicate wells in a 96-well microtitre plate.
5. Add 50 µl of supernatants to wells and incubate with 100 µl of substrate solution.
6. Incubate at 37°C until a yellow colour develops (typically between a few minutes and up to 1 hour). Stop reaction with 100 µl NaOH.
7. Read absorption at 405 nm on spectrophotometric plate reader.
8. Determine protein concentration of supernatants in parallel to AlP according to supplier's protocol.

---

methods give an index of change in collagen synthesis, neither of them gives a true account of total synthesis as much of the pro-collagen synthesized is not ultimately deposited in the insoluble matrix [177]. In addition, both of these methods estimate total collagen and do not distinguish between type I and other collagen types, which is often achieved by measurement of expression of mRNA for alpha I (1) pro-collagen, using Northern blot analyses. Where required, other assays for specific collagen types and detailed kinetic determinations can be done.

Determination of serum osteocalcin has been shown to be a useful marker for the clinical assessment of bone turnover [178]. Because of this there are a number of commercially available radioimmunoassay kits which can be used to measure total osteocalcin in culture supernatants. These assay kits tend to be specific for the assay of human osteocalcin and do not crossreact with (say) rat or mouse osteocalcin, though individual laboratories have prepared other osteocalcin antibodies useful for these species. In carrying out such assays, one of us (FH) has obtained more reliable results by adding the protease inhibitor apropotin (Trazylol, Sigma Chemical Company) to samples on collection, following which samples can be stored frozen at –20°C without measurable loss of activity. Polyclonal antisera to human and rat osteocalcin are also available commercially for immunohistochemical detection of osteocalcin. Further details on the measurement of markers of bone synthesis are described in Chapter 9.

Although there has been much research into the nature of the non-collagenous proteins of bone, in general reagents for their identification are not commercially available. However, samples of appropriate antisera and cDNAs will often be generously given by other workers on request. Notable in this regard are the laboratories of the Bone Research Branch of the National Institute of Dental Research, NIH. This group has been in the forefront of work describing the non-collagenous bone matrix proteins and has an extensive range of polyclonal antisera and cDNAs available for research studies. (For a full list of these reagents, see [179].)

Non-specific binding and high background are sometimes a problem with immunohistochemistry and appropriate controls need to be carefully employed to ensure reliable results, even when using apparently well described protocols supplied with the antisera. Depending solely on the omission of the primary antibody is not always a reliable control because of the amplification that occurs in second antibody and binding detection stages. The use of pre-immune serum or antiserum preabsorbed with its antigen are good controls. In addition it is useful to have a negative control cell line (e.g. skin fibroblasts for osteocalcin) and a positive control (a known positive cell line) wherever possible. Further details of the techniques of immunocytochemistry as applied to bone are provided in Chapter 8.

1.8 CONCLUSIONS

Technically, the culture of osteoblastic cells has become relatively simple and can be readily achieved using methods common to all anchorage-dependent cell cultures. However, the general interpretation of data obtained with these methods can be difficult in view of the many variables described in this chapter. Critical factors that may influence these results include the heterogeneity of cultures, differentiation stages of cells, source and species of cells, culture conditions and, in the case of established cell lines, possible alterations from a 'normal' phenotype. The researcher new to osteoblast cell culture might be bewildered or discouraged by this range of cell culture systems and their obvious imperfections, but this has certainly not been our intention in writing this chapter. Rather, we would hope to give encouragement by the many insights in bone biology that have been obtained using these models, albeit imperfect. In adopting such methods for the testing of hypotheses we would hope that investigators use caution in extrapolating observations from one system to another and the situation *in vivo*.

APPENDIX 1.A COMMERCIAL SUPPLIERS OF REAGENTS

Most reagents described are readily obtainable from many suppliers, and information on suppliers of some specific reagents is provided here. In each case the UK supplier's details are given but most have agents in many countries.

Alpha-modified MEM with nucleotides
Sigma Chemical Company Ltd
Fancy Road
Poole
Dorset BH12 4QH.
Tel. 01800 373731
Fax 01800 378785

or Gibco BRL Ltd
Paisley PA4 9RF

Human osteocalcin radioimmunoassay kit
CIS Bio International
Gif-sur-Yvette
France.

Polyclonal osteocalcin antiserum
Biogenesis Ltd
New Fields
Stinsford Road
Poole
Dorset
BH17 0NF
Tel. 01202 660006
Fax 01202 660020

cAMP radioimmunoassay and enzyme immunoassay kits
Amersham International plc
Amersham Place
Little Chalfont
Buckinghamshire
HP7 9NA
Tel. 01494 544000
Fax 01494 542266

REFERENCES

1. Nijweide. P.J., Burger. E.H. and Feyen, J.H. (1986) Cells of bone: proliferation, differentiation, and hormonal regulation. *Physiological Reviews* **66**, 855–886.
2. Rodan, G.A., Heath, J.K., Yoon, K. *et al.* (1988) Diversity of the osteoblast phenotype, in *Cell and Molecular Biology of Vertebrate Hard Tissues*, (eds D. Evered and S. Harnett), John Wiley & Sons, Chichester, pp. 78–85.
3. Aubin, J.E., Turksen, K. and Heersche, J.N.M. (1993) Osteoblastic cell lineage, in *Cellular and Molecular Biology of Bone*, (ed. M. Noda), Academic Press Inc., San Diego, pp. 1–45.
4. Freshney, R.I. (1983) *Culture of Animal Cells: a Manual of Basic Technique*, Alan R. Liss Inc., New York.
5. Owen, M. (1967) Uptake of [3H] uridine into precursor pools and RNA in osteogenic cells. *Journal of Cell Science* **2**, 39–56.
6. Owen, M. (1970) The origin of bone cells. *International Review of Cytology* **28**, 213–238.

7. Owen, M. and Friedenstein, A.J. (1988) Stromal cells: marrow-derived osteogenic precursors, in *Cell and Molecular Biology of Vertebrate Hard Tissues*, (eds D. Evered and S. Harnett), John Wiley & Sons, Chichester, pp. 42–53.

8. Wolpert, L. (1988) Stem cells: a problem in assymetry. *Journal of Cell Science* **10**, 1–19.

9. Nakahata, T., Gross, A.J. and Ogawa, M. (1982) A stochastic model of self renewal and commitment to differentiation of the primitive hemopoietic stem cells in culture. *Journal of Cellular Physiology* **113**, 455–458.

10. Friedenstein, A.J., Chailalhyan, R.K. and Gerasimov, U.V. (1987) Bone marrow osteogenic stem cells: in vitro cultivation and transplantation in diffusion chambers. *Cell and Tissue Kinetics* **20**, 263–272.

11. Friedenstein, A.J. (1976) Precursor cells of mechanocytes. *International Reviews in Cytology* **47**, 327–359.

12. Beresford, J.N., Joyner, C.J., Devlin, C. and Triffitt, J.T. (1994) The effects of dexamethasone and 1,25-dihydroxyvitamin D_3 on osteogenic differentiation of human marrow stromal cells in vitro. *Archives of Oral Biology* **39**, 941–947.

13. Beresford, J.N., Bennett, J.H., Devlin, C. *et al.* (1992) Evidence for an inverse relationship between the differentiation of adipocytic and osteogenic cells in rat marrow stromal cell cultures. *Journal of Cell Science* **102**, 341–351.

14. Maniatopoulos, C., Sodek, J. and Melcher, A.H. (1988) Bone formation *in vitro* by stromal cells obtained from bone marrow of young adult rats. *Cell Tissue Research* **254**, 317–330.

15. Reznikov, C.A., Brankow, D.W. and Heidelberger, C. (1973) Establishment and characterisation of a cloned line of C3H mouse embryo cells sensitive to postconfluence inhibition of division. *Cancer Research* **33**, 3231–3238.

16. Grigoriadis, A.E., Heersche, J.N.M. and Aubin, J.E. (1988) Differentiation of muscle, fat, cartilage and bone from progenitor cells present in a bone derived clonal cell population: effects of dexamethasone. *Journal of Cell Biology* **106**, 2139-2151.

17. Yamaguchi, A. and Kahn, A.J. (1991) Clonal osteogenic cell lines express myogenic and adipocytic developmental potential. *Calcified Tissue International* **49**, 221–225.

18. Kellermann, O., Buc-Caron, M.H., Marie, P.J. *et al.* (1990) An immortalized osteogenic cell line derived from mouse teratocarcinoma is able to mineralize in vivo and in vitro. *Journal of Cell Biology* **110**, 123–132.

19. Poliard, A., Lamblin, D., Marie, P.J. *et al.* (1993) Commitment of the teratocarcinoma-derived mesodermal clone C1 towards terminal osteogenic differentiation. *Journal of Cell Science* **106**, 503–511.

20. Grigoriadis, A.E., Heersche, J.N. and Aubin, J.E. (1990) Continuously growing bipotential and monopotential myogenic, adipogenic, and chondrogenic subclones isolated from the multipotential RCJ 3.1 clonal cell line. *Developmental Biology* **142**, 313–318.

21. Konieczny, S.F. and Emerson, C.P. Jr (1984) 5-Azacytidine induction of stable mesodermal stem cell lineages from 10T1/2 cells: evidence for regulatory genes controlling determination. *Cell* **38**, 791–800.

22. Turksen, K. and Aubin, J.E. (1991) Positive and negative immunoselection for enrichment of two classes of osteoprogenitor cells. *Journal of Cell Biology* **114**, 373–384.

23. Bellows, C.G., Aubin, J.E. and Heersche, J.N. (1987) Physiological concentrations of glucocorticoids stimulate formation of bone nodules from isolated rat calvaria cells in vitro. *Endocrinology* **121**, 1985–1992.

24. Hughes, F.J., Collyer, J., Stanfield, M. and Goodman, S.A. (1995) The effects of bone morphogenetic protein-2, -4 and -6 on differentiation of rat osteoblast cells *in vitro*. *Endocrinology* **136**, 2671–2677.

25. McCulloch, C.A.G., Strugurescu, M., Hughes, F. *et al.* (1991) Osteogenic progenitor cells in rat bone marrow stromal populations exhibit self-renewal in culture. *Blood* **77**, 1906–1911.

26. Kamalia, N., Mcculloch, C.A.G., Tenebaum, H.C. and Limeback, H. (1992) Dexamethasone recruitment of self-renewing osteoprogenitor cells in chick bone marrow stromal cell cultures. *Blood* **79**, 320–326.

27. Bellows, C.G., Heersche, J.N. and Aubin, J.E. (1990) Determination of the capacity for proliferation and differentiation of osteoprogenitor cells in the presence and absence of dexamethasone. *Developmental Biology* **140**, 132–138.

28. Aubin, J.E., Gupta, A., Zirngibl, R. and Rossant, J. (1995) Bone sialoprotein knockout mice have bone abnormalities. *Bone* **17**, 558 (Abstr. 3a).

29. Ducy, P., Desbois, C., Boyce, B. *et al.* (1996) Increased bone formation in osteocalcin-deficient mice. *Nature* **382**, 448–452.

30. Young, M.F., Ibaraki, K., Kerr, J.M. and Heegaard, A.-M. (1993) Molecular and cellular biology of the major noncollagenous proteins in bone, in *Cellular and Molecular Biology of Bone* (ed. M. Noda), Academic Press Inc., San Diego, pp. 191–234.

31. Fisher, L.W. (1985) The nature of the proteoglycans of bone, in *The Chemistry and Biology of Mineralised Tissues* (ed. W.T. Butler), EBSCO Media, Birmingham, Alabama, pp. 188–196.

32. Noda, M., Yoon, K., Rodan, G.A. and Koppel, D.E. (1987) High lateral mobility of endogenous and transfected alkaline phopshatase: a phosphatidylinositol anchored membrane protein. *Journal of Cell Biology* **105**, 1671–1677.

33. Modrowski, D. and Marie, P.J. (1993) Cells isolated from the endosteal bone surface of adult rats express differentiated osteoblastic characteristics in vitro. *Cell and Tissue Research* **271**, 499–505.

34. Beresford, J.N., Gallagher, J.A., Poser, J.W. and Russell, R.G. (1984) Production of osteocalcin by human bone cells in vitro. Effects of $1,25(OH)_2D_3$, $24,25(OH)_2D_3$, parathyroid hormone, and glucocorticoids. *Metabolic Bone Disease and Related Research* **5**, 229–234.

35. Murray, E., Provvedini, D., Curran, D. *et al.* (1987) Characterization of a human osteoblastic osteosarcoma cell line (SaOS-2) with high bone alkaline phosphatase activity. *Journal of Bone and Mineral Research* **2**, 231–238.

36. Partridge, N.C., Alcorn, D., Michelangeli, V.P. *et al.* (1983) Morphological and biochemical characterization of four clonal osteogenic sarcoma cell lines of rat origin. *Cancer Research* **43**, 4308–4314.

37. Rao, L.G., Ng, B., Brunette, D.M. and Heersche, J.N.M. (1977) Parathyroid hormone and prostaglandin E_1 response in a selected population of bone cells after repeated subculture and storage at –80°C. *Endocrinology* **100**, 1233–1241.

38. Robey, P.G. and Termine, J.D. (1985) Human bone cells in vitro. *Calcified Tissue International* **37**, 453–460.

39. Rodan, S.B., Imai, Y., Thiede, M.A. *et al.* (1987) Characterization of a human osteosarcoma cell line (Saos-2) with osteoblastic properties. *Cancer Research* **47**, 4961–4966.

40. Akatsu, T., Takahashi, N., Udagawa, N. *et al.* (1991) Role of prostaglandins in interleukin-1-induced bone resorption in mice in vitro. *Journal of Bone and Mineral Research* **6**, 183–189.

41. Ellies, L.G. and Aubin, J.E. (1990) Temporal sequence of interleukin 1 alpha-mediated stimulation and inhibition of bone formation by isolated fetal rat calvaria cells in vitro. *Cytokine* **2**, 430–437.

42. Evans, D.B., Thavarajah, M. and Kanis, J.A. (1990) Involvement of prostaglandin E_2 in the inhibition of osteocalcin synthesis by human osteoblast-like cells in response to cytokines and systemic hormones. *Biochemical and Biophysical Research Communications* **167**, 194–202.

43. Saito, S., Ngan, P., Rosol, T. *et al.* (1991) Involvement of PGE synthesis in the effect of intermittent pressure and interleukin-1 beta on bone resorption. *Journal of Dental Research* **70**, 27–33.

44. Schwartz, Z., Dennis, R., Bonewald, L. *et al.* (1992) Differential regulation of prostaglandin E2 synthesis and phospholipase A2 activity by $1,25\text{-}(OH)_2D_3$ in three osteoblast-like cell lines (MC-3T3-E1, ROS 17/2.8, and MG-63). *Bone* **13**, 51–58.

45. Fried, A., Benayahu, D. and Wientroub, S. (1993) Marrow stroma-derived osteogenic clonal cell lines: putative stages in osteoblastic differentiation. *Journal of Cellular Physiology* **155**, 472–482.

46. Kasugai, S., Oida, S., Iimura, T. *et al.* (1995) Expression of prostaglandin E receptor subtypes in bone: expression of EP2 in bone development. *Bone* **17**, 1–4.

47. Scutt, A. and Bertram, P. (1995) Bone marrow cells are targets for the anabolic actions of prostaglandin E2 on bone: induction of a transition from nonadherent to adherent osteoblast precursors. *Journal of Bone and Mineral Research* **10**, 474–487.

48. Evans, D.B., Bunning, R.A., Van Damme, J. and Russell, R.G. (1989) Natural human IL-1 beta exhibits regulatory actions on human bone-derived cells in vitro. *Biochemical and Biophysical Research Communications* **159**, 1242–1248.

49. Evans, D.B., Bunning, R.A. and Russell, R.G. (1990) The effects of recombinant human interleukin-1 beta on cellular proliferation and the production of prostaglandin E2, plasminogen activator, osteocalcin and alkaline phosphatase by osteoblast-like cells derived from human bone. *Biochemical and Biophysical Research Communications* **166**, 208–216.

50. Hanazawa, S., Ohmori, Y., Amano, S. *et al.* (1986) Human purified interleukin-1 inhibits DNA synthesis and cell growth of osteoblastic cell line (MC3T3-E1), but enhances alkaline phosphatase activity in the cells. *FEBS Letters* **203**, 279–284.

51. Ikeda, E., Kusaka, M., Hakeda, Y. *et al.* (1988) Effect of interleukin 1 beta on osteoblastic clone MC3T3-E1 cells. *Calcified Tissue International* **43**, 162–166.

52. Jin, C.H., Miyaura, C., Ishimi, Y. *et al.* (1990) Interleukin 1 regulates the expression of osteopontin mRNA by osteoblasts. *Molecular and Cellular Endocrinology* **74**, 221–228.

53. Gowen, M., Wood, D.D. and Russell, R.G. (1985) Stimulation of the proliferation of human bone cells in vitro by human monocyte products with interleukin-1 activity. *Journal of Clinical Investigation* **75**, 1223–1229.

54. Kuroki, T., Shingu, M., Koshihara, Y. and Nobunaga, M. (1994) Effects of cytokines on alkaline phosphatase and osteocalcin production, calcification and calcium release by human osteoblastic cells. *British Journal of Rheumatology* **33**, 224–230.

55. Rickard, D.J., Gowen, M. and MacDonald, B.R. (1993) Proliferative responses to estradiol, IL-1 alpha and TGF beta by cells expressing alkaline phosphatase in human osteoblast-like cell cultures. *Calcified Tissue International* **52**, 227–233.

56. Stashenko, P., Dewhirst, F.E., Rooney, M.L. *et al.* (1987) Interleukin-1 beta is a potent inhibitor of bone formation in vitro. *Journal of Bone and Mineral Research* **2**, 559–565.

57. Antosz, M.E., Bellows, C.G. and Aubin, J.E. (1989) Effects of transforming growth factor β and epidermal growth factor on cell proliferation and the formation of bone nodules in isolated fetal rat calvaria cells. *Journal of Cellular Physiology* **140**, 386–395.

58. Clover, J. and Gowen, M. (1994) Are MG-63 and HOS TE85 human osteosarcoma cell lines representative models of the osteoblastic phenotype? *Bone* **15**, 585–591.

59. Grzesik, W.J. and Robey, P.G. (1994) Bone matrix RGD glycoproteins: immunolocalization and interaction with human primary osteoblastic bone cells in vitro. *Journal of Bone and Mineral Research* **9**, 487–496.

60. Hughes, D.E., Salter, D.M., Dedhar, S. and Simpson, R. (1993) Integrin expression in human bone. *Journal of Bone and Mineral Research* **8**, 527–533.

61. Hughes, D.E., Salter, D.M. and Simpson, R. (1994) CD44 expression in human bone: a novel marker of osteocytic differentiation. *Journal of Bone and Mineral Research* **9**, 39–44.

62. Horowitz, M.C., Fields, A., DeMeo, D. *et al.* (1994) Expression and regulation of Ly-6 differentiation antigens by murine osteoblasts. *Endocrinology* **135**, 1032–1043.

63. Jamal, H.H. and Aubin, J.E. (1996) CD44 expression in fetal rat bone: in vivo and in vitro analysis. *Experimental Cell Research* **223**, 467–477.

64. Nakamura, H., Kenmotsu, S., Sakai, H. and Ozawa, H. (1995) Localization of CD44, the hyaluronate receptor, on the plasma membrane of osteocytes and osteoclasts in rat tibiae. *Cell and Tissue Research* **280**, 225–233.

65. Pavasant P., Shizari, T.M. and Underhill, C.B. (1994) Distribution of hyaluronan in the epiphysial growth plate: turnover by CD44-expressing osteoprogenitor cells. *Journal of Cell Science* **107**, 2669–2677.

66. Van Vlasselaer, P., Falla, N., Snoeck, H. and Mathieu, E. (1994) Characterization and purification of osteogenic cells from murine bone marrow by two-color cell sorting using anti-Sca-1 monoclonal antibody and wheat germ agglutinin. *Blood* **84**, 753–63.

67. Simmons, P.J. and Torok-Storb, B. (1991) Identification of stromal cell precursors in human bone marrow by a novel monoclonal antibody, STRO-1. *Blood* **78**, 55–62.

68. Gronthos, S., Graves, S.E., Ohta, S. and Simmons, P.J. (1994) The STRO-1+ fraction of adult human bone marrow contains the osteogenic precursors. *Blood* **84**, 4164–4173.

69. Nakamura, T., Gross, M., Yamamuro, T. and Liao, S.-K. (1987) Identification of a human osteosarcoma-associated glycoprotein with monoclonal antibodies: relationship with alkaline phosphatase. *Biochemistry and Cell Biology* **65**, 1091–1097.

70. Bruder, S.P. and Caplan, A.I. (1990) Osteogenic cell lineage analysis is facilitated by organ cultures of embryonic chick periosteum. *Developmental Biology* **141**, 319–329.

71. Turksen, K., Bhargava, U., Moe, H.K. and Aubin, J.E. (1992) Isolation of monoclonal antibodies recognising rat bone-associated molecules in vivo and in vitro. *Journal of Histochemistry and Cytochemistry* **40**, 1339–1352.

72. Yamaguchi, A. and Kahn, A.J. (1993) Monoclonal antibodies that recognize antigens in human osteosarcoma cells and normal fetal osteoblasts. *Bone and Mineral* **22**, 165–176.

73. Aubin, J.E. and Turksen, K. (1996) Monoclonal antibodies as tools for studying the osteoblast lineage. *Microscopy Research and Technique* **33**, 128–140.

74. Nijweide, P.J. and Mulder, R.J. (1986) Identification of osteocytes in osteoblast-like cell cultures using a monoclonal antibody specifically directed against osteocytes. *Histochemistry* **84**, 342–347.

75. Bruder, S.P. and Caplan, A.I. (1989) First bone formation and the dissection of an osteogenic lineage in the embryonic chick tibia is revealed by monoclonal antibodies against osteoblasts. *Bone* **10**, 359– 35.

76. van der Plas, A. and Nijweide, P.J. (1992) Isolation and purification of osteocytes. *Journal of Bone and Mineral Research* **7**, 389–396.

77. van der Plas, A., Aarden, E.M., Feijen, J.H. *et al.* (1994) Characteristics and properties of osteocytes in culture. *Journal of Bone and Mineral Research* **9**, 1697–1704.

78. Wetterwald, A. and Fleisch, H. (1990) A monoclonal antibody recognises a cell membrane antigen present in a subpopulation of rat osteoblasts and in osteocytes. *Calcified Tissue International* **46** (Suppl. 2), A13 (Abstr. 52).

79. Nose, K., Saito, H. and Kuroli, T. (1990) Isolation of a gene sequence induced later by tumor-promoting 12-O-tetradecanolyphorbol-13-acetate im mouse osteoblastic cells (MC3T3-E1) and expressed constitutively in *ras* transformed cells. *Cell Growth and Differentiation* **1**, 511–518.

80. Wetterwald, A., Hofstetter, W., Cecchini, M.G. *et al.* (1996) Characterisation and cloning of the E11 antigen, a marker expressed by rat osteoblasts and osteocytes. *Bone* **18**, 125–132.

81. Sprague, L., Wetterwald, A., Heinzman, U. and Atkinson, M.J. (1996) Phenotypic changes following over-expression of sense of antisense E11 cDNA in ROS 17/2.8 cells. *Journal of Bone and Mineral Research* **11**, S132.

82. Haynesworth, S.E., Baber, M.A. and Caplan, A.I. (1992) Cell surface antigens on human marrow-derived mesenchymal cells are detected by monoclonal antibodies. *Bone* **13**, 69–80.

83. Bellows, C.G., Aubin, J.E. and Heersche, J.N. (1991) Initiation and progression of mineralization of bone nodules formed in vitro: the role of alkaline phosphatase and organic phosphate. *Bone and Mineral* **14**, 27–40.

84. Bellows, C.G., Heersche, J.N. and Aubin, J.E. (1992) Inorganic phosphate added exogenously or released from beta-glycerophosphate initiates mineralization of osteoid nodules in vitro. *Bone and Mineral* **17**, 15–29.

85. Marsh, M.E., Munne, A.M., Vogel, J.J. *et al.* (1995) Mineralization of bone-like extracellular matrix in the absence of functional osteoblasts. *Journal of Bone and Mineral Research* **10**, 1635–1643.

86. Beertsen, W. and Vandenbos, T. (1992) Alkaline phosphatase induces the mineralization of sheets of collagen implanted subcutaneously in the rat. *Journal of Clinical Investigation* **89**, 1974–1980.

87. Khouja, H.I., Bevington, A., Kemp, G.J. and Russell, R.G.G. (1990) Calcium and orthophosphate deposits in vitro do not imply osteoblast mediated mineralization: mineralization by betaglycerophosphate in the absence of osteoblasts. *Bone* **11**, 385–391.

88. Bellows, C.G., Aubin, J.E., Heersche, J.N.M. and Antosz, M.E. (1986) Mineralized bone nodules formed in vitro from enzymatically released rat calvaria cell populations. *Calcified Tissue International* **38,**, 143–154.

89. Bhargava, U., Bar-Lev, M., Bellows, C.G. and Aubin, J.E. (1988) Ultrastructural analysis of bone nodules formed in vitro by isolated fetal rat calvaria cells. *Bone* **9**, 155–163.

90. Ecarot-Charrier, B., Glorieux, F.H., van der Rest, M. and Pereira, G. (1983) Osteoblasts isolated from mouse calvaria initiate matrix mineralization in culture. *Journal of Cell Biology* **96**, 639–643.

91. Nefussi, J.R., Boy-Lefevre, M.L., Boulekbache, H. and Forest, N. (1985) Mineralization in vitro of matrix formed by osteoblasts isolated by collagenase digestion. *Differentiation* **29**, 160–168.

92. Long, M.W., Robinson, J.A., Ashcraft, E.A. and Mann, K.G. (1995) Regulation of human bone marrow-derived osteoprogenitor cells by osteogenic growth factors. *Journal of Clinical Investigation* **95**, 881–887.

93. Lee, K.L., Aubin, J.E. and Heersche, J.N. (1992) Beta-glycerophosphate-induced mineralization of osteoid does not alter expression of extracellular matrix components in fetal rat calvarial cell cultures. *Journal of Bone and Mineral Research* **7**, 1211–1219.

94. Rao, L.G., Murray, T.M. and Heersche, J.N. (1983) Immunohistochemical demonstration of parathyroid hormone binding to specific cell types in fixed rat bone tissue. *Endocrinology* **113**, 805–810.

95. Silve, C.M., Hradek, G.T., Jones, A.L. and Arnaud, C.D. (1982) Parathyroid hormone receptor in intact embryonic chicken bone: characterization and cellular localization. *Journal of Cell Biology* **94**, 379–386.

96. Rouleau, M.F., Mitchell, J. and Goltzman, D. (1988) In vivo distribution of parathyroid hormone receptors in bone: evidence that a predominant osseous target cell is not the mature osteoblast. *Endocrinology* **123**, 187–191.

97. Rouleau, M.F., Mitchell, J. and Goltzman, D. (1990) Characterization of the major parathyroid hormone target cell in the endosteal metaphysis of rat long bones. *Journal of Bone and Mineral Research* **5**, 1043-1-53.

98. Heath, J.K., Rodan, S.B., Yoon, K.G. and Rodan, G.A. (1989) Rat calvaria cell lines immortalised by SV40 large T-antigen – constitutive and retinoic acid-inducible expression of osteoblastic features. *Endocrinology* **124**, 3060–3068.

99. Shalhoub, V., Conlon, D., Tassinari, M. *et al.* (1992) Glucocorticoids promote development of the osteoblast phenotype by selectively modulating expression of cell growth and differentiation associated genes. *Journal of Cellular Biochemistry* **50**, 425–440.

100. Owen, T.A., Aronow, M., Shalhoub, V. *et al.* (1990) Progressive development of the rat osteoblast phenotype in vitro: reciprocal relationships in expression of genes associated with osteoblast proliferation and differentiation during formation of the bone extracellular matrix. *Journal of Cellular Physiology* **143**, 420–430.

101. Owen, T.A., Aronow, M.S., Barone, L.M. *et al.* (1991) Pleiotropic effects of vitamin D on osteoblast gene expression are related to the proliferative and differentiated state of the bone cell phenotype: dependency upon basal levels of gene expression, duration of exposure, and bone matrix competency in normal rat osteoblast cultures. *Endocrinology* **128**, 1496–1504.

102. Aronow, M.A., Gerstenfeld, L.C., Owen, T.A. *et al.* (1990) Factors that promote progressive development of the osteoblast phenotype in cultured fetal rat calvaria cells. *Journal of Cellular Physiology* **143**, 213–221.

103. Liu, F., Malaval, L., Gupta, A.K. and Aubin, J.E. (1994) Simultaneous detection of multiple bone-related mRNAs and protein expression during osteoblast differentiation: polymerase chain reaction and immunocytochemical studies at the single cell level. *Developmental Biology* **166**, 220–234.

104. Breen. E.C., Ignotz, R.A., McCabe, L. *et al.* (1994) TGF beta alters growth and differentiation related gene expression in proliferating osteoblasts in vitro, preventing development of the mature bone phenotype. *Journal of Cellular Physiology* **160**, 323–335.

105. Stein, G.S. and Lian, J.B. (1993) Molecular mechanisms mediating proliferation/differentiation interrelationships during progressive development of the osteoblast phenotype. [Review]. *Endocrine Reviews* **14**, 424–442.

106. Smith, E., Frenkel, B., Schlegel, R. *et al.* (1995) Expression of cell cycle regulatory factors in differentiating osteoblasts: postproliferative up-regulation of cyclins B and E. *Cancer Research* **55**, 5019–5024.

107. Tsuji, T., Hughes, F.J., McCulloch, C.A.G. and Melcher, A.H. (1990) The effect of donor age on the osteogenic cells of rat bone marrow. *Mechanisms of Ageing and Development* **51**, 121–132.

108. Lennon, D.P., Haynesworth, S.E., Young, R.G. *et al.* (1995) A chemically defined medium supports in vitro proliferation and maintains the osteochondral potential of rat marrow-derived mesenchymal stem cells. *Experimental Cell Research* **219**, 211–222.

109. Denis, I., Pointillart, A. and Lieberherr, M. (1994) Cell stage-dependent effects of ascorbic acid on cultured porcine bone cells. *Bone & Mineral* **25**, 149–161.

110. Franceschi, R.T., Iyer, B.S. and Cui, Y. (1994) Effects of ascorbic acid on collagen matrix formation and osteoblast differentiation in murine MC3T3-E1 cells. *Journal of Bone and Mineral Research* **9**, 843–854.

111. Harada, S., Matsumoto, T. and Ogata, E. (1991) Role of ascorbic acid in the regulation of proliferation in osteoblast-like MC3T3-E1 cells. *Journal of Bone and Mineral Research* **6**, 903–908.

112. Malaval, L., Modrowski, D., Gupta, A.K. and Aubin, J.E. (1994) Cellular expression of bone-related proteins during in vitro osteogenesis in rat bone marrow stromal cell cultures. *Journal of Cellular Physiology* **158**, 555–572.

113. Quarles, L.D., Yohay, D.A., Lever, L.W. *et al.* (1992) Distinct proliferative and differentiated stages of murine MC3T3-E1 cells in culture: an in vitro model of osteoblast development. *Journal of Bone and Mineral Research* **7**, 683–692.

114. Torii, Y., Hitomi, K. and Tsukagoshi, N. (1994) L-Ascorbic acid 2-phosphate promotes osteoblastic differentiation of MC3T3-E1 mediated by accumulation of type I collagen. *Journal of Nutritional Science and Vitaminology* **40**, 229–238.

115. Dixon, S.J. and Wilson, J.X. (1992) Adaptive regulation of ascorbate transport in osteoblastic cells. *Journal of Bone and Mineral Research* **7**, 675–681.

116. Franceschi, R.T., Wilson, J.X. and Dixon, S.J. (1995) Requirement for Na(+)-dependent ascorbic acid transport in osteoblast function. *American Journal of Physiology* **268**, C1430–C1439.

117. Boskey, A.L., Stiner, D., Doty, S.B. and Binderman, I. (1991) Requirement of vitamin C for cartilage calcification in a differentiating chick limb-bud mesenchymal cell culture. *Bone* **12**, 277–282.

118. Beresford, J.N., Graves, S.E. and Smoothy, C.A. (1993) Formation of mineralized nodules by bone derived cells in vitro: a model of bone formation? *American Journal of Medical Genetics* **45**, 163–178.

119. McCulloch, C.A.G. and Tenenbaum, H.C. (1986) Dexamethasone induces proliferation and terminal differentiation of osteogenic cells in tissue culture. *Anatomical Record* **215**, 397.

120. Ogata, Y., Yamauchi, M., Kim, R.H. *et al.* (1995) Glucocorticoid regulation of bone sialoprotein (BSP) gene expression. Identification of a glucocorticoid response element in the bone sialoprotein gene promoter. *European Journal of Biochemistry* **230**, 183–192.

121. Majeska, R.J., Nair, B.C. and Rodan, G.A. (1985) Glucocorticoid regulation of alkaline phosphatase in the osteoblastic osteosarcoma cell line ROS 17/2.8. *Endocrinology* **116**, 170–179.

122. Takuwa, Y., Ohse, C., Wang, E.A. *et al.* (1991) Bone morphogenetic protein-2 stimulates alkaline phosphatase activity and collagen synthesis in cultured osteoblastic cells, MC3T3-E1. *Biochemical and Biophysical Research Communications* **174**, 96–101.

123. Falla, N., Van, V., Bierkens, J. *et al.* (1993) Characterization of a 5-fluorouracil-enriched osteoprogenitor population of the murine bone marrow. *Blood* **82**, 3580–3591.

124. Herbertson, A. and Aubin, J.E. (1995) Dexamethasone alters the subpopulation make-up of rat bone marrow stromal cell cultures. *Journal of Bone and Mineral Research* **10**, 285–294.

125. Benayahu, D., Kletter, Y., Zipori, D. and Wientroub, S. (1989) Bone marrow-derived stromal cell line expressing osteoblastic phenotype in vitro and osteogenic capacity in vivo. *Journal of Cellular Physiology* **140**, 1–7.

126. Gundle, R. and Beresford, J.N. (1995) The isolation and culture of cells from explants of human trabecular bone. *Calcified Tissue International* **56**, S8–S10.

127. Whitson, S.W., Harrison, W., Dunlap, M.K. *et al.* (1984) Fetal bovine bone cells synthesize bone-specific matrix proteins. *Journal of Cell Biology* **99**, 607–614.

128. Wong, G.L. and Cohn, D.V. (1975) Target cells in bone for parathormone and calcitonin are different: enrichment for each cell type by sequential digestion of mouse calvaria and selective adhesion to polymeric surfaces. *Proceedings of the National Academy of Sciences USA* **72**, 3167–3171.

129. Nijweide, P.J., van Iperen-van Gent, A.S., Kawilarang-de Haas E.W. *et al.* (1982) Bone formation and calcification by isolated osteoblastlike cells. *Journal of Cell Biology* **93**, 318–323.

130. Bellows, C.G. and Aubin, J.E. (1989) Determination of numbers of osteoprogenitor cells present in isolated fetal rat calvaria cells *in vitro*. *Developmental Biology* **133**, 8–13.

131. Bellows, C.G., Ishida, H., Aubin, J.E. and Heersche, J.N. (1990) Parathyroid hormone reversibly suppresses the differentiation of osteoprogenitor cells into functional osteoblasts. *Endocrinology* **127**, 3111–3116.

132. Harris, S.E., Bonewald, L.F., Harris, M.A. *et al.* (1994) Effects of transforming growth factor beta on bone nodule formation and expression of bone morphogenetic protein 2, osteocalcin, osteopontin, alkaline phosphatase, and type I collagen mRNA in long-term cultures of fetal rat calvarial osteoblasts. *Journal of Bone and Mineral Research* **9**, 855–863.

133. Hughes, F.J. and Howells, G.L. (1993) Interleukin-11 inhibits bone formation in vitro. *Calcified Tissue International* **53**, 362–364.

134. Hughes, F.J. and Howells, G.L. (1993) Interleukin-6 inhibits bone formation in vitro. *Bone and Mineral* **21**, 21–28.

135. Leboy, P.S., Beresford, J.N., Devlin, C. and Owen, M.E. (1991) Dexamethasone induction of osteoblast mRNAs in rat marrow stromal cell cultures. *Journal of Cellular Physiology* **146**, 370–378.

136. Hughes, F.J. and McCulloch, C.A.G. (1991) Stimulation of the differentiation of osteogenic rat bone marrow stromal cells by osteoblast cultures. *Laboratory Investigation* **64**, 617–622.

137. Taylor, S.M. and Jones, P.A. (1979) Multiple new phenotypes induced in 10T1/2 and 3T3 cells treated with 5-azacytidine. *Cell* **17**, 771–779.

138. Taylor, S.M. and Jones, P.A. (1982) Changes in phenotypic expression in embryonic and adult cells treated with 5-azacytidine. *Journal of Cellular Physiology* **111**, 187–194.

139. Wang, E.A., Israel, D.I., Kelly, S. and Luxenberg, D.P. (1993) Bone morphogenetic protein-2 causes commitment and differentiation in C3H10T1/2 and 3T3 cells. *Growth Factors* **9**, 57–71.

140. Ahrens, M., Ankenbauer, T., Schroder, D. *et al.* (1993) Expression of human bone morphogenetic proteins-2 or -4 in murine mesenchymal progenitor C3H10T1/2 cells induces differentiation into distinct mesenchymal cell lineages. *DNA and Cell Biology* **12**, 871–880.

141. Katagiri, T., Yamaguchi, A., Ikeda, T. *et al.* (1990) The non-osteogenic mouse pluripotent cell line, C3H10T1/2, is induced to differentiate into osteoblastic cells by recombinant human bone morphogenetic protein-2. *Biochemical and Biophysical Research Communications* **172**, 295–299.

142. Aubin, J.E., Heersche, J.N.M., Merrilees, M.J. and Sodek, J. (1982) Isolation of bone cell clones with differences in growth, hormone responses, and extracellular matrix production. *Journal of Cell Biology* **92**, 452–461.

143. Gitelman, S.E., Kirk, M., Ye, J.Q. *et al.* (1995) Vgr-1/BMP-6 induces osteoblastic differentiation of pluripotential mesenchymal cells. *Cell Growth & Differentiation* **6**, 827–836.

144. Yamaguchi, A., Katagiri, T., Ikeda, T. *et al.* (1991) Recombinant human bone morphogenetic protein-2 stimulates osteoblastic maturation and inhibits myogenic differentiation in vitro. *Journal of Cell Biology* **113**, 681–687.

145. Poliard, A., Nifuji, A., Lamblin, D. *et al.* (1995) Controlled conversion of an immortalized mesodermal progenitor cell towards osteogenic, chondrogenic, or adipogenic pathways. *Journal of Cell Biology* **130**, 1461–1472.

146. Chentoufi, J., Hott, M., Lamblin, D. *et al.* (1993) Kinetics of in vitro mineralization by an osteogenic clonal cell line (C1) derived from mouse teratocarcinoma. *Differentiation* **53**, 181–189.

147. Asahina, I., Sampath, T.K. and Hauschka, P.V. (1996) Human osteogenic protein-1 induces chondroblastic, osteoblastic, and/or adipocytic differentiation of clonal murine target cells. *Experimental Cell Research* **222**, 38–47.

148. Brennan, T.J., Edmondson, D.G., Li, L. and Olson, E.N. (1991) Transforming growth factor beta represses the actions of myogenin through a mechanism independent of DNA binding. *Proceedings of the National Academy of Sciences USA* **88**, 3822–3826.

149. Kodama, H., Amagai, Y., Sudo, H. *et al.* (1982) Establishment of a clonal osteogenic cell line from newborn mouse calvaria. *Japanese Journal of Oral Biology* **23**, 899–904.

150. Sudo, H., Kodama, H., Amagai, Y. *et al.* (1983) In vitro differentiation and calcification in a new clonal osteogenic cell line derived from newborn mouse calvaria. *Journal of Cell Biology* **96**, 191–198.

151. Benayahu, D., Fried, A., Shamay, A. *et al.* (1994) Differential effects of retinoic acid and growth factors on osteoblastic markers and CD10/NEP activity in stromal-derived osteoblasts. *Journal of Cellular Biochemistry* **56**, 62–73.

152. Benayahu, D., Kompier, R., Shamay, A. *et al.* (1994) Mineralization of marrow-stromal osteoblasts MBA-15 on three-dimensional carriers. *Calcified Tissue International* **55**, 120–127.

153. Benayahu, D., Fried, A. and Wientroub, S. (1995) PTH and 1,25(OH)2 vitamin D priming to growth factors differentially regulates the osteoblastic markers in MBA-15 clonal subpopulations. *Biochemical and Biophysical Research Communications* **210**, 197–204.

154. Sodek, J., Kim, R.H., Ogata, Y. *et al.* (1995) Regulation of bone sialoprotein gene transcription by steroid hormones. *Connective Tissue Research* **32**, 209–217.

155. Yoon, K.G., Rutledge, S.J., Buenaga, R.F. and Rodan, G.A. (1988) Characterization of the rat osteocalcin gene: stimulation of promoter activity by 1,25-dihydroxyvitamin D3. *Biochemistry* **27**, 8521–8526.

156. Franceschi, R.T. and Young, J. (1990) Regulation of alkaline phosphatase by 1,25-dihydroxyvitamin D3 and ascorbic acid in bone-derived cells. *Journal of Bone and Mineral Research* **5**, 1157–1167.

157. Zhou, H., Hammonds, R. Jr, Findlay, D.M. *et al.* (1993) Differential effects of transforming growth factor-beta 1 and bone morphogenetic protein 4 on gene expression and differentiated function of preosteoblasts. *Journal of Cellular Physiology* **155**, 112–119.

158. Ng, K.W., Gummer, P.R., Michelangeli, V.P. *et al.* (1988) Regulation of alkaline phosphatase expression in a neonatal rat clonal calvarial cell strain by retinoic acid. *Journal of Bone and Mineral Research* **3**, 53–61.

159. Forrest, S.M., Ng, K.W., Findlay, D.M. *et al.* (1985) Characterization of an osteoblast-like clonal cell line which responds to both parathyroid hormone and calcitonin. *Calcified Tissue International* **37**, 51–56.

160. Ferrier, J., Ward-Kesthely, A., Heersche, J.N.M. and Aubin, J.E. (1989) Membrane potential changes, cAMP stimulation and contraction in osteoblast-like UMR 106 cells in response to calcitonin and parathyroid hormone. *Bone and Mineral* **4**, 133–145.

161. Randall, J.C., Morris, D.C., Zeiger, S. *et al.* (1989) Presence and activity of alkaline phosphatase in two human osteosarcoma cell lines. *Journal of Histochemistry and Cytochemistry* **37**, 1069–1074.

162. McQuillan, D.J., Richardson, M.D. and Bateman, J.F. (1995) Matrix deposition by a calcifying human osteogenic sarcoma cell line (SaOS-2). *Bone* **16**, 415–426.

163. Farley, J.R., Kyeyune-Nyombi, E., Tarbaux, N.M. *et al.* (1989) Alkaline phosphatase activity from human osteosarcoma cell line SaOS-2: an isoenzyme standard for quantifying skeletal alkaline phosphatase activity in serum. *Clinical Chemistry* **35**, 223–229.

164. Jaaskelainen, T., Pirskanen, A., Ryhanen, S. *et al.* (1994) Functional interference between AP-1 and the vitamin D receptor on osteocalcin gene expression in human osteosarcoma cells. *European Journal of Biochemistry* **224**, 11–20.

165. Billiau, A., Edy, V.G., Heremans, H. *et al.* (1977) Human interferon: mass production in a newly established cell line, MG-63. *Antimicrobial Agents and Chemotherapy* **12**, 11–15.

166. Franceschi, R.T., James, W.M. and Zerlauth, G. (1985) 1 alpha, 25-dihydroxyvitamin D_3 specific regulation of growth, morphology, and fibronectin in a human osteosarcoma cell line. *Journal of Cellular Physiology* **123**, 401–409.

167. Dedhar, S., Mitchell, M.D. and Pierschbacher, M.D. (1989) The osteoblast-like differentiated phenotype of a variant of MG-63 osteosarcoma cell line correlated with altered adhesive properties. *Connective Tissue Research* **20**, 49–61.

168. Lajeunesse, D., Kiebzak, G.M., Frondoza, C. and Sacktor, B. (1991) Regulation of osteocalcin secretion by human primary bone cells and by the human osteosarcoma cell line MG-63. *Bone and Mineral* **14**, 237–250.

169. Johansen, J.S., Williamson, M.K., Rice, J.S. and Price, P.A. (1992) Identification of proteins secreted by human osteoblastic cells in culture. *Journal of Bone and Mineral Research* **7**, 501–512.

170. Jukkola, A., Risteli, L., Melkko, J. and Risteli, J. (1993) Procollagen synthesis and extracellular matrix deposition in MG-63 osteosarcoma cells. *Journal of Bone and Mineral Research* **8**, 651–657.

171. Ali, N.N., Rowe, J. and Teich, N.M. (1996) Constitutive expression of non-bone/liver/kidney alkaline phosphatase in human osteosarcoma cell lines. *Journal of Bone and Mineral Research* **11**, 512–520.

172. Hassager, C., Fitzpatrick, L.A., Spencer, E.M. *et al.* (1992) Basal and regulated secretion of insulin-like growth factor binding proteins in osteoblast-like cells is cell line specific. *Journal of Clinical Endocrinology and Metabolism* **75**, 228–233.

173. Okazaki, R., Conover, C.A., Harris, S.A. *et al.* (1995) Normal human osteoblast-like cells consistently express genes for insulin-like growth factors I and II but transformed human osteoblast cell lines do not. *Journal of Bone and Mineral Research* **10**, 788–795.

174. Keeting. P.E., Scott, R.E., Colvard, D.S. *et al.* (1992) Development and characterization of a rapidly proliferating, well-differentiated cell line derived from normal adult human osteoblast-like cells transfected with SV40 large T antigen. *Journal of Bone and Mineral Research* **7**, 127–136.

175. Harris, S.A., Enger, R.J., Riggs, B.L. and Spelsberg, T.C. (1995) Development and characterization of a conditionally immortalized human fetal osteoblastic cell line. *Journal of Bone and Mineral Research* **10**, 178–186.

176. Walker, K.E., Houghton, A., Russell, R.G.G. and Stringer, B.M.J. (1995) Identification of conditionally immortalised human osteoprogenitor cell lines responsive to estrogen. *Bone* **17**, 565 (Abstr. 30).

177. Fisher, L.W., Gehron Robey, P., Tuross, N. *et al.* (1987) The 24 000 Mr phosphoprotein from developing bone is the *N*-propeptide of the alpha1 chain of type I collagen. *Journal of Biological Chemistry* **262**, 13457–13463.

178. Prince, P.A., Ponthermore, J.G. and Deftos, L.J. (1980) A new biochemical marker for bone metabolism. *Journal of Clinical Investigation* **66**, 878–883.

179. Fisher, L.W., Stubbs, J.T. III and Young, M.F. (1995) Antisera and cDNA probes to human and certain animal model bone matrix noncollagenous proteins. *Acta Orthopedica Scandinavica* (Suppl. 266) **66**, 61–65.

180. Bancroft, J.D. (1990) Enzyme histochemistry, in *Theory and Practice of Histological Techniques*, 3rd edn, (eds J.D. Bancroft and A. Stevens), Churchill Livingstone, London, pp. 379–399.

CHAPTER TWO

In vitro *models for osteoclast recruitment*

Adrienne M. Flanagan and Usha Sarma

2.1 INTRODUCTION

Osteoclasts derive from a mononuclear haematopoietic precursor and are cells that resorb bone. They are rare cells, particularly in physiological states in adult primates, and osteoclasts that exist are difficult to harvest because they are situated on calcified bone surfaces within the medullary cavity. For these reasons the generation of osteoclasts from precursors using *in vitro* models has been of major interest to those involved in studying the biology of bone. The reproducible *in vitro* generation of osteoclasts, using a variety of murine organ and cell culture techniques, is well documented and reviewed and has advanced our knowledge of osteoclast biology [1–10]. The information accrued from these studies has recently led to the formation of cell lines in which a proportion of the cells formed are osteoclasts [8, 11]. Despite considerable effort, analogous studies using human haematopoietic cells have been less successful than those reported for murine systems and this chapter will concentrate on the difficulties involved in generating human osteoclasts *in vitro*.

2.2 ASSESSMENT OF OSTEOCLAST FORMATION *IN VITRO*

If *in vitro* models are used to study osteoclast formation from precursors it is essential to define criteria whereby these cells are classified as osteoclasts in their foreign environment. Bone resorption is recognized as the major role of the osteoclast *in vivo* and the occurrence of this specialized function *in vitro* is accepted as unequivocal evidence that osteoclast formation has taken place. However, characterization of the cell that is responsible for this activity is more difficult. In tissue sections osteoclasts can be reliably identified as multinucleate cells which lie adjacent to bone surfaces, but at sites away from bone other forms of multinucleate cells exist, including macrophage polykaryons and placental syncytiotrophoblasts, both of which are morphologically similar to osteoclasts. Indeed, it is impossible to distinguish between these cell types if they are visualized outside their normal microenvironment, using either phase contrast or transmitted light

Methods in Bone Biology. Edited by Timothy R. Arnett and Brian Henderson.
Published in 1997 by Chapman & Hall, London. ISBN 0 412 75770 2.

microscopy without employing special staining techniques. Histochemistry and immunohistochemistry can be used to differentiate between these cells, but only if they are used critically; for example, labelling of multinucleate cells with antibodies which recognize *c-fms* (the receptor for macrophage-colony stimulating factor) would not be a useful means for selecting osteoclasts from other multinucleate cells, as all express this protein.

2.2.1 Multinuclearity

Tissue culture of peripheral blood and bone marrow cells in the presence of 1,25 dihydroxyvitamin D_3, with and without various cytokines, results in the formation of large numbers of multinucleate 'osteoclast-like' cells and the question therefore arises whether these cells are osteoclasts or macrophage polykaryons. Both derive from haematopoietic precursors and share several surface antigens. However, there are also significant differences between them: macrophages do not resorb bone; macrophages do not express the calcitonin receptor (CTR); and macrophages express certain surface antigens, including CD11a, CD11b, CD11c, CD14 and CD18, which are not expressed on osteoclasts [12]. Hence multinucleation is not a reliable means of identifying osteoclasts *in vitro*.

2.2.2 The calcitonin receptor

Osteoclasts express the CTR *in vivo* [13], as do osteoclasts isolated from rat [14] and fetal bone [15]. It appears to be specific for osteoclasts in the bone marrow since there are no reports that the CTR is expressed on other bone marrow cells *in vivo* in either physiological or pathological conditions. Furthermore, good correlation has generally been found between bone resorption and the number of CTR-positive cells in long-term murine bone marrow cultures [3, 4, 16]. A recent publication by Owens *et al.* [17] provided additional and supportive evidence for the correlation of CTR-positive cells with bone resorption *in vitro* when they showed that formation of CTR-positive cells from haematopoietic precursors could be induced in the presence of spleen stromal cells (this process does not require bone cells) but that these cells lacked the property required to induce the CTR-positive cells to resorb bone. They then demonstrated that bone resorption occurred in cultures in which CTR-positive cells had been formed within 24 hours of plating rat osteoblastic UMR cells.

Cells which label with iodinated salmon calcitonin (^{125}Is/CT) in murine *in vitro* models never label with F4/80 or Moma-2 antibodies, markers for macrophages, indicating that macrophages and osteoclasts are phenotypically different [8]. Analogous studies have not yet been reported using human cells.

2.2.3 Tartrate-resistant acid phosphatase

In tissue sections of bone, tartrate-resistant acid phosphatase (TRAP) is a reliable marker for osteoclasts. The great majority of the osteoclasts that stain with TRAP are multinucleate and are generated by fusion of mononucleate forms, of which only a small number stain with TRAP [18]. This suggests that this enzyme is expressed late in osteoclast development, a finding that is consistent with the *in vitro* studies reported by Takahashi *et al.* [19]. However, expression of this enzyme is not restricted to the osteoclast; TRAP is present in mononucleate and multinucleate macrophages in pathological conditions and can be induced in cells of the mononuclear phagocyte system when grown *in vitro* [20–23]. Hattersley and Chambers [23] found that a poor correlation existed between the number of CTR- and TRAP-positive cells in their murine bone marrow cultures, demonstrating the potential for generating spurious data if this marker were used in isolation. Others have reported a strong correlation between these two markers but it is not clear why such differences exist. This issue is unresolved but since the criteria that are used for osteoclast formation are the basis for analysis and interpretation of the osteoclast generation experiments, it stands to reason that in the event that TRAP is used as a marker for osteoclast formation the responsibility lies with each researcher to demonstrate clearly that a good correlation exists between the numbers of CTR- and TRAP-positive cells in their culture systems. Moreover, since osteoclast formation is inordinately sensitive to small changes in sera, pCO_2 and pH, it is essential that this correlation is constantly validated and is not executed and reported on one occasion only and forever thereafter used as evidence that such correlation exists.

2.2.4 Vitronectin receptor

Monoclonal antibody 23c6 recognizes the $\alpha_v\beta_3$ subunit of the vitronectin receptor on the membrane of human osteoclasts in frozen tissue sections of bone [24], osteoclastoma [25] and osteoclast-rich tumours [26, 27], as well as freshly isolated human osteoclasts from fetal long bones [15]. Strong membrane staining is achieved using the substrates alkaline phosphatase or peroxidase provided the tissue is fixed in either cold acetone or paraformaldehyde but not in formalin, even if antigen-retrieval techniques are used (unpublished observation). Macrophages do not express this antigen [25, 28, 29].

Previously, the emphasis in many studies involving human osteoclast formation was not placed on bone resorption but rather on the formation of 'osteoclast-like' multinucleate cells, and in some cases these cells were labelled with TRAP or antibody 23c6 [30–41]. However, the staining of the 23c6-positive cells in these studies was weak and displayed a diffuse cytoplasmic pattern, unlike that of the strong, largely membrane staining of isolated osteoclasts [15]. Furthermore, the same 'osteoclast-like' cells stained strongly with macrophage markers CD11a, CD11b and CD11c [15], which are known not to be expressed by osteoclasts [12], indicating that these cells were more likely to be macrophage polykaryons than osteoclasts.

Variability in immunohistochemical staining makes interpretation of data unreliable and underscores the necessity to use more than one marker for the identification of osteoclasts in tissue culture experiments. In the virtual absence of bone resorption and assessment of expression of the CTR on osteoclasts, unequivocal evidence does not exist that the so called osteoclast-like cells have the ability to resorb bone (osteoclasts). However, it has been proposed that these cells have failed to differentiate fully into resorbing cells due to the absence of an unknown stimulus lacking in the cultures [42]. With the newly attained knowledge that CTR-positive cells can be formed but which fail to resorb bone in the absence of an additional biological stimulus [17], it will now be possible to test whether the cells described as osteoclast-like in human bone marrow cultures [15, 30, 32,

34–37, 39–41] can be induced to resorb bone. This can be achieved by testing for the induction of bone resorption 24 hours after the addition of UMR cells to cultures containing 'osteoclast-like' cells.

2.2.5 Macrophage-colony stimulating factor and the formation of human osteoclast in human bone marrow cultures

Our experience and that of others is that until recently the formation of bone-resorbing human osteoclasts has been largely unsuccessful [15, 42, 43]. The problem has been solved by finding that the addition of recombinant human M-CSF to human bone marrow cultures results in the reproducible formation of substantial numbers of osteoclasts as assessed by bone resorption, and the generation of 23c6- and CTR-positive cells (Figs 2.1–2.3, Plates 1 and 2) [44]. This finding is consistent with the report that M-CSF is essential for murine osteoclast formation [45–49] and the report that M-CSF increases osteoclast formation in co-cultures of mice embryonic metatarsals and human cord blood [50]. With the knowledge since 1991 that M-CSF played an important role in osteoclast formation [47, 49], one might wonder now why it took so long to identify M-CSF as the missing factor in osteoclast-deficient human bone marrow cultures. The reason is probably because the human experiments were over-influenced by the reports from several groups which showed that addition of M-CSF to murine liquid bone marrow cultures resulted in a reduction in osteoclast numbers and inhibition of bone resorption [4, 51, 52]. It is not clear why this occurs: it could be a difference in the amount of M-CSF being synthezised in the two species; alternatively it could be a difference in sensitivity to M-CSF. In any case, there is an insufficiency of this cytokine in human bone marrow cultures for substantial osteoclast formation.

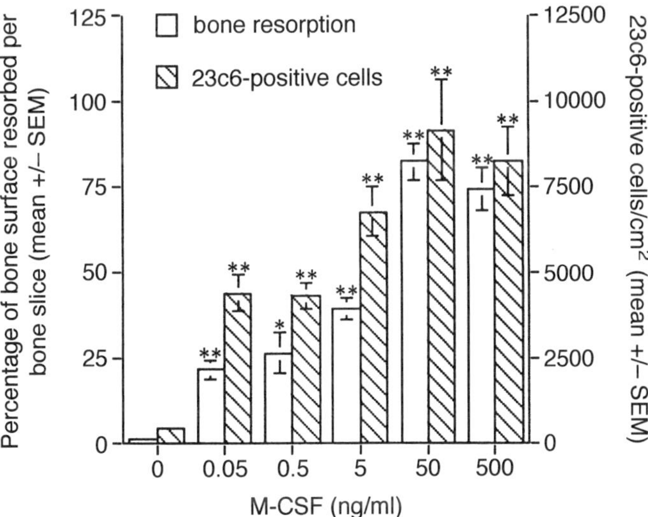

Figure 2.1 Bone resorption and numbers of 23c6-positive cells in human bone marrow cultures in response to increasing concentrations of M-CSF, 14 days after the cells were sedimented onto bone slices. 1,25 Dihydroxyvitamin D_3 was present in all cultures. The area of bone resorption and the number of 23c6-positive cells in the M-CSF-treated cultures were compared with these parameters, respectively, in cultures without M-CSF. *P <0.001; ** P <0.0005.

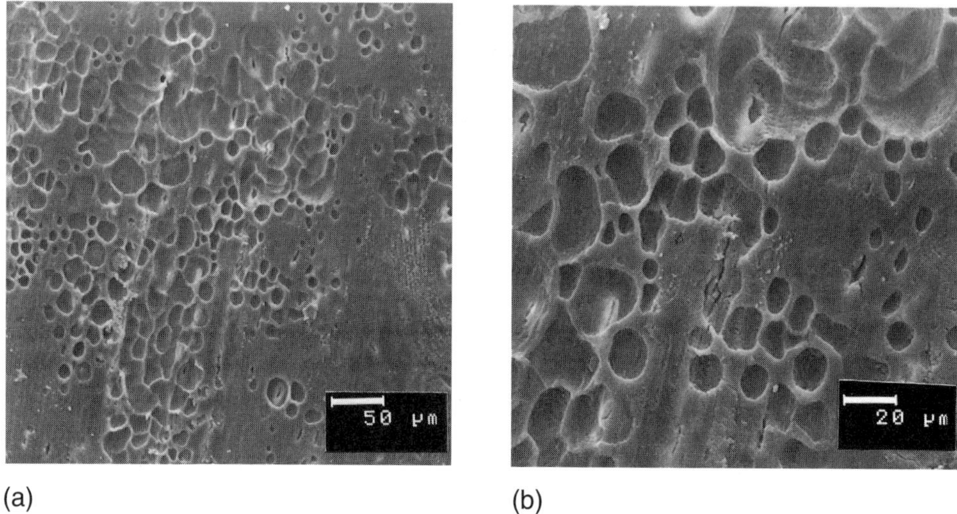

(a)   (b)

Figure 2.2   Scanning electron photomicrographs of bone slices incubated with human bone marrow cells. (a) A low-power magnification shows extensive bone resorption 14 days after the addition of M-CSF (50 ng/ml) and 1,25 dihydroxyvitamin D_3. (b) A higher magnification of the same bone slice shows confluent osteoclastic excavations.

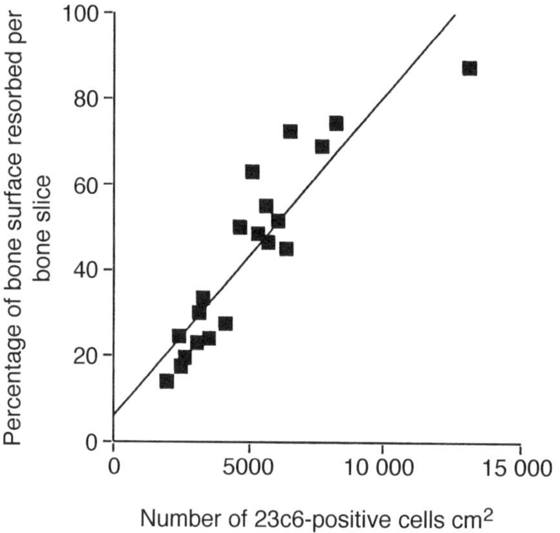

Figure 2.3   Results pooled from 20 human bone marrow experiments in which bone resorption and 23c6-positive cell numbers were measured 14 days after the addition of 1,25 dihydroxyvitamin D_3 in combination with M-CSF 50 ng/ml.

The generation of human osteoclasts in the presence of M-CSF does not depend on the age or sex of the bone marrow donors, and mixing the bone marrow cells from two or more individuals does not alter the osteoclast-formative response of M-CSF. 1,25 Dihydroxyvitamin D_3 is not required for osteoclast formation in these cultures and this is consistent with the knowledge that osteoclast formation occurs in individuals without the 1,25 dihydroxyvitamin D_3 receptor.

The effect of M-CSF on osteoclast formation in human bone marrow cultures is dramatic. A small number of resorption pits and 23c6-positive cells are identified three days after the bone marrow cells are plated on devitalized bone slices in the presence of M-CSF and there is a dramatic increase in both these indices over the next 7 days. In 20 experiments we found that $45.04 \pm 2.23\%$ (mean $\pm$ SEM) of the surfaces of the bone slices were resorbed by day 14. The 23c6-positive cells peaked on day 14 and their numbers strongly correlated with the area of bone resorption (Fig. 2.3). Thereafter the number of 23c6-positive cells declined and by day 21 had virtually disappeared [44]. This 14-day life span is in keeping with the *in vivo* life span of the osteoclast proposed by others [53–55]. Hence our studies suggest that we induce a CTR-negative, 23c6-negative osteoclast precursor in our culture system to differentiate into a bone-resorbing, post-mitotic, CTR-positive, 23c6-positive osteoclast. Optimization of the culture conditions may preserve the osteoclast precursor in culture, permitting continued osteoclast formation, but this may not be feasible if the osteoclast precursors that are present in our cultures are post-mitotic.

From our studies we found that the increase in bone resorption in response to M-CSF was associated with increasing numbers of 23c6- and CTR-positive cells but the area of bone resorbed per cell was also increased in the presence of M-CSF and 1,25 dihydroxyvitamin D_3 compared with that in the presence of 1,25 dihydroxyvitamin D_3 alone [44]. The increased resorption per cell can be accounted for either by increased resorptive activity or by increased life span of the osteoclast population. The latter is favoured on the basis of comparing the areas resorbed per cell per day in time-course experiments between cultures with and without M-CSF; furthermore M-CSF has been shown to increase survival of isolated rat osteoclasts [56]. However, it is probably best to study this effect of M-CSF using isolated human fetal osteoclasts.

Monoclonal antibody 23c6 labelled a population of mononuclear cells or cells with no more than four nuclei in our human bone marrow cultures (Plate 1). These cells had an unusual and striking dendritic appearance (Plate 1), which was similar to that described by Fuller and Chambers in rabbit bone marrow cultures [57]. This characteristic appearance occurred when the cells were formed on either bone slices or plastic coverslips, implying that this morphology did not depend on the type of substrate used. It is noteworthy that we would have failed to recognize that we had generated large numbers of cells that were capable of resorbing bone if the presence of osteoclast-like cells were the sole criterion used for identification of osteoclasts in these cultures.

Labelling of our cultures with antibody 23c6 results in strong and predominantly membrane stain, similar to the staining on isolated osteoclasts as previously reported [25]. The 23c6-positive cells are related to bone resorption spatially as well as in a time-dependent and dose-responsive manner (Fig. 2.1 and Plate 1). Further evidence that these cells are osteoclasts is the finding that 88.5% of 23c6-positive cells also express the CTR (Plate 2).

Although osteoclasts *in vivo* are largely multinucleate, they form by fusion of mononucleate forms [18]. The mononucleate forms in our cultures have the phenotypic and functional characteristics of the multinucleate forms. Hence it is reasonable to conclude that osteoclasts are bone-resorptive cells that are mononucleate or multinucleate depending on various stimuli, the nature of which are not yet fully understood.

The effect of M-CSF on osteoclast formation in human bone marrow cultures appears to be specific. Several other cytokines including interleukin (IL) 1α, IL-1β, IL-3, IL-6, IL-11, leukaemia inhibitory factor (LIF), granulocyte-macrophage colony stimulating factor (GM-CSF), tumour necrosis factor α (TNFα) and transforming growth factor β (TGFβ) were compared with M-CSF but none increased bone resorption above that found in cultures treated with 1,25 dihydroxyvitamin D_3 alone. Indeed several of these factors, including IL-1α, IL-1β, GM-CSF, TNFα and TGFβ, reduced bone resorption and the number of 23c6-positive cells compared with the cultures treated with 1,25 dihydroxyvitamin D_3 alone. LIF and IL-11, however, had no effect on bone resorption but they both showed increased numbers of 23c6-positive cells; this requires further study. The absence of induction of bone resorption by any of these cytokines does not exclude the possibility that they modulate osteoclastic bone resorption, as it may be that these factors are already sufficient in our cultures. It does suggest that M-CSF does not mediate its effect through these factors, nor do these factors induce bone resorption by inducing M-CSF [44].

The effect of M-CSF on the generation of osteoclasts from human bone marrow is serum-dependent. We have found that when human bone marrow cells are cultured in some batches of sera, osteoclast formation fails to occur. It is not obvious what is deficient in these sera or alternatively whether inhibitory factors are present in these batches. Optimization of our culture conditions, particularly serum and cell density, has enabled us to generate human osteoclasts in the absence of M-CSF [58, 59]. On occasions we find that up to 25% of the surface of the bone slices are resorbed in such cultures but more commonly only 1–3% of the bone surfaces are resorbed. The unpredictability of osteoclast formation in cultures without exogenous M-CSF makes such an assay inefficient, particularly when the tissue used for the cultures is so hard to come by. It is not clear why bone marrow cells from certain individuals result in substantial resorptive activity in the absence of M-CSF.

2.2.6 The generation of human osteoclasts from peripheral and cord blood

There has recently been a preliminary report that co-culture of human peripheral blood with the rat osteoblastic UMR cell line in the presence of M-CSF and 1,25 dihydroxyvitamin D_3 results in bone resorption and the formation of cells with an osteoclastic morphology: 23c6- and CTR-positive cells [60]. This cell line synthesizes murine M-CSF and supports osteoclast formation from murine peripheral blood in the absence of exogenous M-CSF [61]. However, as murine M-CSF does not exert a biological effect on human cells, addition of exogenous human M-CSF is required in this culture system for human osteoclast formation. UMR supernatant failed to induce osteoclast formation or bone resorption from peripheral blood, indicating that an osteoclast-inductive factor in addition to M-CSF is required for osteoclast formation and that this factor is membrane-bound.

We have found that osteoclasts are generated from peripheral blood irrespective of either the age or sex of the blood donors. Cells that were 23c6-positive, 88% of which were CTR-positive, appeared in our co-cultures between days 5 and 10 but the greatest increase occurred between days 15 and 21 (Fig. 2.4). This correlates closely with what occurs in human bone marrow system. As in bone marrow cultures, there is a strong correlation between osteoclast numbers

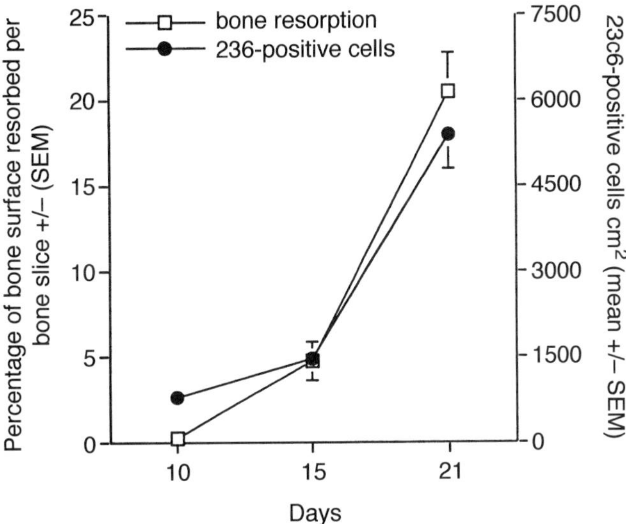

Figure 2.4 Bone resorption and numbers of 23c6- and CTR-positive cells in co-cultures of peripheral blood with UMR rat osteoblastic cells. Data derived from three separate experiments, each containing six bone slices, at each time point.

(23c6- and CTR-positive cells) with bone resorption. Whereas addition of 1,25 dihydroxyvitamin D_3 is not required for the formation of osteoclasts in human bone marrow cultures, it appears to be essential for osteoclast formation using co-cultures of peripheral blood and UMR cells.

2.3 METHODOLOGY

2.3.1 Human bone marrow cultures

Human bone marrow is aspirated under 1% lignocaine local anaesthesia from the posterior iliac crest of healthy male and female volunteers between the ages of 20 and 80 years (approved by St Mary's Hospital medical ethics committee) into preservative-free heparin (CP Pharmaceuticals Ltd, Wrexham, UK). Marrow mononucleate cells are isolated by centrifugation over ficoll/hypaque density gradients, as described in the manufacturer's instructions (Sigma Chemical Co., Poole, Dorset, UK), re-suspended in medium (Sigma Chemical Co.) and plated in tissue culture flasks (Corning Glass Works, Corning, NY) at approximately $10^5/cm^2$ and incubated at 37°C in a humidified atmosphere of 5% CO_2/95% air. We found that osteoclasts are formed in similar numbers in RPMI or minimum essential medium (MEM). The medium is supplemented with 10% heat-inactivated batch-tested serum, L-glutamine, 100 IU benzylpenicillin/ml (Gibco Life Technologies, Paisley, Scotland, UK), 100 µg streptomycin/ml (Gibco) and 10^{-6} M hydrocortisone (Sigma Chemical Co.); hydrocortisone may not be necessary with certain batches of sera. The bone marrow cells are fed twice weekly by replacing half the medium. After 10–14 days, when a heterogeneous population of cells including fibroblasts, adipocytes, macrophages and other haematopoietic cells has formed a confluent layer, the cells are recovered from the flasks using trypsin/EDTA solution (Sigma

Chemical Co.), washed in phosphate buffered saline (PBS) (Sigma Chemical Co.) and re-suspended in medium containing the supplements stated above. The recovered cells are counted using a haemocytometer. The cells are sedimented onto devitalized bone slices ($3 \times 3 \times 0.1$ mm) in a 96-well plate (Corning Glass Works) at 10^5 cells/well. 1,25 Dihydroxyvitamin D_3 (10^{-8} M) and M-CSF, alone or in combination, are then added to the cultures. The cells on the bone slices are fed by replacing half the medium every 3–4 days.

2.3.2 Peripheral blood and UMR co-cultures

The experiments are carried out as previously described [60]. Peripheral blood is aspirated into a heparinized syringe. Using a ficoll/hypaque gradient at room temperature, approximately 2×10^7 mononuclear cells are separated from 10 ml of peripheral blood. These are sedimented at a density of 5×10^4 cells onto UMR cells which have been plated onto bone slices 24 hours earlier at a density of 10^4 cells/well of a 96-well plate. One hour later the bone slices are washed in PBS and transferred into a new well of a 24-well plate. At this point 1,25 dihydroxyvitamin D_3 (10^{-8} M) and M-CSF are added to the cultures. The cultures are fed thrice weekly by semi-depletion and repletion of the medium and reagents. We find that osteoclast formation occurs in both MEM and RPMI. Hydrocortisone (10^{-6} M) is used but this requirement appears to depend on the batch of serum being used.

2.3.3 Immunohistochemistry and quantitation of osteoclast parameters

At the end of the experiments, the bone slices are removed from the wells, washed in PBS, air-dried and fixed in acetone (Analar, Hayman Ltd, Essex, UK). Immunohistochemistry, using monoclonal antibody 23c6 (discussed earlier) kindly provided by Professor M. Horton, London, which preferentially stains osteoclasts [24, 29], is carried out using conventional techniques. The optimal concentration for antibody 23c6 is ascertained by titrating out each new batch and staining slides which have been dabbed with fragments of fresh osteoclastoma or fetal bones. The number of 23c6-positive cells is counted by transmitted light microscopy on 30% of the surface of each of the bone slices and the numbers of mono-, bi- and multi-nucleate (more than three nuclei per cell) cells can also be determined. The total surface of each bone slice is quantified under ×100 or ×200 magnification using a microscope that is linked to a computerized image analysis system (Seescan Ltd, Cambridge, UK). The area of bone resorption on each bone slice is traced with a cursor and at the completion of each bone slice the total area resorbed is automatically calculated as a percentage of the total surface area of each bone slice.

2.3.4 Autoradiography of ^{125}I salmon CT (^{125}IsCT)

Salmon calcitonin (sCT) (kindly provided by Rhône-Poulenc Rorer, PA) is iodinated using a modification of the chloramine T method [62]. Bone-marrow cultures on bone slices are washed in PBS and placed in fresh wells containing 100 µl of 0.1% BSA (bovine serum albumin) in Medium 199 (Gibco). Then 100 µl of ^{125}IsCT (0.2 nM) is added to each well. Non-specific binding is assessed by adding excess unlabelled sCT (300 nM) prior to the addition of the labelled sCT. Following incubation for one hour at 22°C, the bone slices are washed several times in PBS and air dried. They are then fixed in acetone and immunohistochemistry is carried out with antibody 23c6. The bone slices are processed for autoradiography as previously described [14]; they are coated with K5 nuclear emulsion (Ilford, Cheshire, UK). After development, the cultures are counter-stained with toluidine blue. Poor washing of the ^{125}IsCT-labelled bone slices and incubation of the autoradiographs for too long a period before development will lead to a high background level of grains; this can be avoided by setting up extra bone slices in each experiment and developing them at different time points to ascertain the most appropriate time to develop the autoradiographs.

2.4 CONCLUSION

The finding that the addition of M-CSF to human bone marrow cultures results in the reproducible formation of large numbers of osteoclasts *in vitro* will permit the biology of this cell to be studied. Indeed, with respect to elucidating the means whereby 17β-oestradiol reduces bone resorption, exploitation of our human model has proved to be considerably more informative than its murine counterpart. Whereas others have reported that it is difficult to demonstrate an 17β-oestradiol-induced reduction in bone resorption in murine culture systems, we have found that osteoclast numbers, assessed by CTR- and 23c6-positive cells, and bone

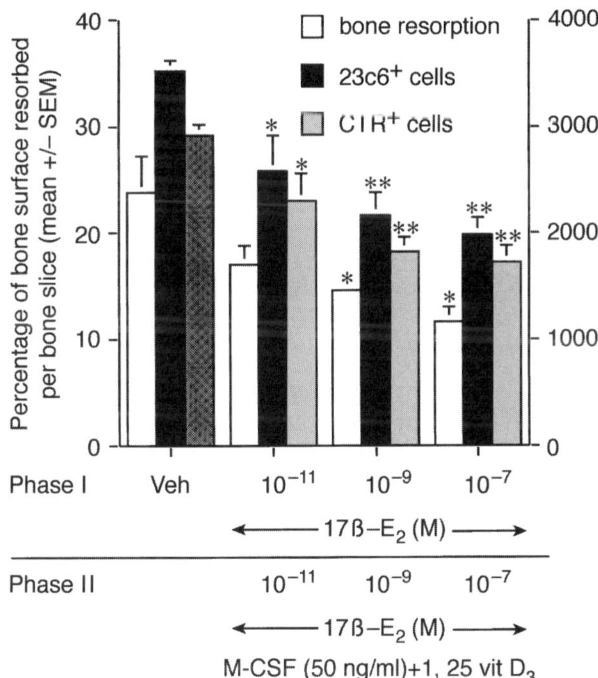

Fig. 2.5 The effect of 17β-oestradiol (17β-E$_2$) on bone resorption, 23c6+ and CTR+ cells in human bone marrow cultures. The cultures were grown in 25 cm^2 flasks, each containing the hormone or vehicle as stated, for phase I of the experiment. Cells were then trypsinized and plated in 96-well plates and cultured in conditions as stated for phase II. Each variable contains six bone slices. *P <0.001; **P <0.0005 vs. relevant parameters in control cultures (untreated with 17β-E$_2$). Veh, vehicle; 1,25-vit D$_3$, 1,25 dihydroxyvitamin D$_3$ (10^{-8} M).

resorption are reduced in a dose-responsive manner (17β-E_2: 10^{-11} M – 10^{-7} M). By exploiting our two-phase system we have found that the effect is mediated through the osteoclast precursor (Fig. 2.5) [63]. Further in-depth analysis of the mechanism by which 17β-E_2 modulates osteoclast formation can now be carried out.

The UMR and human haematopoietic precursor co-culture system in which human osteoclasts are generated will permit detailed analysis of the relationship of the osteoclast and its precursor to other haematopoietic cells. This has been carried out using analogous murine cultures [6], but the availability of cell surface lineage-specific markers to human cells provides considerable advantages over using murine cells.

REFERENCES

1.   Takahashi, N., Akatsu, T., Udagawa, N. *et al.* (1988) Osteoblastic cells are involved in osteoclast formation. *Endocrinology* **123**(5), 2600–2602.

2.   Takahashi, N., Yamama, H., Yoshiki, S. *et al.* (1988) Osteoclast-like cell formation and its regulation by osteotropic hormones in mouse marrow cultures. *Endocrinology* **122**, 1373–1382.

3.   Hattersley, G. and Chambers, T.J. (1989) Calcitonin receptors as markers for osteoclastic differentiation: correlation between generation of bone-resorptive cells and cells that express calcitonin receptors in mouse bone marrow cultures. *Endocrinology* **125**, 1606–1612.

4.   Shinar, D.M., Sato, M. and Rodan, G.A. (1990) The effect of hemopoietic growth factors on the generation of osteoclast-like cells in mouse bone marrow cultures. *Endocrinology* **126**(3), 1728–1735.

5.   Akatsu, T., Tamura, T., Takahashi, N. *et al.* (1992) Preparation and characterization of mouse osteoclast-like multinucleate cell population. *Journal of Bone and Mineral Research* **7**, 1297–1306.

6.   Kerby, J.A., Hattersley, G., Collins, D.A. and Chambers, T.J. (1992) Derivation of osteoclasts from hematopoietic colony-forming cells in culture. *Journal of Bone and Mineral Research* **7**(3), 353–362.

7.   Tanaka, S., Takahashi, N., Udagawa, N. *et al.* (1993) Macrophage colony-stimulating factor is indispensable for both proliferation and differentiation of osteoclast progenitors. *Journal of Clinical Investigation* **91**, 257–263.

8.   Chambers, T.J., Owens, J.M., Hattersley, G. *et al.* (1993) Generation of osteoclast-inductive and osteoclastogenic cell lines from the *H-2K^b^tsA58* transgenic mouse. *Proceedings of the National Academy of Sciences USA* **90**, 5578–5582.

9.   Quinn, J.M., McGee, J.O. and Athanasou, N.A. (1994) Cellular and hormonal factors influencing monocyte differentiation to osteoclast bone resorbing cells. *Endocrinology* **134**, 2416–2223.

10. Wesolowski, G., Duong, L.T., Lakkakorpi, P.T. *et al.* (1995) Isolation and characterization of highly enriched, prefusion mouse osteoclastic cells. *Experimental Cell Research* **219**, 679–686.

11. Ho Shin, J., Kukita, A., Ohki, K.P. *et al.* (1995) In vitro differentiation of the murine macrophage cell line BDM-1 into osteoclast-like cells. *Endocrinology* **136**, 4285–4292.

12. Athanasou, N.A. and Quinn, J. (1990) Immunophenotypic differences between osteoclasts and macrophage polykaryons: immunohistochemical distinction and implications for osteoclast ontogeny and function. *Journal of Clinical Pathology* **43**, 997–1003.

13. Warshawsky, J., Goltzman, D., Rouleau, M.F. and Bergeron, J.J.M. (1980) Direct *in-vivo* demonstration by autoradiography of specific binding sites for calcitonin in skeletal and renal tissues of the rat. *Journal of Cell Biology* **85**, 682–694.

14. Nicholson, G.C., Moseley, J.M., Sexton, P.M. *et al.* (1986) Abundant calcitonin receptors in isolated rat osteoclasts. *Journal of Clinical Investigation* **78**, 355–360.

15. Flanagan, A.M., Horton, M.A., Dorey, E.L. *et al.* (1992) An assessment of the ability of human bone marrow cultures to generate osteoclasts. *International Journal of Experimental Pathology* **73**, 387–401.

16. Shinar, D.M. and Rodan, G.A. (1990) Biphasic effects of transforming growth factor β on the production of osteoclast-like cells in mouse bone marrow cultures: the role of prostaglandins in the generation of these cells. *Endocrinology* **126**(6), 3153–3158.

17. Owens, J.M., Gallagher, A.C., Chambers, T.J. (1996) Bone cells required for osteoclastic resorption but not for osteoclastic differentiation. *Biochemical and Biophysical Research Communications* **222**, 225–229.

18. Kaye, M. (1984) When is it an osteoclast? *Journal of Clinical Pathology* **37**, 398–400.

19. Takahashi, N., Udagawa, N., Tanaka, S. *et al.* (1994) Postmitotic osteoclast precursors are mononuclear cells which express macrophage-associated phenotypes. *Developmental Biology* **163**, 212–221.

20. Bianco, P., Costantini, M., Dearden, L.C. and Bonucci, E. (1987) Expression of tartrate-resistant acid phosphatase in bone marrow macrophages. *Basic Applied Histochemistry* **31**, 433–440.

21. Snipes, R.G., Lam, K.W., Dodd, R.C. *et al.* (1986) Acid phosphatase in mononuclear phagocytes and the U937 cell line: monocyte-derived macrophages express tartrate-resistant acid phosphatase. *Blood* **67**, 729–734.

22. Yan, L.T., Li, C.Y. and Finkel, H.E. (1972) Leukemic reticuloendotheliosis. The role of tartrate-resistant acid phosphatase in diagnosis and splenectomy in treatment. *Archives of Internal Medicine* **130**, 248–256.

23. Hattersley, G. and Chambers, T.J. (1989) Generation of osteoclastic function in mouse bone marrow cultures: multinuclearity and tartrate-resistant acid phosphatase are unreliable markers for osteoclastic differentiation. *Endocrinology* **124**, 1689–1696.

24. Horton, M.A., Rimmer, E.F., Lewis, D *et al.* (1984) Cell surface phenotype of the human osteoclast: phenotypic relationship to other bone marrow derived cell types. *Journal of Pathology* **144**, 281–294.

25. Horton, M.A., Lewis, D., McNulty, K. *et al.* (1985) Monoclonal antibodies to osteoclastomas (giant cell bone tumours): definition of osteoclast-specific cellular antigens. *Cancer Research* **45**, 5663–5669.

26. Flanagan, A.M., Tinkler, S.M.B., Horton, M.A. *et al.* (1988) The multinucleate cells in giant cell granulomas of the jaw are osteoclasts. *Cancer* **62**, 1139–1145.

27. Flanagan, A.M. and Chambers, T.J. (1989) Osteoclasts are present in the giant cell variant of malignant fibrous histiocytoma. *Journal of Pathology* **159**, 53–57.

28. Horton, M.A., Lewis, D., McNulty, K. *et al.* (1985) Human fetal osteoclasts fail to express macrophage antigens. *British Journal of Experimental Pathology* **66**, 103–108.

29. Davies, J., Warwick, J., Tutty, N. *et al.* (1989) The osteoclast functional antigen, implicated in the regulation of bone resorption, is biochemically related to the vitronectin receptor. *Journal of Cell Biology* **109**, 1817–1826.

30. MacDonald, B.R., Mundy, G.R., Clark, S. *et al.* (1986) Effects of human recombinant CSF-GM and highly purified CSF-1 on the formation of multinucleated cells with osteoclast characteristics in long-term bone marrow cultures. *Journal of Bone and Mineral Research* **1**, 227–233.

31. Chenu, C., Pfeilschifter, J., Mundy, G.R. and Roodman, G.D. (1988) Transforming growth factor β inhibits formation of osteoclast-like cells in long-term human marrow cultures. *Proceedings of the National Academy of Sciences USA* **85**, 5683–5687.

32. Hughes, D.E., MacDonald, B.R., Russell, R.G.G. and Gowen, M. (1989) Inhibition of osteoclast-like cell formation by bisphosphonates in long-term cultures of human bone marrow. *Journal of Clinical Investigation* **83**, 1930–1935.

33. Kukita, A., Bouewald, L., Rosen, D. *et al.* (1990) Osteoinductive factor inhibits formation of human osteoclast-like cells. *Proceedings of the National Academy of Sciences USA* **87**, 3023–3026.

34. Kukita, T. and Roodman, G.D. (1989) Development of a monoclonal antibody to osteoclasts formed *in vitro* which recognizes mononuclear osteoclast precursors in the marrow. *Endocrinology* **125**, 630–637.

35. Kukita, T., McManus, L.M., Civin, C. and Roodman, G.D. (1989) Osteoclast-like cells formed in long-term human bone marrow cultures express a similar surface phenotype as authentic osteoclasts. *Laboratory Investigation* **60**, 532–538.

36. Takahashi, N., Kukita, T., MacDonald, B.R. *et al.* (1989) Osteoclast-like cells form in long-term human bone marrow but not in peripheral blood. *Journal of Clinical Investigation* **84**, 543–550.

37. Chenu, C., Kurihara, N., Mundy, G.R. and Roodman, G.D. (1990) Prostaglandin E_2 inhibits formation of osteoclast-like cells in long-term human marrow cultures but is not a mediator of the inhibitory effects of transforming growth factor β. *Journal of Bone and Mineral Research* **5**, 677–687.

38. Kurihara, N., Gluck, S. and Roodman, G.D. (1990) Sequential expression of phenotype markers for osteoclasts during differentiation of precursors for multinucleated cells formed in long term human marrow cultures. *Endocrinology* **127**(6), 3215–3221.

39. Kurihara, N., Bertolini, D., Suda, T. *et al.* (1990) IL-6 stimulated osteoclast-like multinucleated cell formation in long-term human cultures by inducing IL-1 release. *Journal of Immunology* **144**, 4226–4230.

40. Kurihara, N., Chenu, C., Miller, M. *et al.* (1990) Identification of committed mononuclear precursors for osteoclast-like cells formed in long term human marrow cultures. *Endocrinology* **126**(5), 2733–2741.

41. Thavarajah, M., Evans, D.B., Binderup, L. and Kanis, J.A. (1990) 1,25(OH)$_2$D$_3$ and calcipotriol (MC903) have similar effects on the induction of osteoclast-like cell formation in human bone marrow cultures. *Endocrinology* **125**, 630–637.

42. Roodman, G.D. (1995) Application of bone marrow cultures to the study of osteoclast formation and osteoclast precursors in man. *Calcified Tissue International* **56** (Suppl. 1), S22–S23.

43. Pacifici, R. (1995) Cytokines and osteoclast activity. *Calcified Tissue International* **56**, S27–S28.

44. Sarma, U. and Flanagan, A.M. (1996) Macrophage-colony stimulating factor (M-CSF) induces substantial osteoclast formation in human bone-marrow cultures. *Blood* **88**, 2531–2540.

45. Wiktor-Jedrzejczak, W., Bartocci, A., Ferrante, A.W. Jr *et al.* (1990) Total absence of colony-stimulating factor 1 in the macrophage-deficient osteopetrotic (op/op) mouse. *Proceedings of the National Academy of Sciences USA* **87**, 4828–4832.

46. Felix, R., Cecchini, M.G., Hofstetter, W. *et al.* (1990) Impairment of macrophage colony-stimulating factor production and lack of resident bone marrow macrophages in the osteopetrotic op/op mouse. *Journal of Bone and Mineral Research* **5**, 781–789.

47. Felix, R., Cecchini, M.G. and Fleisch, H. (1990) Macrophage colony stimulating factor restores *in vivo* bone resorption in the *op/op* osteopetrotic mouse. *Endocrinology* **127**(6), 2592–2594.

48. Kodama, H., Yamasaki, A., Abe, M. *et al.* (1993) Transient recruitment of osteoclasts and expression of their function in osteopetrotic *(op/op)* mice by a single injection of macrophage colony-stimulating factor. *Journal of Bone and Mineral Research* **8**, 45–50.

49. Yoshida, H., Hayashi, S.-I., Kunisada, T. *et al.* (1990) The murine mutation osteopetrosis is in the coding region of the macrophage colony stimulating factor gene. *Nature* **345**, 442–444.

50. Mbalaviele, G., Orcel, P., Morieux, C. *et al.* (1995) Osteoclast formation from human cord blood mononuclear cells co-cultured with mice embryonic metatarsals in the presence of M-CSF. *Bone* **16**, 171–177.

51. Takahashi, N., Udagawa, N., Akatsu, T. *et al.* (1991) Role of colony-stimulating factors in osteoclast development. *Journal of Bone and Mineral Research* **6**, 977–985.

52. Hattersley, G. and Chambers, T.J. (1990) Effects of interleukin 3 and of granulocyte-macrophage and macrophage colony stimulating factors on osteoclast differentiation from mouse hemopoietic tissue. *Journal of Cellular Physiology* **142**, 201–209.

53. Jaworski, Z.F.G., Duck, B. and Sekaly, G. (1981) Kinetics of osteoclasts and their nuclei in evolving secondary Haversian systems. *Journal of Anatomy* **133**, 397–405.

54. Loutit, J.F. and Townsend, K.M.S. (1982) Longevity of osteoclasts in radiation chimeras of osteopetrotic beige and normal mice. *British Journal of Experimental Pathology* **63**, 221–223.

55. Marks, S.C. Jr and Schneider, G.B. (1982) Transformation of osteoclast phenotype in rats cured of congenital osteopetrosis. *Journal of Morphology* **174**, 141–147.

56. Fuller, K., Owens, J.M., Jagger, C.J. *et al.* (1993) Macrophage colony-stimulating factor stimulates survival and chemotactic behavior in isolated osteoclasts. *Journal of Experimental Medicine* **178**, 1733–1744.

57. Fuller, K. and Chambers, T.J. (1987) Generation of osteoclasts in cultures of rabbit bone marrow and spleen cells. *Journal of Cellular Physiology* **132**, 441–452.

58. Flanagan, A.M., Stow, M.D. and Williams, R. (1995) The effect of interleukin-6 and soluble interleukin-6 receptor protein on the bone resorptive activity of human osteo-clasts generated *in vitro*. *Journal of Pathology* **176**, 289–297.

59. Flanagan, A.M., Stow, M.D., Kendall, N. and Brace, W. (1995) The role of dihydroxy-vitamin D3 and prostaglandin E2 in the regulation of human osteoclastic bone resorption. *International Journal of Experimental Pathology* **76**, 37–42.

60. Fujikawa, Y., Quinn, J.M.W., Sabokbar, A. *et al.* (1996) The human osteoclast precursor circulates in the monocyte fraction. *Endocrinology* **137**, 4058–4060.

61. Quinn, J.M.W., McGee, J.O. and Athanasou, N.A. (1994) Cellular and hormonal factors influencing monocyte differentiation to osteoclastic bone-resorbing cells. *Endocrinology* **134**, 2416–2423.

62. Hunter, W.M. and Greenwood, F.C. (1962) Preparation of iodine 131 labelled human growth hormone of high specific activity. *Nature* **194**, 495.

63. Saema, U., Edwards, M., Motoyoshi, K. and Flanagan, A.M. (1997) Inhibition of bone resorption by 17β-estradiol in human bone-marrow cultures. *Journal of Cellular Physiology*. In press.

CHAPTER THREE

Isolation and culture of osteoclasts and osteoclast resorption assays

Richard J. Murrills, David W. Dempster and Timothy R. Arnett

3.1 INTRODUCTION

The central role that bone resorption plays in the pathogenesis of human disorders such as osteoporosis, rheumatoid arthritis, Paget's disease, hypercalcaemia of malignancy and osteopetrosis has long prompted the scientific community to inquire into the biology of the osteoclast. Over time, the investigative tools have changed. In recent years, the advances seen first in *in vitro* organ culture of bone, then in the isolation, purification and culture of selected cell types and ultimately in the molecular biology of osteoclasts have led to a rapidly improving understanding of the molecular engines comprising the osteoclast and its interactions with neighbouring and distant cell types. This chapter will begin by outlining the techniques that have been developed for isolating and purifying osteoclasts before concentrating on the technique with which we have most of our personal experience, the assay of bone resorption using isolated osteoclasts on bone slices, and finally closing with a summary of the advances that have been possible using these techniques. Our definition of an osteoclast is based upon function, namely that the ability to resorb a pit under conducive conditions is an essential requirement.

3.2 HISTORY AND SURVEY OF TECHNIQUES

3.2.1 Organ culture assays of bone resorption

Organ culture assays of bone resorption, using whole fetal or neonatal bones (usually minus the cartilaginous epiphyses), were the main method by which the functional ability of the osteoclast was investigated *in vitro* for 20 years following their inception in the early 1960s [1, 2] and continue to be in wide use today. A wide range of hormones, cytokines, growth factors, arachidonic acid metabolites, inhibitors of the mechanism of bone resorption, regulators of intracellular signalling, prospective anti-resorptive drugs and inorganic ions have been tested in these assays and they have provided valuable information on the mechanism and regulation of bone resorption. These techniques have the advantage that the

Methods in Bone Biology. Edited by Timothy R. Arnett and Brian Henderson.
Published in 1997 by Chapman & Hall, London. ISBN 0 412 75770 2.

bone is separate from all circulating and most mechanical influences, making these controllable by the investigator. Also, the bone cells remain in a lifelike spatial arrangement, increasing the likelihood that cellular interactions are representative of those that occur in the body.

The disadvantages of these systems are that the target cells of a particular hormone or other agent can never be adequately identified: the agent could act on bone resorption either by a direct action on the osteoclast or the osteoclast precursor, or through an indirect action on any of the other cells present in the bone (osteoblasts, marrow stromal cells). Hence, the conclusions from this assay are limited because the relevant cell populations cannot be isolated or manipulated. Similarly, biochemical or genetic studies undertaken using these models are ambiguous as to the cell type being studied. A further limitation is that measurement of bone resorption is usually by means of measuring calcium release, whether measured by radioactive techniques or not. This has been shown to correlate with osteoclast number [3], but calcium measurement itself is not a direct measurement of bone resorption and provides no information concerning the number and size of the resorption lacunae.

3.2.2 Isolation of osteoclasts

The reasons for isolating and purifying osteoclasts and for developing isolated osteoclast assays were many. To perform biochemical and genetic studies, large numbers of purified osteoclasts were required. Techniques had to be developed to study the functional ability of the isolated osteoclast to resorb bone *in vitro*, to enable qualitative study of the osteoclast as it resorbed bone in a controlled environment and to provide a system in which the different cellular components of bone could be studied in isolation and their relationships manipulated.

A first step to achieving this was the isolation of osteoclasts. From the start, the osteoclast presents special problems. First, it is a rare cell, making up approximately 1% or less of the cells associated with bone or bone marrow, making the recovery of large numbers of osteoclasts extremely difficult. Second, it is terminally differentiated, being incapable of dividing within the body or *in vitro*, making long-term culture of primary osteoclast cell lines impossible. Third, it is associated with a tough, mineralized material, making recovery of the osteoclasts difficult and exposing the cells to potential damage. Over the last 15 years, a range of techniques has been developed to overcome these problems.

The first step in isolating osteoclasts is choice of source. Large numbers of differentiated osteoclasts are only routinely obtainable from chickens, particularly laying hens or sometimes younger animals, that have been fed a low calcium diet. Similar numbers of human osteoclasts can be obtained from giant cell tumours of bone, though these will not be routinely available to an investigator. Apart from these two models, the yield diminishes and the choice narrows to fetal or neonatal laboratory animals – rats, rabbits, mice or chicks. In the case of mice, reasonably large numbers of osteoclasts have been obtained from 2 h sedimentations of the marrow from 13-day-old animals, with pitting analysed over a period of several days [4, 5]. Osteoclasts have also been recovered from human fetal material [6, 7] but this, again, is not routinely available. To obtain osteoclasts from young adult or adult material, or simply to increase the numbers of osteoclasts, a source of osteoclast precursors can be used that can be induced to mature into differentiated osteoclasts *in vitro*. This approach has been used with chickens, with great success in purification [8, 9], and also with mouse [10] and rat [11]. Recent publications suggest that this should also be possible with human osteoclasts [12, 13].

3.2.3 Recovery of osteoclasts: enzymatic, hormonal and mechanical detachment

Three approaches are available for removing differentiated osteoclasts from the surface of bone: enzymatic release, induced detachment and mechanical release. The traditional tissue culture methods of enzymatic release were among the first to be applied to the recovery of osteoclasts. Attempts to recover osteoclasts by sequential collagenase digestion of mouse calvaria succeeded in isolating cells that had some of the characteristics of osteoclasts – namely, calcitonin responsiveness and some ability to release calcium from devitalized calvariae when incubated for 72 h [14]. The identity of these cells still remains uncertain. Initially mononuclear, they formed some multinuclear cells upon culture with devitalized bone in the above experiment and may represent a mixed population of cells that contains some osteoclast precursors. More recent techniques for recovering *bona fide* osteoclasts from the medullary bones of hypocalcaemic chickens have used collagenase, trypsin [15] and dispase, which has also been used to recover human osteoclasts from giant cell tumour of bone [16].

Early observations that calcitonin caused avian osteoclasts to retract from the bone [17] prompted some investigators to use this as a means of detaching osteoclasts from the bone surface. A modification of this approach has recently been used to recover mouse osteoclast precursors from *in vitro* marrow cultures using echistatin to detach the cells [18]. Such approaches have the potential both for removing the cell from the substrate with minimal damage and also, if sufficiently selective agents are used, for generating a purified population.

Mechanical release has been the main technique by which osteoclasts have been recovered. This originated in the early work of Boyde and Jones [19] and Chambers *et al.* [20] in the recovery of osteoclasts from neonatal rabbits, rats and embryonic chicks. Osteoclasts are removed from bone either by mincing the tissue of fetal or neonatal animals or by scraping the exposed endosteal surfaces of longitudinally split long bones. These simple techniques remain in wide use today.

More recently, osteoclasts have been recovered from marrow cultures, offering the possibility of handling much larger numbers of osteoclasts from mammalian sources [10]. The recovery of osteoclast precursors from mammalian marrow cultures or avian sources is a technique that allows considerable yield and purity [8, 9, 18]. Avian osteoclasts that form from precursors are capable of causing release of calcium from bone particles, but mammalian osteoclast precursors appear to require co-culture with osteoblastic cells to acquire bone resorptive function, at least as measured by the bone slice assay [18].

3.2.4 Purification of osteoclasts: unit gravity sedimentation, density gradient separation, immunomagnetic techniques, adhesion

It is at present unlikely that any preparation of primary mature osteoclasts will be 100% pure. Unit gravity sedimentation of chicken osteoclasts was the first technique to yield large numbers of osteoclasts [21] and has since been used by other laboratories for the purification of avian osteoclasts [22, 23]. Density gradient separation has been employed to generate bone marrow mononuclear cells containing osteoclast precursors [24] and also to further purify avian osteoclasts recovered by enzymatic digestion [15, 25]. Immunomagnetic techniques using antibody 121F have been used on chicken and human osteoclasts [15, 16, 25] and antibody to mouse vitronectin receptor has been used to recover osteoclasts immunomagnetically from mouse marrow cultures (Hong, personal communication) grown on collagen gels and then treated with collagenase to recover cells. The osteoclast's ability to adhere rapidly to plastic or bony substrates has also been widely used as a simple technique in partially purifying osteoclasts from mixed cell suspensions and obtaining 'functionally purified' osteoclasts incapable of responding to parathyroid hormone, IL-1, 1,25 dihydroxyvitamin D_3 and TNF [26–30]. However, the degree of purification obtained by this method is not as impressive as by other methods. Evely *et al.* [31] quote purification factors of approximately two-fold, with osteoclasts still very much in the minority. Longer adhesion times have been successfully used to remove mature osteoclasts from mixed bone marrow suspensions, leaving only relatively undifferentiated precursors [11]. Enzymatic treatments that rely upon the osteoclast's ability to adhere tightly to surfaces have been used with success to remove non-osteoclastic cells from *in vitro* cultures of rabbit osteoclasts with pronase and EDTA [32] and from co-cultures of CFU-enriched marrow with ST2 cells with collagenase [33]. Recently, May and Gay [34] have explored methods for the purification of avian osteoclasts by selective adhesion to vitronectin-coated surfaces following migration from endosteal surfaces.

3.2.5 Culture of osteoclasts

Differentiated osteoclasts survive only a few days at most *in vitro*, with time-course studies showing a steady decline in mature osteoclasts in the pitting assay in the absence of conditions under which new osteoclasts can be formed from precursors [35]. Four to five days has been used as a time during which existing osteoclasts will die or detach from the culture dish and so leave a preparation of marrow cells depleted in mature osteoclasts for differentiation studies [4, 5, 36]. The culture of differentiated osteoclasts is therefore initially a question of maintenance of mature osteoclasts. In this regard, very little has been studied, although M-CSF and IL-1α have been shown to prolong the survival of differentiated osteoclasts *in vitro* [37, 38]. If adequate numbers of osteoclast precursors are present in the preparation or continue to be generated, then longer term culture and passaging becomes possible. Jones *et al.* [39] have maintained chick marrow preparations for 6 weeks with continued resorption. Chick marrow cultures contain large numbers of TRAP+ mononucleated osteoclast precursors that can be shown to differentiate into functional osteoclasts [40]. Recently, Zambonin-Zallone and colleagues have obtained a giant cell tumour sample that appears to be capable of continually generating osteoclast precursors and forming osteoclasts even after several passages [41].

3.2.6 The search for osteoclast cell lines

An osteoclast-like cell line would clearly be invaluable for biochemical, genetic and cell–cell interaction studies and has been sought for nearly two decades. As yet, a satisfactory cell line remains elusive, but several close approximations do exist.

HL-60 cells are a human promyelocytic cell line that has shown the ability to differentiate into multinucleate cells capable of excavating pits in bone slices, binding calcitonin and expressing tartrate-resistant acid phosphatase [42]. However, the procedure for differentiating these cells is lengthy and complex, involving culture in methylcellulose, subcloning, incubation with two different types of conditioned media and immune panning. Furthermore, the cells only form pits in bone slices at the rate of approximately 10 pits/10 000 cells per 7 days. Hence, although these studies show that differentiation along the osteoclast lineage is possible for HL-60 cells, they do not appear to represent, at least at present, a valuable source of human osteoclasts for *in vitro* use.

FLG 29.1 cells are human leukaemic cells that have been reported to differentiate into cells resembling osteoclasts following exposure to phorbol ester [43]. The differentiated cells display an antigenic profile similar to human fetal osteoclasts, with some multinuclearity, a cAMP response to calcitonin (albeit weak) and the ability to resorb bone particles. However, FLG 29.1 cells have not been demonstrated to resorb pits in bone slices. Hence, although these cells display some of the characteristics of osteoclasts, they cannot be said to be an osteoclast cell line. However, they may be sufficiently close to an osteoclast to provide useful biochemical information about human osteoclasts that is not available by other means and are being used in a growing number of studies by their originators.

BDM-1 cells are a mouse macrophage-like cell line, an F4/80 antigen subclone of which can differentiate into cells showing some osteoclast characteristics, including the ability to form pits on bone slices [44]. An assessment of the usefulness of this subclone as a model of osteoclast biology must await further study by other laboratories.

Zambonin-Zallone and co-workers have derived cell populations from human giant cell tumour of bone (osteoclastoma) that retain the ability to form multinucleate osteoclast-like cells with passaging *in vitro* [41]. These show a weak cAMP response to calcitonin and can resorb bone from bone particles and bone slices. Autoradiography detects calcitonin receptors on both mononuclear and multinuclear cells in the cultures. These cells have been used in a number of cell signalling studies.

Chambers *et al.* [45] used the H-2KbtsA58 transgenic mouse to create cell lines capable of differentiating into osteoclasts. The cells from this transgenic mouse can be conditionally immortalized by exposing them to interferon at 33°C. This system requires the presence of similarly immortalized stromal cells to support differentiation of the osteoclastogenic cell line. Boyce *et al.* [46] have also attempted to create an osteoclast cell line using transgenic technology, by targeting the SV40 T-antigen to the tartrate-resistant acid phosphatase (TRAP) promoter. However, despite the existence of osteoclast tumours and sheets of osteoclasts ('tumourlets') and mitotic osteoclasts in the mice and the ability of isolated osteoclasts to resorb bone *in vitro*, no cell line could be developed from this transformation.

Despite this recent progress in the identification of osteoclast-like cell lines, there are as yet no published reports of their routine use in assays of bone resorption. In some cases, this may be due to a limited ability of some of the cells to make pits in bone slices. However, a *bona fide* osteoclastic cell line remains an attractive idea and one that is the subject of active research.

3.2.7 Particle-based osteoclast resorption assays

Bone particle-based assays were the first bone resorption assays to be used with isolated cell populations and initially employed monocytes and macrophages as surrogate osteoclasts [47–49]. Bone resorption in these assays is detected by the release of radiolabelled calcium or proline from the particles, which have usually been sieved to achieve a particle size of greater than 10 μM to avoid the possibility of degradation by phagocytosis, a phenomenon not thought to be representative of physiological bone resorption. Avian osteoclasts (either mature osteoclasts or osteoclast precursors allowed to differentiate into mature osteoclasts *in vitro*) have more recently been used with this approach [22, 50], as have mammalian osteoclasts derived from bone marrow cultures [51]. Osteoclast-to-bone contact and clear zones and ruffled borders have been noted in this system [52] and bone particle assays continue to be used today. The disadvantage of this system is that cells that cannot excavate pits in bone slices (e.g. macrophages, monocytes and preosteoclasts such as FLG 29.1 cells) are capable of releasing calcium and proline from bone particles. Hence, there is some concern that at least part of the 'resorption' seen in this assay represents a process that is different from the excavation of Howship's lacunae and may be of restricted physiological relevance. A similar technology, using devitalized calvaria prelabelled with radioactive calcium and proline, has been used to quantify resorption by chicken osteoclasts [53]. In this assay, ruffled borders are also demonstrable and resorption is inhibitable by calcitonin and associated with acidification of the resorption lacuna.

3.2.8 Bone slice-based osteoclast resorption assays

Bone slice-based assays were developed in the early 1980s by Boyde and Jones [19] and Chambers *et al.* [20]. These consist of a bone (or dentine) slice upon which is incubated a population of cells containing osteoclasts. It has clear advantages over the organ culture assays of resorption, because isolated cells can now be used and the cell population can be manipulated, and over the particle assays, because measurement of resorption is by direct quantitation of osteoclastic resorption lacunae, the *in vitro* version of Howship's lacunae. A detailed description of the way in which this assay is performed in our laboratories is presented below, followed by a summary of the advances in knowledge that have been gained from it.

3.3 THE DISAGGREGATED OSTEOCLAST RESORPTION ASSAY

3.3.1 Our laboratory methods

This section is intended simply to set out the procedures used in our laboratories, with minimal discussion of variations used by other workers or the issues surrounding some of the procedures. These will be dealt with in sections 3.3.2 and 3.3.3, respectively. In this way, a newcomer to the technique can see the details of our methods clearly and in somewhat greater detail than has been presented in our publications. In doing so, we are not necessarily saying that this is the only way to perform the assay, simply that this is what we do and for several years it has worked satisfactorily. Over such a period of time, it is inevitable that each of us has introduced modifications to the assay: we have attempted to incorporate these into our description without creating too much confusion.

(a) Preparation of bone/dentine slices

Whole bovine femurs can be obtained either from animal tissue suppliers or from local butchers. Confiscated dentine (mainly elephant) may often be obtained on request from customs, fisheries and wildlife agencies. We initially obtained whole bovine femurs from slaughterhouses as soon as possible after killing the animal and these were shipped frozen or on ice before being stored in a –80°C freezer to minimize post-mortem degradation. However, some leeway may be permissible in the immediate post-mortem protocol – bones obtained free from local butchers appear to work just as well. Cylinders (50–75 mm) are cut from the diaphyseal region where cortical bone is thickest, using a hacksaw. Surplus adherent tissue is stripped off the periosteal surface using a scalpel. The cylinders are then cut radially to create 3–5 'arcs' of bone, which are then 'defatted' by sonicating for two 30-minute periods in a dilute solution of 7X detergent (ICN, Costa Mesa, CA) in distilled or Milli-Q water. The arcs are then clamped into a Buehler low speed saw equipped with a diamond wafering blade and arc-shaped slices 600 μm thick are cut using 10% ethanol as wetting solution. One complete turn of the micrometer on the Buehler saw is roughly equivalent to 600 μM. The slices are transferred to Milli-Q water on ice in a 60 mm plastic tissue culture Petri dish as they are cut and stacked in register. To make bone squares, batches of 10–20 arc-shaped slices are then clamped in such a way that 4 mm strips can be cut (Fig. 3.1) while grasping the unclamped part of the arcs firmly between fine forceps. Seven complete turns of the micrometer will produce approximately 4 mm strips. The strips are then turned around, clamped and cut into squares, again grasping the unclamped end firmly between forceps. In this way, large numbers of slices can be generated at one time – the use of more than one saw at a time to generate the arcs also increases the rate at which slices can be made. One of our laboratories (RJM) is currently using three at one time.

Stacks of slices are placed in plastic tissue culture Petri dishes as they are made and, when sufficient have been collected, a batch of slices is washed and cleaned in three changes of Milli-Q water with sonication before being air-dried in a laminar flow hood. Slices can be stored dry either in the refrigerator or frozen. Bone and dentine slices can also be prepared as 5 mm disks using a standard paper hole punch on wet slices – these fit very well into the wells of 96-well plates.

Prior to an experiment, the slices are numbered on one side with a pencil and each slice is placed in one of the wells of a 96-well plate, number side down. Slices can be autoclaved or sterilized using ethanol and dried in a laminar flow cabinet if sterility is a problem or long-term incubations (longer than 2 days) are being used. Each slice is then pre-wetted with 100 μl of settling medium (Medium 199/Hanks) at 37°C. If settling of cells is to be performed in a water bath, the plate containing the slices is placed on a frame (in our case a plastic-coated wire test-tube rack) in a water bath such that the bottom of the plate is completely contacted by water at 37°C.

(b) Preparation of reagents

The culture medium used for the incubation phase of the assay is Medium 199 with Earle's salts containing 0.7 g $NaHCO_3$/l (to ensure a pH of 6.9 or below on

Figure 3.1 Manufacture of bone slices using a Buehler low-speed saw equipped with a diamond wafering blade.

incubation in a 5% or 10% CO_2 environment) with 10 mM HEPES buffer and 10% heat-inactivated fetal calf serum. Powdered medium containing no $NaHCO_3$ is purchased from Gibco (Grand Island, NY) and bicarbonate is added to a final concentration of 0.7 g/l before HEPES and antibiotics (penicillin/streptomycin or amphotericin) are added and the medium is filtered with a 0.2 µm filter and stored in tissue culture flasks. Milli-Q water or water purchased from tissue culture suppliers can be used to dilute the powder. Alternative methods of lowering pH are: adding 10 mEq HCl/l to standard bottled Medium 199; or incubating standard Medium 199/Earle's in 10% CO_2. In our experience, low bicarbonate appears to reduce pit size somewhat and the best resorption is obtained using 10% CO_2. The possible toxicity of HEPES should also be considered; see also section 3.3.3 (*f*).

Dilution of compounds, peptides or other agents to be tested is performed under the laminar flow hood in culture medium in 15 ml centrifuge tubes and 0.4–1.0 ml aliquots of each of the final solutions are placed in the wells of a 24-well plate. This is usually done either prior to commencing dissection or while cells are settling. Time taken is 30–60 minutes, depending on solubility and number of different vehicles used. The plates are then incubated in the 37°C/10% CO_2 incubator until it is time to commence the transfer of slices, by which time they will have equilibrated to a pH level of approximately 6.9. Reagents of questionable stability can be added in minimal volume to the wells once slices have been transferred to the 'treatment' wells of the 24-well plate. The pH of the equilibrated culture medium is most accurately measured using a blood gas analyser in which the CO_2 concentration is measured simultaneously, thus permitting a correction for the pH drift (upwards) that occurs as CO_2 is lost as soon as the medium is removed from the incubator.

(c) Source of osteoclasts

We have routinely used osteoclasts from neonatal rats (aged 2–6 days, usually 4 days) and embryonic chicks (aged 16–21 days, usually 17 days). Neonatal rats are routinely available in most laboratories and we have used both Sprague-Dawley and Wistar with no obvious strain differences. Embryonic chicks are obtained by purchasing fertilized eggs from Spafas (Preston, CT) and incubating in an egg incubator until required. Both neonatal rats and embryonic chicks are killed by decapitation, or by cervical dislocation if decapitation is not permitted. Five animals are used per experiment. In terms of resorptive efficiency, embryonic chick osteoclast preparations resorb more bone (1–3 pits/osteoclast per day) than neonatal rat osteoclast preparations (0.2–0.5 pits/osteoclast per day). To date, it seems that purified osteoclasts are almost incapable of resorbing pits at a comparable level in this assay – purified mouse osteoclasts resorb very little unless osteoblasts are added [54] and purified chicken osteoclasts resorb at a rate of one pit per 20 000 preosteoclasts [55].

(d) Dissection of bones

All long bones (humeri, radii, ulnae, femora, tibiae, fibulae) are removed and freed of adherent muscle and their epiphyses are cut off with a scalpel at the level of the growth plate and discarded (Fig. 3.2). This dissection is performed on a dissecting board using a scalpel and fairly fine forceps to hold the bones. Once clean, the bones are then transferred to 35 mm plastic tissue culture Petri dishes containing 3.5 ml medium 199/Hanks salts without serum, on ice. We routinely have two people dissecting animals – time taken for dissection of bones from five animals (10 bones per animal, ignoring fibulae) is approximately 30 minutes.

(e) Preparation of cell suspension and plating

Bones are then minced to create a cell suspension, using a scalpel and holding the bones with forceps. Mincing typically takes about 3–5 minutes per animal. The bone debris is then repeatedly sucked in and out (20–50 times) of a transfer pipette that has had its tip cut back such that the opening is approximately 5 mm. Refinements introduced into this aspect of the assay that may increase yield include mincing in non-tissue culture treated dishes and vortexing the final matrix/cell mixture for 20 seconds. The Petri dish is then tilted at a slight angle so as to allow the bulk of the bone matrix to settle for about 10 seconds into the corner of the dish and the cell suspension is then taken up gently from the top of the liquid into a (unaltered) transfer pipette, trying to avoid any large pieces of bone matrix debris. This is then plated into a V-bottomed plastic trough and the suspension mixed gently by sucking in and out of the transfer pipette. The cell suspension is then taken up into the tips of an 8-tip pipette and 100 µl is plated onto each of the bone

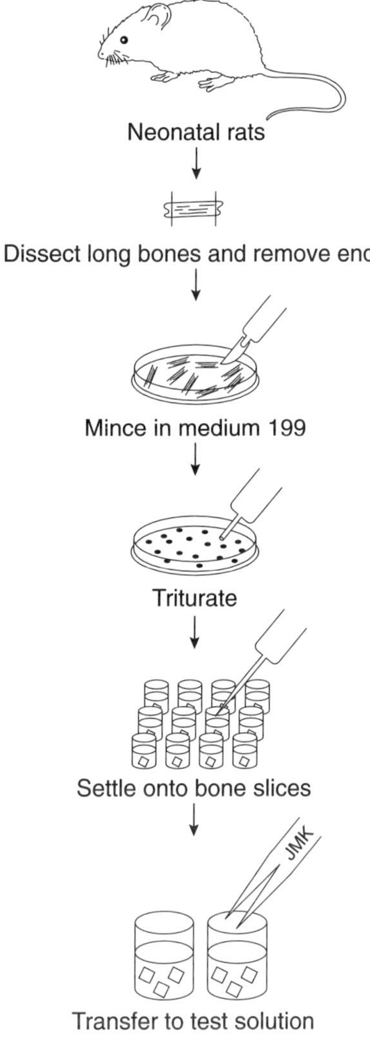

Figure 3.2 Dissection of long bones, preparation of cell suspension and plating of cells onto bone slices prior to incubation. Neonatal rats are shown as the source of osteoclasts but we have also routinely used embryonic chick as a source of bones.

slices that have been placed in the wells of the 96-well plate and prewetted with 100 µl of Medium 199/Hanks salts. During settling, the 96-well plate is maintained at 37°C using a water bath (see above) or by transferring it, as soon as cells have been added, to a 37°C humidified air incubator if PBS or medium 199/Hanks is the settling medium, or to a CO_2 incubator if using MEM/Earle's. The slices are arranged such that each dispensing from the 8-tip repeating pipette (Eppendorf with attachment) goes onto one slice from each of the treatments to be tested, in an attempt to avoid any apparent treatment effects that are in fact plating error (Fig. 3.3). If an 8-tip pipette is not used, a single-tip pipette or repeating pipette can be used. In this case, plating error is minimized by plating successive 100 µl

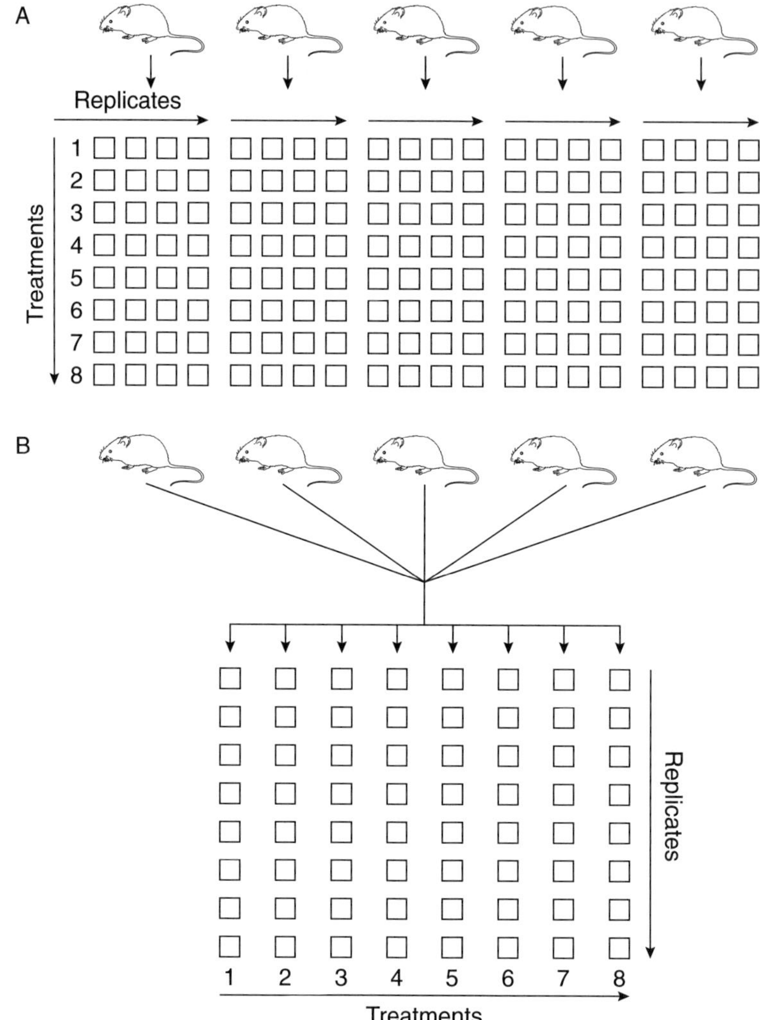

Figure 3.3   Design of experiment using the pitting assay. Each square represents a bone slice in a well of a 96-well plate. In Design A, cells from individual animals are kept separate. In Design B, cells from five animals are pooled to make a single cell suspension.

doses in an up-and-down pattern (Fig. 3.4). When a single-tip Eppendorf repeating pipette with Combitip (VWR, Plainfield, NJ/Brinkmann Instruments, Westbury, NY) or Gilson Pipetman single-shot pipette (Rainin, Woburn, MA) has been used in our laboratories, the tips have been cut back to minimize blockage and possible damage to the cells. However, when we adopted the 8-tip repeating pipette, this practice was abandoned with no obvious adverse effects. When cells from five animals are pooled before being plated (see experimental Design B in section (k) below), the procedure changes slightly. Instead of mincing in 2–4 ml of medium, the cell suspension is prepared in 20 ml. Time taken to prepare cell suspension and plate by Design A is 15–20 minutes and for Design B is 10–15 minutes. We routinely have one person mincing bones and another preparing the cell suspension and plating when using Design A, but one person is sufficient to perform Design B.

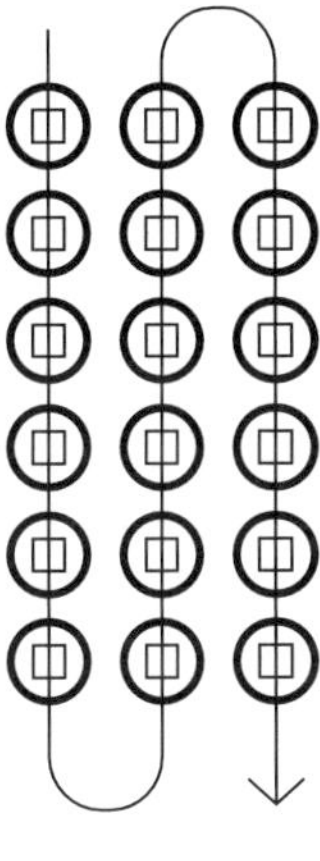

Plating

Figure 3.4 Up-and-down method of plating the cell suspension using single-shot pipettes. This method minimizes plating error between treatment groups, as each treatment receives cells from non-consecutive stages of the plating process.

(f) Settling onto bone slices, rinsing and transfer into culture wells
We settle osteoclasts for 25–35 minutes prior to removing them from the wells (still keeping the plate on the water bath) using fine forceps and rinsing them in 5 ml of prewarmed (in an air incubator) Medium 199/Hanks to remove non-adherent cells. As a rule, we attempt to standardize the rinsing by doing 10 side-to-side movements per slice. The slices are then transferred to the wells of 24-well plates that hold medium containing the various treatments; these are also resting on the frame in the water bath. The forceps are rinsed in fresh medium after each transfer to avoid cross-contamination of treatments. Slices are transferred in the up-and-down pattern previously described for plating (Fig. 3.4). All this is done on the bench, not under the laminar flow hood. Time taken to transfer is approximately 30 minutes using Design A and approximately 10 minutes for Design B.

(g) Stopping the experiment
At the end of the incubation period (see below), the experiment can be stopped either by fixation or by simply removing the cells by sonication. We have successfully used glutaraldehyde fixation (2% glutaraldehyde in 0.1 M cacodylate buffer) for 30 minutes at 4°C for staining pits with toluidine blue and for histochemical staining of osteoclasts with TRAP. In our experience, simply removing the cells by sonication without fixation is also sufficient for staining pits with toluidine blue. For double-staining cells for TRAP and alkaline phosphatase, a brief (5 minutes) fix with citrate/acetone/formaldehyde (the fixative from the Sigma alkaline phosphatase kit) prior to staining first for alkaline phosphatase and then for TRAP results in good double staining, with osteoclasts showing red and osteoblasts blue (Plate 3).

(h) Staining pits
Pits are stained on slices that have had the cells removed by sonication for two 15-second bursts in 0.25 M NH_4OH followed by one 15-second burst in distilled

water and two 15-second bursts in acetone. The cells are then air-dried and the numbering on the reverse of the slice is refreshed with a pencil. Slices are stained by floating approximately 10 slices at a time, resorbed side down, on a small pool of 1% toluidine blue in 1% borax in a 35 mm Petri dish for 4 minutes. The slices are then rinsed by dropping them into a large volume (approximately 300 ml) of distilled water in a Tri-pour beaker, pouring off the water once all the slices are in and filling up again with water before draining and air-drying the slices. The slices are then ready for counting.

(i) Quantitation of resorption

Our routine method of quantifying resorption is simply to count the pits 'blind', revealing the pencilled number only after the count is performed. Counting is performed by placing eight slices at a time on a microscope slide in a line and counting each in turn. Each slice takes about 3 minutes to count – longer if there are numerous (> 50) pits. We find that applying strips of tape to the slide to create a 'channel' for the slices helps in maintaining their position and orientation. Slices are scanned up and down using a ×16 to ×25 objective and an eyepiece reticle to mark the side limits of the scanned track and thus ensure that pits are not counted twice. Pits that are completely separated by an area of unresorbed bone are counted as separate pits and are recorded with an electronic counter. If the resorbed area is continuous, no matter how large the area or how thin the connection, it is counted as one pit. If a pit is recognizable as such, it is counted, no matter how small. In training observers to recognize pits, it is first important to become familiar with the background morphology of the bone slice as it appears after staining and then to look for stained areas that have some circularity or at least some clearly defined rounded margin. Smaller pits tend to be round, echoing the ring shape of the clear zone, but larger pits can be complex because they may have been formed by multiple resorptive episodes superimposed on one another as the osteoclast moves on the bone surface. Elongated serpentine pits that we refer to as 'snail trails' are presumably formed by an osteoclast continuing to resorb as it moves across the surface of the bone slice. Larger clear zones tend to deviate from a perfect ring, part of the margin bending in towards the centre to become kidney-shaped in some cases [35] – clear zones such as this are presumably responsible for kidney-shaped pits.

In our experience, staining of pits with toluidine blue is satisfactory but can be uneven, some pits staining strongly and others weakly. In addition there can be different staining intensities within one pit. A more difficult problem are slices that 'overstain' with toluidine blue, even though they have only been exposed for 4 minutes. The background bone, instead of being white or only lightly stained, is a rich purple color, with the pits showing as areas of light blue, making it much more difficult to recognize and spot the pits. This appears to be due to the stainability of the slice, not to the stain or the staining time. These can be counted, albeit with some difficulty, but use of reflected light may circumvent this problem. Alternatively, some destaining can be achieved using 70% ethanol. The problem of background staining and the presence of morphological features such as osteons and Volkmann's canals can be avoided by the use of dentine or ivory as the mineralized substrate.

In some experiments, and when we were initially establishing the assay, we quantified resorption by measuring the plan area of the pits and expressing bone resorption as the total plan area resorbed per slice. Area measurement can be performed using any modern morphometry program and takes about four to five times as long as counting the pits (see discussion below).

(j) Cell counts

Osteoclasts are best counted following staining for TRAP. We follow the definition of an osteoclast as a TRAP+ cell having three or more recognizable nuclei (see below) and count all the osteoclasts on all slices (DWD) or a subsample of slices (section (k) below). Osteoblasts are quantitated (primarily to gain information on possible cytotoxicity of agents being tested and also to check for mitogenicity) by counting cells in two random fields defined by an eyepiece reticle using a ×16 or ×25 objective per slice. These can be converted to an estimate of the number of osteoblasts per slice or per unit area, if desired.

(k) Experimental design

We have used two basic experimental designs (Fig. 3.3), each one usually using cells from five animals. Design A, with which we have had most experience, keeps the cells from each animal separate and treats each animal as a 'block' in a block design, using three or four replicates per treatment per animal (a total of 15–20 slices per treatment), whereas Design B pools cells from the five animals and uses eight replicates of each treatment.

Design A   The cell suspension from each animal is prepared separately and sequentially and applied to four replicate slices for each treatment. The four replicates can either be allocated so that three are used for pit counts and one for cell counts, or so that TRAP staining is performed on all slices, followed by pit counts, with expression of resorption as pits/osteoclast per slice. After losses of medium to bone matrix and plasticware, 3.5 ml of medium is required to make sufficient cell suspension from all of the long bones of one animal to plate eight treatments, including control.

Design B   A cell suspension is prepared from all the bones from the five animals in 20 ml of medium and plated onto eight replicate slices per treatment, ensuring that plating is performed such that each 'squirt' from the repeating 8-channel pipette is applied equally to all treatments to minimize the effects of possible plating error on the experiment.

(l) Statistics

Design A is analysed by two-way analysis of variance and differences from control are detected by either the method of least significant difference (LSD) or Dunnett's test, which is more conservative. Design B is analysed by one-way analysis of variance, again followed by LSD or Dunnett's. Log or root transformation is often required to bring the data into the normal distribution and homogeneous variances that allow ANOVA to perform at its best. Non-parametric tests such as Mann–Whitney or Wilcoxon can be used if the requirements of normality and homogeneous variances are still not satisfied even after transformation. This is particularly true for pit areas that are typically distributed in a highly skewed fashion, there being far more small pits than large ones.

(m) Troubleshooting

The most serious problem that is encountered in this assay is a lack of pits. In our experience, there is considerable variability in the mean control pit numbers over time and often from experiment to experiment. We have even noted a marked difference between an assay performed one morning and one performed that same afternoon using the same materials and pups from the same litter. When such inter-experiment variability in the basal level of resorption is present and the overall average level of resorption for the laboratory is somewhat low (10 pits per slice or less), deviation below the average level can result in assays in which resorption is so low (less than 5 pits per slice) that meaningful conclusions are all but impossible, because even large (50%) inhibitory effects do not produce statistically significant differences. Low pit numbers can be doubly frustrating as they appear to have no obvious cause. In many cases, there are apparently sufficient osteoclasts on the slice to produce a reasonable number of pits but activity (pits/osteoclast) remains low.

Despite our many years of collective experience, we have not come up with any reliable way of dealing with this. In these circumstances, we check basic tissue technique – namely, the freshness of the medium, the CO_2 concentration of the incubator (using a Fyrite) and the pH of the medium – and if none of these is obviously a problem, we 'change everything' in the hope that this will rectify the problem. We have empirically examined different batches of serum and the presence or absence of glutamine in the medium and found neither of these to have large effects on performance. In addition we have examined age of pups (within the range 2–6 days), different batches of bone slices, fasting of pups before dissection and presence or absence of HEPES and again found none of these to explain or correct poor performance.

Problems can occur in the visualization of pits. With toluidine blue staining, the background bone can overstain such that pits are difficult to distinguish and this can be solved by destaining in ethanol (section (i) above). Pits close to the edge of the slice can be difficult to see if using transmitted light but this can be solved by using reflected light microscopy (section 3.3.2(f)).

3.3.2 Variations reported by other laboratories

(a) Choice of substrate

The work of Jones *et al.* [56] has demonstrated that osteoclasts are likely to be able to resorb pits in almost any mineralized matrix. Dentine has been most routinely used, particularly from sperm whale and also ivory. Both of these have advantages; they are:

- free from much of the complicated morphology of bone that can create problems if image analysis is used;
- much easier to cut into squares or disks;
- of more uniform mineralization – the osteons present in bovine cortical bone are of different ages and are mineralized to varying degrees and could theoretically have greater or lesser resorbability.

The disadvantages of these alternative substrates are:

- availability, particularly whale dentine;
- the fact that they are not the physiological substrate of interest for most osteoclast researchers.

However, where agents have been evaluated on bovine cortical bone as well as an alternative substrate, similar results have been obtained with either substrate [57].

(b) Source of osteoclasts
Just over half of the published literature using this assay has used rat osteoclasts; rabbit and avian assays together represent about a third, with mouse and human currently making up the remainder in roughly equal proportions. In our laboratory, in addition to the rat assay described above, we have routinely used embryonic chick and occasionally human fetal osteoclasts. Embryonic chick osteoclasts are prepared in exactly the same way except that cells are allowed to adhere to bone slices for 90 minutes instead of 25–35 minutes. We found a large difference in yield and resorptive capacity between different batches of human fetal osteoclasts [7] and so little can be said with confidence concerning their suitability for assay. Recovery of human fetal osteoclasts was either by mincing (if possible) or by recovering the marrow by flushing with medium if bones were sufficiently calcified. Rabbit [57–59], deer [60] and mouse [61] osteoclasts have also been used by other laboratories.

The use of purified chicken osteoclasts [62] or osteoclasts recovered from marrow cultures [10] offers the possibility of using much larger numbers of osteoclasts per slice, with a consequently much larger area of bone resorbed and the potential to analyse resorption on each slice by image analysis [4] or point-counting of a part of the bone slice, a technique first applied to embryonic chick osteoclasts [63]. It should be noted that image analysis has been successfully applied to pits formed on bone slices by rat osteoclasts and stained with wheat germ agglutinin [64] and the values obtained have been shown to correlate with SEM measurements.

Recently, reports on the use of human peripheral blood and marrow to generate large numbers of bone-resorbing osteoclasts have offered the promise of using human osteoclasts routinely in resorption-based assays [12, 13].

(c) Settling time for osteoclast suspension
We have used 25–35 minutes to settle our osteoclast suspensions. This settling time was determined by the design of our experiments (Design A) and the procedure we use to get slices into their respective treatment wells – individual transfer of 120 slices from their settling wells to their treatment wells takes about 25–35 minutes. Thus, to ensure that each slice has cells attaching to it for approximately the same time, a 25–35-minute settling period is required. Design B, with fewer total slices in an experiment, offers the possibility that settling time can be reduced.

In Chambers' laboratory osteoclasts from rats and rabbits are routinely settled for shorter periods (10–15 minutes) and this appears to be the crucial factor in removing the osteoclast's ability to respond to parathyroid hormone and in obtaining a 'functionally isolated' preparation of osteoclasts [26–30]. Such shorter settling times are achieved after placing all the slices from one animal in one well of a square-well Petri dish for the settlement phase of the assay.

Other laboratories settle their osteoclast suspensions for longer, e.g. 1 hour [57]. This may ensure the maximum number of osteoclasts adhere and may increase the resorptive capacity of the preparation. However, in our experience, overlong settling times (2.5 h) induced a stimulatory response to indomethacin, suggesting that inhibitory prostaglandins were being released that also appeared to kill cells

(Murrills and Dempster, unpublished data). In some cases, osteoclasts may be under tonic inhibition by prostaglandins released in these cultures even after shorter settling times [65].

(d) Choice of culture medium

A survey of the literature shows that the most popular medium for use in the disaggregated osteoclast pitting assay is minimal essential medium (MEM) or some variation of it. The choice of medium may have important consequences for the time-course of the assay – we have found that, for neonatal rat osteoclasts, incubation at low pH (below 6.6) tends to favour a rapid onset of resorption, with numerous pits, that slows after approximately 6 h [35].

As pH increases, resorption may take longer to commence but then may continue for well over 24 hours [35]. Choice of medium can therefore be important if one is studying long- or short-term effects or agents that have only a relatively short life in culture. Embryonic chick osteoclasts appear to be less sensitive to pH [35, 63].

A number of laboratories have reduced the amount of serum in their cultures. Raynal *et al.* [57] have successfully used 2% FCS in α-MEM pH 7.0 for 24 h incubation of rabbit osteoclasts; and serum-free medium has been employed by Hall *et al.* [66] and Chowdhury *et al.* [67].

(e) Incubation period

Our standard incubation period is 24 h. However, we and other workers have presented data from or conducted entire studies using different incubation times, ranging from 1.5 h for time-course studies [35] through 6 h [68] and 18 h [58, 69–71] up to 4 days [61]. The choice of incubation period is sometimes determined by the nature of the study: the pH of the culture medium can be adjusted accordingly (section (d) above). The incubation period has been manipulated to permit distinctions between agents acting on the early phase of bone resorption in this assay and those acting on later phases. For example, bone sialoprotein and amiloride are each more effective in the earlier stages of osteoclast resorption [57, 72]. In addition, the assay has been performed over periods of 6 days or more, using models of osteoclast formation, to test simultaneously the action of certain agents on osteoclast differentiation and activity [57, 73, 74]. Longer periods (2–6 weeks) such as those used by Jones *et al.* [39] have not been used as a routine assay.

(f) Visualizing pits

Scanning electron microscopy (SEM) was the first technique used but this has largely been replaced by light microscope (LM) analysis, which is less cumbersome and more accessible to most laboratories. An advantage of SEM, compared with a standard LM at the usual magnifications and numerical aperture, is that pit depth is more clearly appreciated. However, it has been demonstrated that pits can be visualized equally well using toluidine blue staining and that the results obtained in terms of plan area and number of pits are essentially identical [75, 76]. In addition, modern confocal microscopes or high numerical aperture objectives can permit not only an appreciation of depth but accurate measurement as well. Pit depth has also been measured using transmission electron microscopy of bone slices [4], an approach that is only practical when a large percentage of the slice has been resorbed and pits are consequently easy to locate.

A variety of alternative stains have been used to visualize pits: Coomassie brilliant blue [77], acid haematoxylin [74] or a combination of acid haematoxylin and toluidine blue [57]. One of us (TRA) has modified the usual staining procedure using toluidine blue by rubbing the stain in with gloved fingers, which apparently improves the intensity with which pits are stained. Fluorescein isothiocyanate (FITC)-conjugated lectins have also been shown to stain the bottom of the pits and chondroitinase ABC digestion has been shown to reveal matrix components stainable with a variety of antibodies [78]. Immunohistochemical techniques, using antibodies to Type I collagen present at the base of the pit, have also been used successfully [79].

Reflected light microscopy can provide a high quality image and can be used on unstained specimens as well as stained. In making 'before and after' comparisons while looking at a toluidine blue stained pit, reflected light offers a clearly superior image in terms of defining the boundaries of the pit, but it has a quite different overall appearance that can take a little getting used to. Reflected light allows perfect visualization of pits at the edge of the slices/disks, a region that can be problematic with transmitted light. The choice of using reflected light rather than toluidine blue staining in our laboratory has always been left to the operator, and has proved to be a matter of personal preference with no consistent favourite. Metallurgical objectives with no coverslip correction can improve reflected light images. Tangential light illumination has also been used [80].

(g) Quantitation of resorption
The three basic options for quantifying resorption are morphometrically to count the number of pits, measure the area of pits or measure the volume of pits. The two most commonly used techniques are pit counting and area measurement. Areas can be measured either by tracing around the border of a pit using a digitized morphometry programme or by performing image analysis detecting the color that the pit is stained. We have only used the tracing approach. This can be applied to either SEM or LM images. Area can also be estimated by point counting or some variation of it [57, 63, 74], a technique that is particularly useful when a large amount of resorption is present.

Volume measurement undoubtedly offers the best quantitation of the total amount of bone resorbed on a given slice and provides additional information on the depth of pits (this can also be measured separately without a volume estimation). However, the equipment required to measure volume is expensive and specialized: it can be performed only on the SEM [81, 82] or using video rate laser confocal microscopy [83]. The total cost of a confocal system to measure pit volume may be over US$100 000. Approximate but nonetheless useful depth measurements, by contrast, can be obtained fairly cheaply using reflected light and high numerical aperture objectives (metallurgical) to focus on the bone surface and the bottom of the pit and recording the distance travelled by the stage. Depth measurement offers the only means of studying an aspect of osteoclast biology that may possibly be crucial in the pathogenesis of osteoporosis – namely, the perforation of a trabecular plate by so-called 'killer' osteoclasts.

(h) Statistics
Student's *t* test is often used to test for significant differences in this assay. However, with the multiple comparisons usually employed, this approach runs a high risk of a type I error and ANOVA is the method that we would recommend.

3.3.3 Issues concerning the methods used in the disaggregated osteoclast resorption assay

(a) Functional isolation

The early experiments of Chambers and colleagues demonstrated that if osteoclasts were settled on bone slices for only a limited period (10–15 minutes) prior to rinsing off the non-adherent cells, these preparations could not be stimulated to resorb more bone by a single dose of parathyroid hormone, interleukin-1, 1,25 dihydroxy-vitamin D_3 and tumour necrosis factor-α (TNFα) [26–30]. However, co-culture of these short sediment preparations with osteoblasts restored the ability to respond to these agents. Although the short settlement cultures contained some osteoblasts, their inability to respond to these hormones and cytokines without the presence of additional osteoblasts (or supernatants from osteoblasts treated with these agents) led them to be called functionally purified osteoclasts or functionally isolated osteoclasts. These experiments have been replicated by at least three other laboratories [31, 80, 84], at least for PTH and PTHrP at a single dose, and the concept of inability to respond to PTH has been used as a screen for suitably puri-fied osteoclasts [85, 86]. A difference in the response of 'functionally purified osteoclasts' from osteoclasts obtained from long sediment cultures or co-cultures with osteoblasts has also been described by Chambers' group for oestradiol and hepatocyte growth factor [86, 87] and by Holloway *et al.* for zinc [88].

In our experience, the PTH response is very difficult to abolish in this assay and was still detectable when we attempted to separate osteoclasts from osteoblasts by plating only 25 μl of cells instead of 100 μl on each slice with a 25–30-minute settling period [89]. It is possible that the response was carried by osteoblasts that still remained attached or in close proximity to osteoclasts in these low-density cultures. Alternatively, and consistent with our observation that PTH induced the forma-tion of smaller pits in these cultures, the response could conceivably have been at least partly due to PTH acting on preosteoclasts which may possess PTH recep-tors [90]. When viewing an entire dose–response experiment involving five or more animals, we have also noted that animals appear to differ in their responsiveness to PTH and that this appears to be unrelated to overall osteoblast number on bone slices in those animals. One of us (TRA) has also noted a strong response to 1,25 dihydroxyvitamin D_3 in low density cultures. In our opinion, the concept of 'func-tional purification' or 'functional isolation' is therefore one that is desirable in terms of defining direct or indirect mechanism of action but which can be difficult to achieve because of the narrow window for settling to become 'too long' and the contribution from preosteoclasts that may possess PTH receptors. In the long run, it is likely that studies using highly purified mammalian osteoclasts and preosteo-clasts with or without co-culture with osteoblasts capable of influencing osteoclast activity will provide more definitive systems in which to address the questions raised by the concept of functional purification or isolation.

(b) Quantitation of bone resorption

The method of quantitation of bone resorption in this assay has been a thorny issue for several years and we have stated our position on this previously [76]. The debate arises because the most logical method (measuring volume of bone resorbed) is also the most specialized, the most expensive and the most time-consuming. In justifying the easiest, cheapest and quickest method (pit counting using the light microscope) we have drawn from a wealth of published information and data generated in our own laboratory which indicates that there is a striking parallelism between results obtained by measuring area (and where data are available, volume) and counting pits [76]. This parallelism and correlation is seen not only with our own embryonic chicken and neonatal rat preparations but also with purified laying hen osteoclasts [91]. Some agents have been shown to influence pit size – for example, PTH or bicarbonate, (a reduction) [89, 92], and CO_2, (an increase) [92] – and it cannot be denied that in those cases in which inhibitory agents cause a reduction in pit size alongside a reduction in pit number, measuring area provides an additional sensitivity [70, 71, 93], though this may be small. We are not aware of any instance where counting pits gave a misleading or false negative result as compared with the absolute gold standard, i.e. volume measurements. One possible exception might be expected when studying agents that interfere with the mecha-nism of matrix degradation, such as those that inhibit collagen-degrading enzymes. Such inhibitors appear to influence the depth to which collagen is degraded, and hence the volume of the pit, with little (if any) effect on its area [94]. Delaisse *et al.* [95] had also noted this but interestingly found that the inhibitor in question (E-64) nonetheless caused a reduction in the numbers of pits. Hence, in this instance at least, counting pits still detected an effect of the inhibitors. In a different research area, Yoneda *et al.* [96] saw no change in pit number but a reduction in pit area with 10 and 100 ng of the tyrosine kinase inhibitor Herbimycin A/ml in PTH-stimulated cultures, suggesting that pit counting might fail to detect inhibition by tyrosine kinase inhibitors. However, Hall *et al.* [97] noted the typical parallel reduction of basal resorption on both plan area and number of pits with similar doses of Herbimycin A, and one of us (RJM) has also seen a 50% and greater reduction in pit number at doses of 10 ng/ml and 100 ng/ml. Thus this possible exception to usefulness of counting pits does not appear to be reproduced by other laboratories.

In summarizing our views, we feel that pit counting is a useful assay of *in vitro* bone resorption but one that has to be taken for what it is – a single aspect of the process of bone resorption. More sophisticated measurements, such as volume and depth, are clearly desirable when practical, as volume represents the absolute amount of bone resorbed and the depth of a pit may be a crucial factor in the pathogenesis of osteoporosis.

(c) Variability

The first impression that a newcomer using Design A has of the assay is likely to be its high variability. This is due to:

- an intrinsic variability in the number of pits per slice between replicates;
- variability in the basal level and responsiveness of individual animals;
- observer variability.

The first is alarming (see sample data in Table 3.1) and can be reduced somewhat by correcting for the differing numbers of osteoclasts that may be on each slice. This is done by first counting TRAP+ osteoclasts and then removing the cells and stain-ing the slice for pit counting. Although this reduces variability somewhat, it does not by any means eliminate it. In a subsample of our experiments on file, the CV between replicates in a treatment group was reduced from approximately 40% to approximately 33% by correcting for osteoclast number (using either pit counts or

Table 3.1 Sample data from a typical dose–response pitting assay using Design A and three replicate slices per treatment per animal

Treatment	Rat 1	Rat 2	Rat 3	Rat 4	Rat 5	Mean + SEM
Control	10 8 10	34 17 13	19 32 41	16 26 18	25 37 21	
	(9.3)	(21.3)	(30.7)	(20.0)	(27.7)	21.8 ± 3.7
Low dose	0 0 0	4 18 9	9 25 12	22 14 6	5 3 8	
	(0)	(10.3)	(15.3)	(14.0)	(5.3)	8.9 ± 2.8
Mid dose	0 2 0	1 1 4	0 10 4	13 11 0	1 5 *	
	(0.7)	(2.0)	(4.7)	(8.0)	(3.0)	3.7 ± 1.3
High dose	0 2 0	0 0 0	0 0 0	0 0 0	0 0 0	
	(0.7)	(0)	(0)	(0)	(0)	0.1 ± 0.1

Note the variability between replicate slices and our method of calculating means ± SEM for each treatment. Resorption was analysed by counting pits on each of three replicates per treatment per animal. The mean for each group appears in parentheses and the grand mean of these means was calculated and appears in the final column ± SEM for $n = 5$ animals. * represents a slice that was lost. Only four treatment groups from the eight in this experiment are shown for the sake of clarity.

plan area as a measure of bone resorption). A similarly small or negligible improvement in variability using this technique has been noted by McSheehy *et al.* [98]. The effect of variability between basal levels of animals can be reduced alternatively by:

- expressing data as percent of control (though this needs to be used with care as then, for example, stimulatory effects tend to become inversely proportional to basal level of resorption, which may be a mathematical rather than a biological effect);
- pooling cells from all animals to create one 'artificial' animal, as in Design B;
- performing two-way analysis of variance in a block design with each animal as one of the blocks.

Observer variability appears to be low enough in this assay such that it is no impediment to separate observers (even novice ones) producing essentially identical interpretations of the same experiment.

Variability in the assay is a major problem if the assay is to be used for detailed or sophisticated applications. For example, when calculating $IC_{50}s$ or $ED_{50}s$, it is not uncommon for confidence intervals of the estimates to span an order of magnitude, making comparisons between related agents extremely difficult unless there are huge differences in potency.

Tamura *et al.* [10] have reported that the mouse osteoclast assay generated *in vitro* shows lower variability than the isolated rat osteoclast assay, though the reason for this is not yet clear. It does not appear, at least from our point of view, to be due simply to the larger area of bone resorbed using this approach, as analyses of our isolated chick osteoclast cultures in which large numbers of pits are formed show just as much variability as those with fewer pits (Murrills, unpublished observations). However, it is apparent that mouse marrow-based assays often have encouragingly low variability [4].

(d) Osteoclast counting

It is clear from serial electron microscope sections that mononucleate cells can perform the definitive function of an osteoclast – namely, to excavate a resorption lacuna in bone [99]. However, when we count osteoclasts on bone slices, we routinely count only cells with three or more nuclei. This is partly for historical reasons, as it was initially felt that a multinuclearity was a defining feature of an osteoclast and that a binucleate TRAP+ cell could be a dividing precursor. Although this approach now seems contradictory, its validity is based on the observation that mononucleate and binucleate cells form the majority of the cells present and it is likely that only a small proportion of them will actually be capable of resorption. Hence, counting only the larger cells may give a more accurate estimate of the numbers of cells that are actually capable of resorbing bone.

(e) pH effects on osteoclasts
The deleterious effects of acidosis on the skeletal tissues have long been recognized [100–102]. We became aware of the influence of extracellular pH on bone resorption whilst trying to set up the disaggregated rat osteoclast resorption assay. Our initial attempts to induce rat osteoclasts to resorb bone *in vitro*, using methods that had recently been reported [20, 58], were not successful. We investigated, at some length, the effect of changing the more obvious parameters we thought might affect osteoclast function, such as calcium and phosphate concentration, serum type, substrate, time in culture, etc. The cells remained viable and healthy under most conditions, but resorption pits were never formed. As a final measure before abandoning the model, we tried increasing the incubator CO_2 concentration from the usual 5% to about 10%. Against expectations, not only did the bone cell cultures tolerate the resulting acid conditions, but also resorption pits were formed. Subsequently, we learned that the laboratory whose methods we were following had used a culture medium containing a very low bicarbonate concentration (Hank's salts), intended for use with atmospheric air; when used in a tissue culture incubator containing 5% CO_2, such a medium becomes strongly acidic (see below).

To investigate the influence of extracellular hydrogen ion concentration on rat osteoclast function as simply as possible, we formulated culture media without bicarbonate, buffered only with HEPES and adjusted to varying pH levels. Little or no resorption pit formation was observed at pH 7.4, but progressive acidification to pH 6.8 resulted in stepwise increases in resorption in 24-hour cultures. The primary effect of decreasing extracellular pH was to increase the number of resorption pits formed [75].

Further work confirmed that in media buffered physiologically with HCO_3^-/CO_2, resorption pit formation by rat osteoclasts is stimulated when pH is reduced either by increasing pCO_2 or by decreasing HCO_3^- concentration [92]. This study also showed that the size of resorption pits decreases as HCO_3^- concentration is reduced. In HCO_3^--free media, pits resorbed by rat osteoclasts are typically $2\ \mu m$ deep [75], a value which increased up to about $4\ \mu m$ when HCO_3^- concentration was normal.

Recent investigation of the effect of very small pH changes has demonstrated the remarkable proton sensitivity of rat osteoclasts cultured in HCO_3^-/CO_2-buffered media. The pH response occurs within a relatively limited range, such that shifts in the range pH 7.25–7.05 act as an 'off/on switch' for resorptive activity, with the major change associated with a pH difference of as little as 0.1 unit (Fig. 3.5). Such a pH difference corresponds to an absolute change in H^+ concentration of less than 1.3-fold.

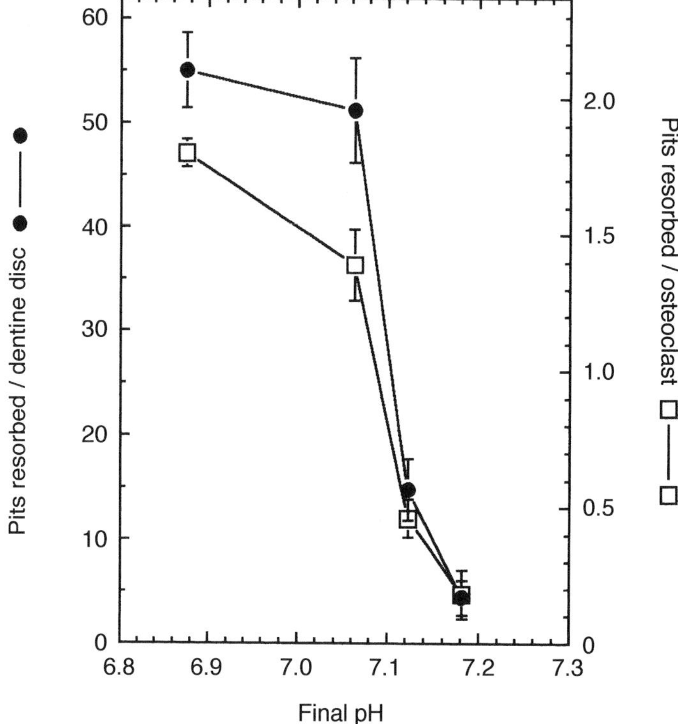

Figure 3.5 Effect of 0, 2.5, 5 and 10 mEq added H^+/l (as HCl) on resorption pit formation by rat osteoclasts in 24-hour cultures. Values are means $\pm$ SEM (n = 5).

The key to these experiments was the use of a clinical blood gas analyser, with rigorous precautions to minimize CO_2 loss from culture media before measurement. Time-course studies over 30 hours indicated that exposure to low pH conditions (pH 7.06) does not activate rat osteoclasts to continue resorbing after transfer to more alkaline media (pH 7.34), and that resorption pit formation is closely related to time spent at low pH (Fig. 3.6) [103].

Interestingly, whereas pit formation by disaggregated rat osteoclasts is strongly stimulated by both HCO_3^- and CO_2 acidosis, in cultured mouse calvarial bones, CO_2 acidosis is much less effective than HCO_3^- acidosis in stimulating osteoclast-mediated calcium release, at any given pH [104; Arnett *et al.*, unpublished data]. The reasons for this disparity are not clear. The response curves of these two resorption systems to HCO_3^- acidosis are closely similar, however [105, 106].

Clearly, the effects of pH on osteoclasts interact with those of other osteotropic agents. Acid-activated resorption pit formation is blocked by calcitonin [58, 75] or can be further stimulated by, for example, 1,25 dihydroxyvitamin D_3 (Fig. 3.7), pertussis toxin [107] or cyclooxygenase inhibitors [65]. It is unclear to what extent the effects of these varying modes of stimulation are additive, or what the upper limit of resorptive efficiency for an osteoclast might be.

The stimulatory effect of low pH has been observed in osteoclasts derived from numerous species, including mouse [10], deer (Arnett, unpublished), human [108], rabbit [59] and embryonic chick [63]. Chick osteoclasts, however, seem able to form resorption pits in somewhat more alkaline conditions (pH 7.4) than mammalian

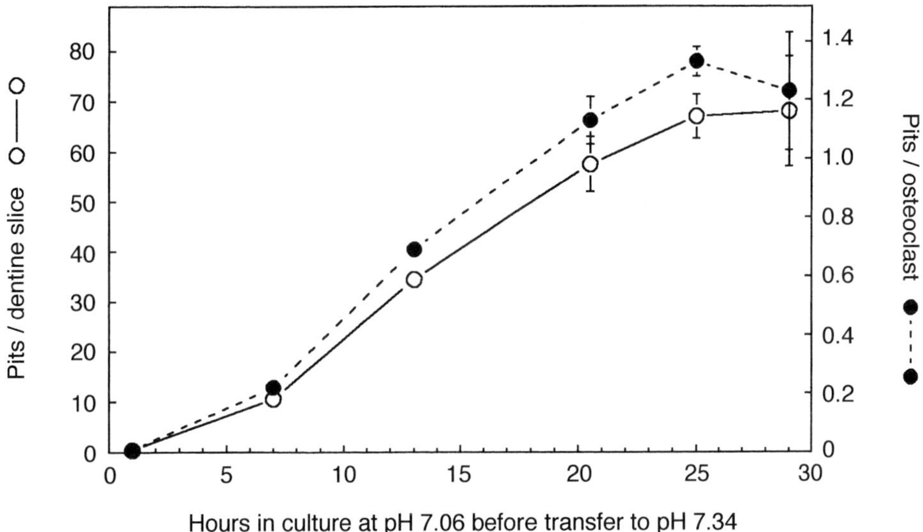

Figure 3.6 Effect of pre-incubation in pH 7.06 medium for varying times on resorption pit formation by osteoclast-containing rat bone cell populations subsequently cultured at pH 7.34. Values are means ± SEM ($n = 5$).

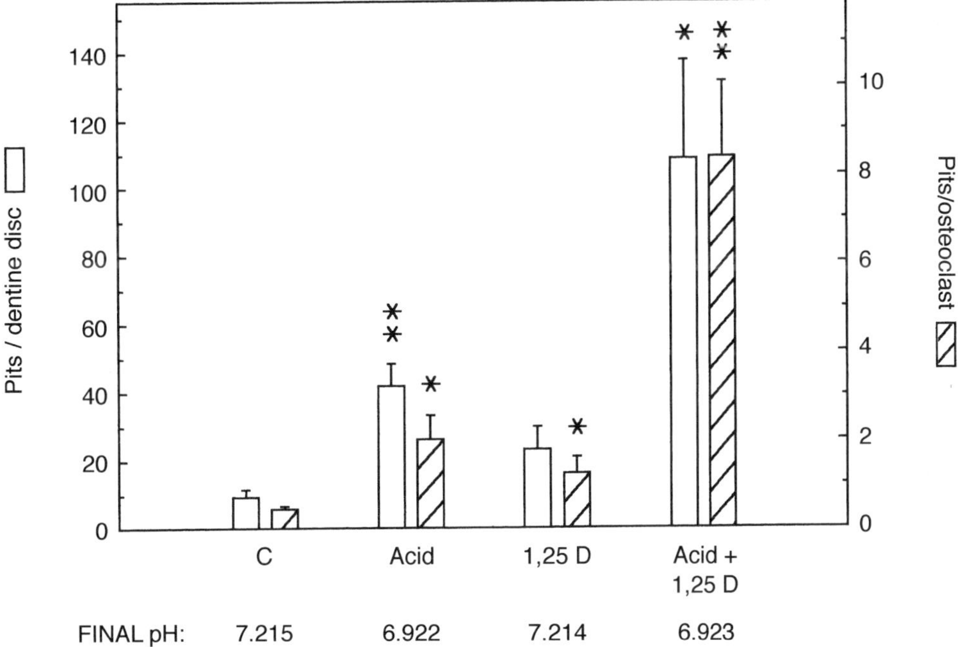

Figure 3.7 Interaction of the effects of pH and 1,25 dihydroxyvitamin D_3 (1,25 D; 10 nM) on resorption pit formation by rat osteoclasts cultured on dentine discs for 24 hours. Values are means ± SEM ($n = 5$); *P <0.05; **P <0.01 vs. respective control. (Reproduced by courtesy of M. Morrison, Department of Anatomy and Developmental Biology, University College, London.)

osteoclasts [63]. Perhaps because of the rapid remodelling that must accompany the very high rates of bone growth seen in developing birds, chick osteoclasts would appear to be 'constitutively activated', and can resorb mineralized substrates at prodigious rates.

In relatively short-term experiments, the effects of extracellular pH on mammalian osteoclastic resorption seem clear cut. In longer term cultures, there are conflicting data. Murrills *et al.* [35] showed that the relative stimulatory effect of low pH on resorption by rat osteoclasts diminished over several days in culture; it should be noted, though, that these experiments studied the effects of severely acid media (below pH 6.7), which can reduce osteoclast survival, as discussed below. Other results (Arnett and Spowage, unpublished) indicated that the stimulatory effect of low pH on pit formation by rat osteoclasts showed no tachyphylaxis after 5 days in culture (pH 7.03 vs. pH 7.23). Recently, we have studied the effect of extracellular pH in 10-day mouse marrow cultures. Paradoxically, TRAP-positive osteoclast formation in this system appears to be stimulated by alkaline conditions and is inhibited below pH 7.0 (Morrison and Arnett, unpublished). Other experiments with 3-day cultures of mouse calvaria have shown that although extracellular protons can stimulate osteoclastic resorption as powerfully as maximal doses of 1,25 dihydroxy-vitamin D_3, parathyroid hormone or prostaglandin E2 [105], acidification may be accompanied by reductions in the number of TRAP+ osteoclasts visible in whole mount preparations (Morrison and Arnett, unpublished).

(f) Buffering and pH: theory and practice for osteoclast culture

The principles governing HCO_3^-/CO_2 buffering of culture media and blood are not always well understood by tissue culturists. Given the unique sensitivity of mammalian osteoclasts to extracellular pH, this consideration is disregarded at the investigator's peril.

In physiologically buffered media the following equilibrium exists:

$$CO_2 + H_2O \leftrightarrow H_2CO_3 \leftrightarrow H^+ + HCO_3^-$$

For the equilibrium constant, K_a:

$$K_a = [H^+] \times [HCO_3^-] \ / \ [CO_2]$$

and:

$$pH = pK_a + \log [HCO_3^-] \ / \ pCO_2 \times \alpha$$

(the Henderson–Hasselbalch equation), where α is the solubility coefficient of CO_2 in water.

Tissue culture media formulated with Earle's salts, containing 2.2 g $NaHCO_3$/l, are normally used in conjunction with a 5% CO_2 atmosphere (equivalent to a partial pressure of 42 mm Hg) to give an operating pH at equilibrium of about 7.25. Addition of protons to culture medium reduces the HCO_3^- concentration and the equilibrium operating pH at constant pCO_2, thus mimicking *in vivo* metabolic acidosis. Conversely, increasing pCO_2 reduces pH, whilst HCO_3^- concentration remains more or less constant, a model of *in vivo* respiratory acidosis. Figure 3.8 shows the relationship between pH and pCO_2 for unmodified and reduced $[HCO_3^-]$ tissue culture media.

For mammalian osteoclast culture, a reduced operating pH is conveniently achieved by adding concentrated HCl directly to standard liquid medium (e.g.

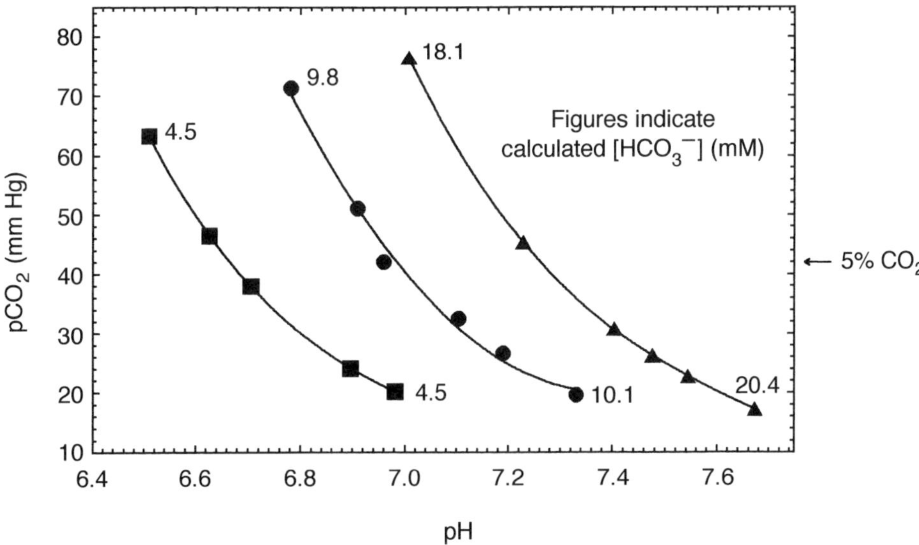

Figure 3.8 Relationship between pH, pCO_2 and $[HCO_3^-]$ in tissue culture media.

MEM) containing Earle's salts plus 10% fetal bovine serum. Concentrated (12 M) HCl has the advantage of being self-sterilizing; the addition of 82 μl of 12 M HCl to 100 ml of culture medium provides 10 mEq H^+/l and results in an optimal operating pH of about 7.0 in a 5% CO_2 atmosphere. Alternatively, a similar operating pH can be achieved with unmodified medium by increasing incubator CO_2 to about 10% (Fig. 3.8).

Accurate measurement of the operating pH of mammalian osteoclast cultures is desirable in most, if not all, experimental situations and is necessary for meaningful comparison of results from different laboratories. Our recent experience has shown that, for HCO_3^-/CO_2-buffered media, this can only be achieved by the use of a properly standardized clinical blood gas analyser. From Fig. 3.5, it can be seen that extracellular pH changes of the order of a few hundredths of one unit may be sufficient to alter osteoclast pit formation significantly. Culture medium pH can be altered directly by acidic or alkaline compounds being tested in the osteoclast assay at high concentrations (generally > 10 μM) or as a result of the metabolic activity of cells. The latter effect can confound the interpretation of results where osteoclast preparations are co-cultured with additional cells or when the effect of cell-conditioned medium is being tested. Furthermore, osteoblast mitogens such as parathyroid hormone (PTH), transforming growth factor-β (TGFβ), insulin-like growth factor (IGF) or interleukin 1 (IL-1) could conceivably stimulate osteoclasts simply via enhanced medium acidification, in addition to any other mechanisms that may be involved. It has been argued that if osteoclast cultures are maintained at a sufficiently acid pH, i.e. at the top of the proton activation curve (below pH

6.7), then small, uncontrolled fluctuations of pH are likely to be of little consequence. The main drawback to this approach would seem to be that in such acidic conditions, osteoclast survival in culture may be impaired [35]; it should also be borne in mind that the buffering power of the HCO_3^-/CO_2 system at very low pH is considerably diminished.

The following procedure has been developed for controlling and measuring the pH of osteoclast cultures. Assays are performed in multiwell plates placed in the bottom of half open, humidified plastic lunchboxes. To prevent CO_2 loss, opening of the incubator door during the last few hours of experiments is avoided. At the end of experiments, culture plates are sealed in the lunchboxes as soon as the incubator door is open, then taken directly to the blood gas analyser (Radiometer ABL4). Immediately, the lunchbox is unsealed, a sample (approximately 200 μl) is analysed to obtain a pH and pCO_2 reading for medium in one well (usually a control). This initial pCO_2 reading is assumed to be the same in all wells and to reflect the actual pCO_2 during the 24-hour incubation. Readings of pH and pCO_2 are then made for all the other wells (treatment groups). The cycle time of most blood gas analysers is 3–4 minutes and the measured pCO_2 will drop by several mm Hg for each successive reading, causing measured pH values to rise accordingly. These pH readings for each well are then back-corrected to the pH value associated with the initially measured pCO_2 value, using calibration curves previously measured for culture medium with differing bicarbonate concentrations (Fig. 3.8). Clearly, it is difficult or impossible to monitor pH during the course of incubations, though pCO_2 can be checked (±0.5%) using a Fyrite combustion kit. 'Initial' pH is measured using cell-free media incubated with or without mineralized substrate slices. To avoid significant modification of culture medium pH by cells in osteoclast resorption assays, it is desirable to use reasonably large volumes (e.g. approximately 1 ml/5 mm bone or dentine disc per 24 hours).

3.4 USE OF THE DISAGGREGATED OSTEOCLAST RESORPTION ASSAY

3.4.1 Identifying the target cells for hormone, cytokine and drug action

The first achievement of the disaggregated osteoclast resorption assay was the demonstration by Chambers and his colleagues that many hormones and cytokines such as PTH, IL-1, TNF and 1,25 dihydroxyvitamin D_3 do not act directly upon osteoclasts but instead act via receptors on the osteoblast which then releases a stimulator, or stimulators, of osteoclasts [26–30]. This phenomenon of osteoblastic influence on osteoclast response has been confirmed in several other labs [109, 110] (see also 3.3.3) and appears to apply to a range of regulatory agents, both stimulatory and inhibitory, including oestradiol [86], bisphosphonates [80], thyroid hormone [111], hepatocyte growth factor [87] and zinc [88]. It should be noted, however, that evidence also exists for PTH receptors and hence a direct action of PTH on osteoclasts [112–114].

3.4.2 Identifying osteoblastic mediators of bone resorption

The identity of such osteoblastic regulators of bone resorption has been the target of much research since but, despite the importance of this factor (or factors), publications characterizing the factor have been infrequent – only one or two papers per year, with the majority of studies using the pitting assay to address the topic.

What has emerged from these studies is a general finding that osteoblasts can and do regulate osteoclastic bone resorption activity, at least *in vitro*. Both stimulators [26–30, 54, 115–121] and inhibitors [121, 122] have been described, usually from rat osteoblasts [26–30, 115, 116, 118, 121] but also from mouse [54, 119, 120] and human [123] and in some cases partially characterized in terms of molecular weight [27, 115, 118, 123]. Modulating factors have been noted to be diffusible [27, 28, 30, 109, 115, 118, 119, 121, 123], contact-mediated/short-range [29] or to be deposited in the extracellular matrix or on the cell surface [116, 122]. Some of these are constitutively produced by osteoblasts [118, 121] whereas the remainder of those quoted here are induced by calciotropic hormones. Most are described as acting on the mature osteoclast, while some [118, 121] appear to act primarily on osteoclast recruitment. Known cytokines such as IL-6 and LIF have also been implicated [124–126]. Partial characterization of factors found in osteoblast supernatants has revealed that most laboratories find heat labile factors, though the molecular weights of these factors vary from 0.5–1 kDa [27], through 9 kDa [123] to greater than 10 kDa [118], or 70 and 110 kDa [115] for diffusible activators and 1–10 kDa for diffusible inhibitors [121]. Where tested, factors have been found not to be prostaglandins [30, 115]. In sum, these studies suggest that there may be several osteoblast-derived products that modify osteoclast activity and some authors have suggested that two or more factors may be required simultaneously for detectable activity [115]. It should be borne in mind that the conditions for detecting activity must be carefully controlled for identifying such factors. In some instances, studies have been performed on cell preparations that are clearly not functionally isolated [115, 123] and in other cases may be at the borderline of functional isolation [118, 121, 122], which generally requires very short settling times (10–15 minutes). Factors identified via systems that are not fully isolated could theoretically be identifying factors that act indirectly on the osteoclast.

The 10 years since the concepts of osteoblast regulation of the osteoclast were first substantiated in isolated systems have therefore produced no general consensus on the osteoclast activator(s). In part this is no doubt due to the difficulties inherent in the osteoclast pitting assay, notably variability, concerns about functional isolation and low throughput. Different approaches, such as differential display, may provide alternative routes of tackling this important topic but it is nonetheless likely that a purified osteoclast preparation (or cell line) unequivocally resorbing the matrix of bone slices will be required to prove finally the activating qualities of the substance(s).

3.4.3 Testing of regulators of bone resorption

A multitude of potential regulators of bone resorption have also been tested in this assay, detailed in Table 3.2. In this mode, the assay has been used not necessarily

Note added in proof: recently, collagen fragments released from human dentine slices by mouse osteoclasts have been found to correlate with morphometric measurements of bone resorption. This technique offers an exciting alternative to morphometry that needs to be explored further. Reference: Apone, S., Lee, M.Y. and Eyre, D.R. (1997) Osteoclasts generate cross-linked collagen N-telopeptides (N-Tx) but not free pyridinolines when cultured on human bone. *Bone* **21**, 129–136.

Table 3.2 Published papers using osteoclast assays, arranged according to substance or topic investigated

Class	Substance/topic	Reference (first author only)
Calcium regulating hormones	Calcitonin	Ali (1984) *Anat Embryol* **170**:51
	"	Arnett 1986 *Endocrinology* **119**:119
	"	Arnett 1987 *Endocrinology* **120**:602
	"	Athanasou 1991 *Br J Canc* **63**:527
	"	Britto 1994 *Endocrinology* **134**:169
	"	Chambers 1984 *J Cell Sci* **66**:383
	"	Chambers 1985 *J Pathol* **145**:297
	"	Chambers 1985 *Endocrinology* **116**:234
	"	Cudd 1995 *J Pharm Sci* **84**:717
		Dempster 1987 *JBMR* **2**:443
	"	Fenton 1993 *J Cell Physiol* **155**:1
	"	Fenton 1991 *Endocrinology* **129**:1762
	"	Fenton 1994 *JBMR* **9**:515
	"	Flanagan 1988 *Cancer* **62**:1139
	"	Fuller 1989 *JBMR* **4**:209
	"	Hall 1994 *BBRC* **199**:1237
	"	Iida 1996 *CTI* **59**:100
	"	Kaji 1993 *Horm Met Res* **25**:421
	"	Kitamura 1993 *Bone* **14**:829
	"	McIntyre 1991 *PNAS* **88**:2936
	"	McSheehy 1995 *Bone* **16**:435
	"	Murrills 1995 *CTI* **57**:47
	"	Murrills 1989 *JBMR* **4**:259
	"	Notoya 1993 *CTI* **53**:206
	"	Notoya 1992 *CTI* **51** Suppl 1:S3
	"	Rathod 1994 *J Pathol* **174**:293
	"	Ries 1992 *JBMR* **7**:931
	"	Sundquist 1990 *BBRC* **168**:309
	"	Su 1992 *Endocrinology* **131**:1497
	"	Takada 1992 *Bone Min* **17**:347
	"	Tamura 1993 *JBMR* **8**:931
	"	Wada 1996 *Endocrinology* **137**:1042
	"	Zaidi 1987 *J Endocrinol* **115**:511
	"	Zaidi 1987 *CTI* **40**:149
	"	Zaidi 1994 *Exp Physiol* **79**:387
	"	Zaidi 1990 *J Endocrinol Invest* **13**:119
	"	Zaidi 1990 *Exp Physiol* **75**:529
	Parathyroid hormone	Allain 1992 *J Endocrinol* **133**:327
	"	Arnett 1986 *Endocrinology* **119**:119
	"	Athanasou 1991 *Br J Canc* **63**:527
	"	Bax 1992 *BBRC* **183**:1153

It is unlikely that this is a complete list and we extend our apologies to authors whose papers have been inadvertently omitted. We have limited this to papers in which some quantitation of osteoclast activity is reported, and have omitted a large number of marrow culture studies in which resorption is presented as a confirmation of the osteoclast nature of the cells counted or as an assay of osteoclastogenesis. Non-conventional abbreviations of journals used: *BBA – Biochimica et Biophysica Acta BBRC – Biochemical and Biophysical Research Communications   CTI – Calcified Tissue International JBMR – Journal of Bone and Mineral Research   JCI – Journal of Clinical Investigation PNAS – Proceedings of the National Academy of Sciences of the United States of America*

Table 3.2 *Continued*

Class	Substance/topic	Reference (first author only)
	"	Britto 1994 *Endocrinology* **134**:169
	"	Chambers 1985 *J Pathol* **145**:297
	"	Chambers 1985 *J Cell Sci* **76**:155
		Chambers 1985 *Endocrinology* **116**:234
	"	Chowdury 1991 *CTI* **49**:275
	"	Evely 1991 *J Bone Min Res* **6**:85
	"	Fenton 1991 *J Cell Physiol* **155**:1
	"	Fenton 1994 J Bone Min Res 9:515
	"	Holloway 1996 *Bone* **19**:137
	"	Ishii 1993 *BBRC* **191**:495
	"	Kaji 1993 *Horm Met Res* **25**:421
	"	Kaji 1992 *BBRC* **182**:1356
	"	Martin 1992 *Polyfunctional Cytokines*
	"	Matsumoto 1992 *Proc Soc Exp Biol Med* **200**:161
	"	McSheehy 1986 *Endocrinology* **19**:1654
	"	McSheehy 1986 *Endocrinology* **118**:824
	"	Murrills 1990 *Endocrinology* **127**:2648
	"	Takada 1992 *Bone Min* **17**:347
	"	Yu 1996 *Bone* **19**:339
Parathyroid hormone-related protein	N-terminal PTHrP	Evely 1991 *JBMR* **6**:85
	N-terminal PTHrP	Kaji 1993 *Horm Met Res* **25**:421
	N-terminal PTHrP	Murrills 1990 *Endocrinology* **127**:2648
	N-terminal PTHrP	Yoneda 1995 *Cancer Res* **55**:1989
	C-terminal PTHrP	Fenton 1993 *J Cell Physiol* **155**:1
	C-terminal PTHrP	Fenton 1991 *Endocrinology* **129**.1762
	C-terminal PTHrP	Fenton 1994 *JBMR* **9**:515
	C-terminal PTHrP	Kaji 1994 *Endocrinology* **134**:1897
	C-terminal PTHrP	Kaji 1995 *Endocrinology* **136**:842
	C-terminal PTHrP	Murrills 1995 *CTI:* **57**:47
Other hormones	CGRP, amylin	Alam 1993 *Exp Physiol* **78**:183
	CGRP	Zaidi 1987 *J Endocrinol* **115**:511
	"	Zaidi 1987 *CTIP* **40**:149
	CGRP and CGRP fragments	Zaidi 1990 *Biochem J* **269**:775
	Amylin	Zaidi 1990 *Exp Physiol* **75**:529
	Insulin, somatostatin, glucagon, corticotropin, substance P, LHRH, corticotropin releasing factor, atriopeptin III	Mentioned in Zaidi 1987 *J Endocrinol* **115**:511
Prostaglandins and other arachidonic acid derivatives	PGE$_2$	Arnett 1987 *Endocrinology* **120**:602
	PGE$_2$	Athanasou 1991 *Br J Canc* **63**:527
	PGE$_1$, PGE$_2$, PGI$_2$	Chambers 1985 *Endocrinology* **116**:234
	PGE$_1$	Chambers 1985 *J Pathol* **145**:297
	PGs, leukotrienes, HETES	Fuller 1989 *JBMR* **4**:209
	5-Lipoxygenase metabolites	Gallwitz 1993 *J Biol Chem* **268**:10087

Table 3.2   *Continued*

Class	Substance/topic	Reference (first author only)
	Peptido-leukotrienes	Garcia 1996 *JBMR* **11**:521
	Leukotriene B-4	Garcia 1996 *JBMR* **11**:1619
	PGE_2	Kaji 1996 *JBMR* **11**:62
	PGE_2	Okuda 1989 *Bone Min* **7**:255
Lipids	PAF	Zheng 1993 *Am J Physiol* **264**:E74
Inorganics/ion transport	pH	Arnett 1986 *Endocrinology* **119**:119
	HCO_3/CO_2	Arnett 1994 *J Bone Min Res* **9**:375
	pH	Arnett 1994 *J Bone Min Res* **9**:375
	H_2O_2	Bax 1992 *BBRC* **183**:1153
	Ni	Bax 1993 *Exp Physiol* **78**:517
	pH	Collin 1992 *Endocrinology* 131:1181
	Cl/HCO_3 inhibitor	Hall 1989 *CTI* **45**:378
	Reactive oxygen intermediates	Hall 1995 *BBRC* **207**:280
	Zinc	Holloway 1996 *Bone* **19**:137
	NO	Kasten 1994 *PNAS* **91**:3569
	NO	MacIntyre 1991 *PNAS* **88**:2936
	Ca, Perchlorate, SCN	Moonga 1991 *BBRC* **76**:923
	Zinc	Moonga 1995 *JBMR* **10**:453
	pH	Murrills 1993 *J Cell Physiol* **154**:511
	Superoxide	Ries 1992 *JBMR* **7**:931
	pH	Shibutani 1993 *JBMR* **8**:331
	Monensin	Tagami 1994 *FEBS Letters* **342**:308
	pH	Tamura 1993 *JBMR* **8**:953
	pH	Walsh 1990 *JBMR* **5**:1243
	Cd	Wilson 1996 *Toxicol Appl Pharmacol* **140**:451
	Zinc	Yamaguchi 1996 *Mol Cell Biochem* **158**:171
	PO_4	Yates 1991 *JBMR* **6**:473
	Perchlorate	Zaidi 1990 *Br J Rheumatol* **29**:406
Intracellular signalling	dbcAMP	Arnett 1987 *Endocrinology* **120**:602
	dbcAMP	Chambers 1985 *J Pathol* **145**:297
	dbcAMP	Chambers 1985 *Endocrinology* **116**:234
	PKC	Fenton 1991 *Endocrinology* **129**:1762
	Calcium	Hall 1994 *BBRC* **202**:456
	c-src Inhibitors	Hall 1994 *BBRC* **199**:1237
	PI3-kinase	Hall 1995 *CTI* **56**:336
	cAMP	Kaji 1992 *BBRC* **182**:1356
	PKA/PKC	Kaji 1993 *Horm Metab Res* **25**:421
	PKA	Kaji 1996 *JBMR* **11**:62
	Calcium, Mg	Kaji 1996 *JBMR* **11**:912
	DAG, PKC	Moonga 1996 *CTI* **59**:105
	G-protein	Moonga 1993 *BBRC* **190**:496
	PKC	Murrills 1992 *JBMR* E**7**:415
	cAMP	Murrills 1990 *Bone* **11**:333
	Calcium	Rathod 1994 *J Pathol* **174**:293
	Calcium	Ritchie 1994 *Endocrinology* **135**:996
	PTPase inhibitors	Schmidt 1996 *PNAS* **93**:3068

Table 3.2 *Continued*

Class	Substance/topic	Reference (first author only)
	PTC, cAMP	Su 1992 *Endocrinology* **131**:1497
	c-src Antisense, c-cbl antisense	Tanaka 1996 *Nature* **383**:528
	c-src Inhibitor (Herbimycin)	Yoneda 1993 *JCI* **91**:2791
	Calcium	Zaidi 1989 *BBRC* **163**:1461
	G-protein	Zhang 1995 *J Cell Sci* **108**:2285
Gene expression	mRNA synthesis, protein synthesis	Collin 1992 *Endocrinology* **131**:1181
	mRNA synthesis, protein synthesis	Hall 1993 *BBRC* **195**:1245
	mRNA synthesis, protein synthesis	McSheehy 1987 *JCI* **80**:425
	c-fos	Udagawa 1996 *Bone* **18**:511
Vitamin D	1,25 dihydroxyvitamin D_3	Athanasou 1991 *Br J Canc* **63**:527
	1,25 dihydroxyvitamin D_3	Chambers 1985 *Endocrinology* **116**:234
	1,25 dihydroxyvitamin D_3	Fuller 1991 *BBRC* **181**:67
	1,25 dihydroxyvitamin D_3	Ishii 1993 *BBRC* **191**:495
	1,25 dihydroxyvitamin D_3, 24,25(OH)D_3	Matsumoto 1992 *Proc Soc Exp Biol Med* **200**:161
	1,25 dihydroxyvitamin D_3	McSheehy 1987 *JCI* **80**:425
	1,25 dihydroxyvitamin D_3	Takada 1992 *Bone Min* **17**:347
	1,25 D_3; 24R, 25D_3	Yamato 1993 *CTI* **52**:255
Thyroid hormone	T3	Allain 1992 *J Endocrinol* **133**:327
	T3	Britto 1994 *Endocrinology* **134**:169
	T4	Britto 1994 *Endocrinology* **134**:169
Retinoic acid	Retinoic acid	O'Neill 1992 *Bone* **13**:23
	Retinoic acid	Saneshige 1995 *Biochem J* **309**:721
Sex hormones	Oestradiol, tamoxifen	Arnett 1996 *J Endocrinol* **149**:503
	Centchroman, raloxifene, oestradiol	Hall 1995 *BBRC* **216**:662
	Extracts from oestradiol-treated osteoblasts	Ishii 1993 *BBRC* **191**:495
	Oestradiol, tamoxifen	Oursler 1993 *Endocrinology* **132**:1373
	Oestradiol	Oursler 1994 *PNAS* **91**:5227
	Oestradiol	Tamura 1993 *JBMR* **8**:953
	E_2, DHT, PG, TMX	Tobias 1991 *Acta Endocrinol* **124**:121
Glucocorticoids	Hydrocortisone, dexamethasone	Tobias 1989 *Endocrinology* **125**:1290
		Dempster 1997 *J Endocrinol* **154**:397
Cytokines, growth factors, morphogens	IL-1α	Athanasou 1991 *Br J Canc* **63**:527
	TGF-β, PDGF, IGF-1, IGF-2, αFGF, bFGF	Fuller 1991 *J Cell Physiol* **147**:208
	TGF-β	Hattersley 1991 *JBMR* **6**:165
	OP-1	Hentunen 1995 *BBRC* **209**:433
	IGF-1, IGF-2	Hill 1995 *Endocrinology* **136**:124
	IL-1α, IL-6	Ishii 1993 *BBRC* **191**:495
	IGF-1	Kaji 1996 *JBMR* **11**:62
	BMP-2	Kanatani 1995 *JBMR* **10**:1681
	MSP	Kurihara 1996 *Blood* **87**:3704

Table 3.2   *Continued*

Class	Substance/topic	Reference (first author only)
	IGF-1, IGF-2	Mochizuki 1992 *Endocrinology* **131**:1075
	TGF-β, OIF	Oreffo 1990 *Endocrinology* **126**:3069
	IL-10	Owens 1996 *J Immunol* **157**:936
	M-CSF, GM-CSF, IL-1β, IL-3, IL-6, TNFα, TGF-β, LIF, IL-11	Sarma 1996 *Blood* **88**:2531
	IL-1	Thomson 1986 *J Med Chem*
	TNFα	Thomson 1987 *J Immunol* **138**:775
Integrins and extracellular matrix	Thrombospondin, fibronectin	Carron 1995 *BBRC* **213**:1017
	Vitronectin receptor, antibody	Chambers 1986 *Bone Min* **1**:127
	Chondroitin sulfates, dextran sulfate	Chowdhury 1992 *JBMR* **7**:771
	Fibronectin	Fuller 1991 *J Cell Physiol* **147**:208
	Laminin	Fuller 1991 *J Cell Physiol* **147**:208
	Echistatin, integrin antibodies	Helfrich 1996 Bone **19**:317
	BORA, a bone matrix protein	Hentunen 1994 *Bone Min* **25**:183
	Bone matrix proteins	Hill 1994 *BBA* **1201**:193
	Vitronectin receptor, antibody, GRGDSP, GRGESP	Horton 1991 *Exp Cell Res* **195**:368
	Kistrin	King 1994 *JBMR* **9**:381
	GRGDS, GRGES, RGD	Lakkakorpi 1991 *J Cell Biol* **115**:1179
	BSP, osteopontin, vitronectin, GRGDS, vitronectin receptor, antibody, Poly Glu, Poly Asp	Ryanal 1996 *Endocrinology* **37**:2347
	Echistatin, Ala[24] echistatin, RGDS	Sato 1990 *J Cell Biol* **111**:1713
	BAG75, osteopontin	Sato 1992 *FASEB J* **6**:2966
	Echistatin, echistatin analogues RGDS, RGDF, RGDW, GPenGHRGD-LRCA	Sato 1994 *JBMR* **9**:1441
Osteoblast-derived modulators of bone resorption	Stimulator	Collin (1992) *Endocrinology* **131**:1181
	Stimulator	Fuller 1991 *J Cell Physiol* **147**: 208
	Stimulator	Fuller 1991 *BBRC* **181**:67
	Stimulator	Ishii 1993 *BBRC* **191**:495
	Inhibitor	
	Stimulator	Jimi 1996 *Endocrinology* **137**:2187
	Stimulator	Jimi 1996 *Endocrinology* **137**:3446
	Stimulator	Katoh 1995 *Bone* **16**:97
	Stimulator	Kuroki 1994 *Clin Exp Immunol* **95**:536
	Stimulator	Perry 1989 *Endocrinology* **125**:2075
	Stimulator	Vitte 1996 *Endocrinology* **137**:2324
	Inhibitor	
	Genetic defect	Sundquist 1995 *Tissue Cell* **27**:569
Other cell-derived modulators	Breast cancer cells supernatants	Clohisy 1996 *J Orthopaed Res* **14**:396
	Supernatants from phagocytosing synovial cells	Kim 1996 *J Biomedical Mater Res* **32**:3

Table 3.2   *Continued*

Class	Substance/topic	Reference (first author only)
Pharmaceuticals	Bisphosphonates	Azuma 1995 *Bone* **16**:235
	Bisphosphonates	Flanagan 1991 *CTI* **49**:407
	Bisphosphonates	Flanagan 1989 *Bone Min* **6**:33
	Bisphosphonates	Murakami 1995 *Bone* **17**:137
	Bisphosphonates	Piper 1994 *CTI* **54**:56
	Bisphosphonates	Sahni 1993 *JCI* **91**:2004
	Bisphosphonates	Sato 1990 *JBMR* **5**:31
	Bisphosphonates	Sato 1991 *JCI* **88**:2095
	Bisphosphonates	Schmidt 1996 *PNAS* **93**:3068
	Bisphosphonates	Selander 1994 *CTI* **55**:368
	Bisphosphonates	Vitte 1996 *Endocrinology* **137**:2324
	Bisphosphonates	Yu 1996 *Bone* **19**:339
	Buprenorphine, naloxone	Hall 1996 *Inflamm Res* **45**:299
	Cyclosporine A	Chowdhury 1991 *CTA* **49**:275
	Cyclosporine A	Chowdhury 1992 *JBMR* **7**:771
	Fluoride	Taylor 1990 *JBMR* **5** Suppl.:S121
	Fluoride	Taylor 1989 *Anat Embryol* **180**:427
	Gallium nitrate	Hall 1990 *Bone Min* **8**:211
	Hydrochlorothiazide	Hall 1994 *CTI* **55**:266
	Ipriflavone	Azria 1993 *CTI* **52**:16
	Ipriflavone	Notoya 1992 *CTI* **51** Suppl 1:S3
	Ipriflavone	Notoya 1993 *CTI* **53**:206
	Phenothiazine, chlorpromazine, amitryptiline, phenazine, phenoxazine	Hall 1996 *Gen Pharmacol* **27**:845
	Promethazine	Hall 1994 *CTI* **55**:68
	Suramin	Yoneda 1995 *Cancer Res* **55**:1989
	Taxol	Hall 1995 *CTI* **57**:463
Cytoskeleton	Cytochalasin B	Chambers 1984 *J Cell Sci* **66**:383
	Cytochalasin D	Sasaki 1993 *CTI* **53**:217
	Antibodies to myosin	Sato 1990 *Cell Motility and the Cytoskeleton* **17**:250
Mechanism of matrix degradation	Tetracycline	Chowdhury 1993 *Agents Actions* **40**:124
	Leupeptin	Chowdhury 1992 *JBMR* **7**:771
	Cysteine proteinase inhibitors	Debari 1995 *CTI* **56**:566
	Collagenase inhibitors	Delaisse 1987 *Bone* **8**:305
	Cysteine proteinase inhibitors	Delaisse 1987 *Bone* **8**:305
	Acid transport	Hall 1990 *J Cell Physiol* **142**:420
	Acid transport	Hall 1991 *CTI* **49**:328
	Acid transport	Hall 1992 *BBRC* **188**:1097
	Acid transport	Hall 1994 *Bone Min* **27**:159
	Cysteine proteinase inhibitors	Hill 1994 *J Cell Biochem* **56**:118
	Cysteine proteinase inhibitors	Kakegawa 1993 *FEBS Letters* **321**:247

Table 3.2   *Continued*

Class	Substance/topic	Reference (first author only)
	Cysteine proteinase inhibitors	Kakegawa 1995 *FEBS Letters* **370**:78
	Acid transport	Laitala 1994 *JCI* **93**:2311
	Tripeptidyl peptidase inhibitor	Page 1993 *Arch Biochem Biophys* **306**:354
	Acid transport	Sarges 1993 *J Med Chem* **36**:2828
	Acid transport	Sundquist 1990 *BBRC* **168**:309
	Acid transport	Sundquist 1994 *JBMR* **9**:1575
	Acid transport	Tamura 1993 *JBMR* **8**:953
	Cathepsin L inhibitor	Woo 1996 *Eur J Pharmacol* **300**:131
Microbial agents	*P. gingivalis* fimbriae	Kawata 1994 *Inf Immun* **62**:3012
	Paramyxovirus infection	Shepard 1996 *J Pathol* **179**:448
	LPS	Simsey-Durrant 1987 *Archs Oral Biol* **32**:911
Miscellaneous	Ethanol	Cheung 1995 *Bone* **16**:143
	Heparin, protamine	Chowdhury 1992 *JBMR* **7**:771
	Acid phosphatase inhibitors	Moonga 1990 *J Physiol* **429**:29
	Zeolite A	Schutze 1995 *J Cell Biochem* **58**:39
	Acid phosphatase inhibitors	Zaidi 1989 *BBRC* **159**:68

Non-conventional abbreviations of journals used:
BBA – Biochimica et Biophysica Acta
BBRC – Biochemical and Biophysical Research Communications
CTI – Calcified Tissue International
JBMR – Journal of Bone and Mineral Research
JCI – Journal of Clinical Investigation
PNAS – Proceedings of the National Academy of Sciences of the United States of America

to provide information concerning the mode of action or target cell involved in a response but to provide an assay of bone resorption that is based upon direct quantitation of bone resorbed.

3.4.4 Confirming the identity of osteoclasts generated in marrow and spleen cultures

The ability to excavate pits in bone slices *in vitro* has become the gold standard by which the differentiation of bona fide osteoclasts is confirmed following *in vitro* marrow or spleen haematopoietic blast cell culture [74]. However, other cell types that are not typically called osteoclasts, such as giant cells from non-osseous sites, can also dig pits [127], and this raises semantic questions as to the definition of an osteoclast. In several studies, pitting has been quantified in parallel with osteoclastogenesis, but it is not clear from these studies whether the osteoclast phenotype or osteoclast activity is being quantified (such studies have been omitted from our table of pitting experiments).

3.4.5 Elucidating the role of the cytoskeleton in osteoclastic polarization and resorption

The work of Vaananen's group has provided novel insights into the structure and role of the cytoskeleton in osteoclasts. Using the bone slice assay, they correlated cytoskeletal structure with resorption pits and identified a double ring structure of vinculin that was associated with osteoclast excavation; they proposed a cycle of cytoskeletal changes that would be consistent with the resorptive cycle of an osteoclast [64, 128–130]. Recent confocal microscope studies using the bone slice assay have extended this and provided information on the membrane domains of the osteoclast [131].

ACKNOWLEDGEMENTS

The authors are grateful to Justine M. Kilb for producing the line drawings and Brian Yarborough for photographic assistance and Chanda Taylor for preparing the cells for plate 3. This work was supported in part by NIH grants AR39191 and AR41331.

REFERENCES

1. Raisz, L.G. (1963) Stimulation of bone resorption by parathyroid hormone in tissue culture. *Nature (London)* **197**, 1015–1016
2. Raisz, L.G. (1965) Bone resorption in tissue culture. Factors influencing the response to parathyroid hormone. *Journal of Clinical Investigation* **44**, 103–116.
3. Abramson, E.C., Chang, J., Mayer, M. *et al.* (1988) Effects of cisplatin on parathyroid hormone- and human lung tumor-induced bone resorption. *Journal of Bone and Mineral Research* **3**, 541–546.
4. Takada, Y., Kusuda, M., Hiura, K. *et al.* (1992) A simple method to assess osteoclast-mediated bone resorption using unfractionated bone cells. *Bone and Mineral* **17**, 347–359.
5. Mochizuki, H., Hakeda, Y., Wakatsuki, N. *et al.* (1992) Insulin-like growth factor-1 supports formation and activation of osteoclasts. *Endocrinology* **131**, 1075–1080.
6. Helfrich, M.H. and Mieremet, R.H.P. (1988) A morphological study of osteoclasts isolated from osteopetrotic microphthalmic (mi/mi) mouse and human fetal long bones using an instrument permitting combination of light and scanning electron microscopy. *Bone* **9**, 113–119.
7. Murrills, R.J., Shane, E., Lindsay, R. *et al.* (1989) Bone resorption by isolated human osteoclasts *in vitro*: effects of calcitonin. *Journal of Bone and Mineral Research* **4**, 259–268.
8. Alvarez, J.I., Teitelbaum, S.L., Blair, H.C. *et al.* (1991) Generation of avian cells resembling osteoclasts from mononuclear phagocytes. *Endocrinology* **128**, 2324–2335.
9. Alvarez, J.I., Ross, F.P., Athanasou, N.A. *et al.* (1992) Osteoclast precursors circulate in avian blood. *Calcified Tissue International* **51**, 48–53.
10. Tamura, T., Takahashi, N., Akatsu, T. *et al.* (1993) New resorption assay with mouse osteoclast-like multinucleated cells formed in vitro. *Journal of Bone and Mineral Research* **8**, 953–960.
11. Hata, K., Kukita, T., Akamine, A. *et al.* (1992) Trypsinized osteoclast-like cells formed in rat bone marrow cultures efficiently form resorption lacunae on dentine. *Bone* **13**, 139–146.
12. Fujikawa, Y., Quinn, J.M.W., Sabokbar, A. *et al.* (1996) The human osteoclast precursor circulates in the monocyte fraction. *Endocrinology* **137**, 4058–4060.

13. Sarma, U. and Flanagan, A.M. (1996) Macrophage colony-stimulating factor induces substantial osteoclast generation and bone resorption in human bone marrow cultures. *Blood* **88**, 2531–2540.

14. Luben, R.A., Wong, G.L. and Cohn, D.V. (1977) Parathormone-stimulated resorption of devitalized bone by cultured osteoclast-type cells. *Nature* **265,** 629–630.

15. Collin-Osdoby, P., Oursler, M.J., Webber, D. and Osdoby, P. (1991) Osteoclast-specific monoclonal antibodies coupled to magnetic beads provide a rapid and efficient method of purifying avian osteoclasts. *Journal of Bone and Mineral Research* **6**, 1353–1365.

16. Oursler, M.J., Pederson, L., Fitzpatrick, L. *et al.* (1994) Human giant cell tumors of the bone (osteoclastomas) are estrogen target cells. *Proceedings of the National Academy of Sciences USA* **91**, 5227–5231.

17. Kallio, D.M., Garant, P.R. and Minkin, C. (1972) Ultrastructural effects of calcitonin on osteoclasts in tissue culture. *Journal of Ultrastructure Research* **39**, 205–216.

18. Wesolowski, G., Duong, L.T., Lakkakorpi, P.T. *et al.* (1995) Isolation and characterization of highly enriched, prefusion mouse osteoclastic cells. *Experimental Cell Research* **219**, 679–686.

19. Boyde, A. and Jones, S.J. (1984) Resorption of dentine by isolated osteoclasts in vitro. *British Dental Journal* **156**, 216–220.

20. Chambers, T.J., Revell, P.A., Fuller, K. *et al.* (1984) Resorption of bone by isolated rabbit osteoclasts. *Journal of Cell Science* **66**, 383–399.

21. Zambonin-Zallone, A., Teti, A. and Primavera, M.V. (1982) Isolated osteoclasts in primary culture; first observations on structure and survival in culture media. *Anatomy and Embryology* **165**, 405–413.

22. Blair, H.C., Kahn, A., Crouch, E.C. *et al.* (1986) Isolated osteoclasts resorb the organic and inorganic components of bone. *Journal of Cell Biology* **102**, 1164–1172.

23. David, P. and Baron, R. (1994) The catalytic cycle of the vacuolar H^+-ATPase. *Journal of Biological Chemistry* **269**, 30158–30163.

24. Demulder, A., Suggs, S.V., Zsebo, K.M. *et al.* (1992) Effects of stem cell factor on osteoclast-like cell formation on long term human marrow cultures. *Journal of Bone and Mineral Research* **7**, 1337–1344.

25. Oursler, M.J., Collin-Osdoby, P., Anderson, F. *et al.* (1991) Isolation of avian osteoclasts: improved techniques to preferentially purify viable cells. *Journal of Bone and Mineral Research* **6**, 375–385.

26. McSheehy, P.M.J. and Chambers T.J. (1986) Osteoblastic cells mediate osteoclastic responsiveness to parathyroid hormone. *Endocrinology* **118**, 824–828.

27. McSheehy, P.M.J. and Chambers, T.J. (1986) Osteoblast-like cells in the presence of parathyroid hormone release soluble factor that stimulates osteoclastic bone resorption. *Endocrinology* **119**, 1654–1659.

28. McSheehy, P.M.J. and Chambers, T.J. (1987) 1.25-Dihydroxyvitamin D_3 stimulates rat osteoclasts to release a soluble factor that increases osteoclastic bone resorption. *Journal of Clinical Investigation* **80**, 425–429.

29. Thomson, B.M., Saklatvala, J. and Chambers, T.J. (1986) Osteoblasts mediate interleukin 1 stimulation of bone resorption by rat osteoclasts. *Journal of Experimental Medicine* **164**, 104–112.

30. Thomson, B.M., Mundy, G.R. and Chambers, T.J. (1987) Tumor necrosis factors a and β induce osteoblastic cells to stimulate osteoclastic bone resorption. *Journal of Immunology* **138**, 775–779.

31. Evely, R.S., Bonomo, A., Schneider, H.-G *et al.* (1991) Structural requirements for the action of parathyroid hormone-related protein (PTHrP) on bone resorption by isolated osteoclasts. *Journal of Bone and Mineral Research* **6**, 85–93.

32. Tezuka, K.-I., Sato, T., Kamioka, H. *et al.* (1992) Identification of osteopontin in isolated rabbit osteoclasts. *Biochemical and Biophysical Research Communications* **186**, 911–917.

33. Shioi, A., Ross, F.P. and Teitelbaum, S.L. (1994) Enrichment of generated murine osteo-clasts. *Calcified Tissue International* **55,** 387–394.

34. May, L.G. and Gay, C.V. (1996) Development of a new method for obtaining osteo-clasts from endosteal surfaces. *In Vitro Cellular & Developmental Biology* **32,** 269–278.

35. Murrills, R.J., Stein, L.S. and Dempster, D.W. (1993) Stimulation of bone resorption and osteoclast clear zone formation by low pH: a time-course study. *Journal of Cellular Physiology* **154,** 511–518.

36. Kaji, H., Sugimoto, T., Kanatani, M. *et al.* (1996) Estrogen blocks parathyroid hormone (PTH)-stimulated osteoclast-like cell formation by selectively affecting PTH-responsive cyclic adenosine monophosphate pathway. *Endocrinology* **137,** 2217–2224.

37. Fuller, K., Owens, J.M., Jagger, C.J. *et al.* (1993) Macrophage colony-stimulating factor stimulates survival and chemotactic behavior in isolated osteoclasts. *Journal of Experimental Medicine* **178,** 1733–1744.

38. Jimi, E., Shuto, T. and Koga, T. (1995) Macrophage colony-stimulating factor and inter-leukin-1a maintain the survival of osteoclast-like cells. *Endocrinology* **136,** 808–811.

39. Jones, S.J., Ali, N.N. and Boyde, A. (1986) Survival and resorptive activity of chick osteoclasts in culture. *Anatomy and Embryology* **174,** 265–275.

40. Prallet, B., Male, P., Neff, L. *et al.* (1992) Identification of a functional mononuclear precursor of the osteoclast in chicken medullary bone marrow cultures. *Journal of Bone and Mineral Research* **7,** 405–414.

41. Grano, M., Colucci, S., De Bellis, M. *et al.* (1994) New model for bone resorption study in vitro: human osteoclast-like cells from giant cell tumors of bone. *Journal of Bone and Mineral Research* **9,** 1013–1020.

42. Yoneda, T., Alsina, M.M., Garcia, J.L. *et al.* (1991) Differentiation of HL-60 cells into cells with the osteoclast phenotype. *Endocrinology* **129,** 683–689.

43. Gattei, V., Bernabei, P.A., Pinto, A. *et al.* (1992) Phorbol ester induced osteoclast-like differentiation of a novel human leukemic cell line (FLG 29.1). *Journal of Cell Biology* **116,** 437–447.

44. Shin, J.H., Kukita, A., Ohki, K. *et al.* (1995) In vitro differentiation of the murine macrophage cell line BDM-1 into osteoclast-like cells. *Endocrinology* **136,** 4285–4292.

45. Chambers, T.J, Owens, J.M., Hattersley, G. *et al.* (1993) Generation of osteoclast-induc-tive and osteoclastogenic cell lines from the H-2KbtsA58 transgenic mouse. *Proceedings of the National Academy of Sciences USA* **90,** 5578–5582.

46. Boyce, B.F., Wright, K., Reddy, S.V. *et al.* (1995) Targeting simian virus 40 T antigen to the osteoclast in transgenic mice causes osteoclast tumors and transformation and apoptosis of osteoclasts. *Endocrinology* **136,** 5751–5759.

47. Kahn, A.J., Stewart, C.C. and Teitelbaum, S.L. (1978) Contact-mediated bone resorp-tion by human monocytes. *Science* **199,** 988–990.

48. Teitelbaum, S.L., Stewart, C.C. and Kahn, A.J. (1979) Rodent peritoneal macrophages as bone resorbing cells. *Calcified Tissue International* **27,** 255–261.

49. Chambers, T.J. (1980) Diphosphonates inhibit bone resorption by macrophages in vitro. *Journal of Pathology* **132,** 255–62.

50. Ross, F.P., Chappel, J., Alvarez, J.I. *et al.* (1993) Interactions between the bone matrix proteins osteopontin and bone sialoprotein and the osteoclast integrin $\alpha_v\beta_3$ potentiate bone resorption. *Journal of Biological Chemistry* **268,** 9901–9907.

51. Bizzarri, C., Shioi, A., Teitelbaum, S.L. *et al.* (1994) Interleukin-4 inhibits bone resorp-tion and acutely increases cytosolic Ca^{2+} in murine osteoclasts. *Journal of Biological Chemistry* **269,** 13817–13824.

52. Zambonin-Zallone, A., Teti, A. and Primavera, M.V. (1984) Resorption of vital or devi-talized bone by isolated osteoclasts in vitro. *Cell and Tissue Research* **235,** 561–564.

53. de Vernejoul, M.-C., Horowitz, M., Demignon, J. *et al.* (1988) Bone resorption by isolated chick osteoclasts in culture is stimulated by murine spleen cell supernatant fluids (osteoclast-activating factor) and inhibited by calcitonin and prostaglandin E_2. *Journal of Bone and Mineral Research* **3,** 69–80.

54. Jimi, E., Nakamura, I., Amano, H. *et al.* (1996) Osteoclast function is activated by osteoblastic cells through a mechanism involving cell-to-cell contact. *Endocrinology* **137**, 2187–2190.

55. Garcia, C., Qiao, M., Chen, D. *et al.* (1996) Effects of synthetic peptido-leukotrienes on bone resorption in vitro. *Journal of Bone and Mineral Research* **11**, 521–529.

56. Jones, S.J., Boyde, A. and Ali, N.N. (1984) The resorption of biological and non-biological substrates by cultured avian and mammalian osteoclasts. *Anatomy and Embryology* **170**, 247–256.

57. Raynal, C., Delmas, P. and Chenu, C. (1996) Bone sialoprotein stimulates *in vitro* bone resorption. *Endocrinology* **137**, 2347–2354.

58. Chambers, T.J., McSheehy, P.M.J., Thomson, B.M. *et al.* (1985) The effect of calcium-regulating hormones and prostaglandins on bone resorption by osteoclasts disaggregated from neonatal rabbit bones. *Endocrinology* **116**, 234–239.

59. Shibutani, T., and Heersche, J.N.M. (1993) Effect of medium pH on osteoclast activity and osteoclast formation in cultures of dispersed rabbit osteoclasts. *Journal of Bone and Mineral Research* **8**, 331–336.

60. Gray, M., Taylor, M.L., Horton, M.A. *et al.* (1991) Studies on cells derived from growing deer antler, in *The Biology of Deer*, (ed. R.D. Brown), Springer-Verlag, New York, 511–519.

61. Kaji, H., Sugimoto, T., Fukase, M. *et al.* (1993) Role of dual signal transduction systems in the stimulation of bone resorption by parathyroid hormone-related peptide. *Hormone and Metabolic Research* **25**, 421–424.

62. Oursler, M.J., Pederson, L., Pyfferoen, J. *et al.* (1993) Estrogen modulation of avian osteoclast lysosomal gene expression. *Endocrinology* **132**, 1373–1380.

63. Morrison, M.S. and Arnett, T.R. (1997) Effect of extracellular pH on resorption pit formation by chick osteoclasts. *Journal of Bone and Mineral Research,* **12**, S 290 (Abstr.).

64. Selander, K., Lehenkari, P. and Vaananen, H.K. (1994) The effects of bisphosphonates on the resorption cycle of isolated osteoclasts. *Calcified Tissue International* **55**, 368–375.

65. Morrison, M.S. and Arnett, T.R. (1996) Indomethacin and ibuprofen stimulate pit formation by rat osteoclasts. *Journal of Bone and Mineral Research* **11**, 1825 (Abstr. P38).

66. Hall, T.J., Jeker, H. and Schaueblin, M. (1995) Wortmannin, a potent inhibitor of phosphatidylinositol 3-kinase inhibits osteoclastic bone resorption *in vitro*. *Calcified Tissue International* **56**, 336–338.

67. Chowdhury, M.H., Hamada, C. and Dempster, D.W. (1992) Effects of heparin on osteoclast activity. *Journal of Bone and Mineral Research* **7**, 771–777.

68. Moonga, B.S., Stein, L.S., Kilb, J.M. *et al.* (1996) Effect of diacylglycerols on osteoclastic bone resorption. *Calcified Tissue International* **59**, 105–108.

69. Chambers, T.J., Fuller, K., McSheehy, P.M.J. *et al.* (1985) The effects of calcium regulating hormones on bone resorption by isolated human osteoclastoma cells. *Journal of Pathology* **145**, 297–305.

70. Zaidi, M., Fuller, K., Bevis, P.J.R. *et al.* (1987) Calcitonin gene-related peptide inhibits osteoclastic bone resorption: a comparative study. *Calcified Tissue International* **40**, 149–154.

71. Zaidi, M., Chambers, T.J., Gaines Das, R.E. *et al.* (1987) A direct action of human calcitonin gene-related peptide on isolated osteoclasts. *Journal of Endocrinology* **115**, 511–518.

72. Hall, T.J., Schaeublin, M. and Chambers, T.J. (1992) Na^+/H^+-antiporter activity is essential for the induction, but not the maintenance of osteoclastic bone resorption and cytoplasmic spreading. *Biochemical and Biophysical Research Communications* **188**, 1097–1103.

73. Hattersley, G. and Chambers, T.J. (1991) Effects of transforming growth factor beta 1 on the regulation of osteoclastic development and function. *Journal of Bone and Mineral Research* **6**, 165–172.

74. Sato, T., Hakeda, Y., Yamaguchi, Y. *et al.* (1995) Hepatocyte growth factor is involved in formation of osteoclast-like cells mediated by clonal stromal cells (MC3T3-G2/PA6). *Journal of Cellular Physiology* **164**, 197–204.

75. Arnett, T.R. and Dempster, D.W. (1986) Effect of pH on bone resorption by rat osteoclasts in vitro. *Endocrinology* **119**, 119–124.

76. Murrills, R.J. and Dempster, D.W. (1990) The effects of stimulators of intracellular cyclic AMP on rat and chick osteoclasts *in vitro*: validation of a simplified light microscope assay of bone resorption. *Bone* **11**, 333–344.

77. Kitamura, K., Katoh, M., Komiyama, O. *et al.* (1993) Establishment of a rapid bone resorption in vitro assay using previously frozen mouse unfractionated bone cells pretreated with parathyroid hormone. *Bone* **14**, 829–834.

78. Kanehisa, J., Shibutani, T., Iwayama, Y. *et al.* (1990) Lectin-binding glycoconjugates and immunoreactive proteoglycans in resorption pits freshly excavated by isolated rabbit osteoclasts. *Archives of Oral Biology* **35**, 771–774.

79. King, K.L., D'Anza, J.J., Bodary, S. *et al.* (1994) Effects of kistrin on bone resorption and serum calcium in vivo. *Journal of Bone and Mineral Research* **9**, 381–387.

80. Sahni, M., Guenther, H.L., Fleisch, H. *et al.* (1993) Bisphosphonates act on rat bone resorption through the mediation of osteoblasts. *Journal of Clinical Investigation* **91**, 2004–2011.

81. Jones, S.J., Boyde, A., Ali, N.N. *et al.* (1986) Variation in the sizes of resorption lacunae made in vitro. *Scanning Electron Microscopy* **IV**, 1571–1580.

82. Sasaki, T., Debari, K. and Hasemi, M. (1993) Measurement of Howship's resorption lacunae by a scanning probe microscope system. *Journal of Electron Microscopy* **42**, 356–359.

83. Jones, S.J., Boyde, A., Piper, A. *et al.* (1992) Confocal microscope mapping of osteoclastic resorption. *Microscopy and Analysis* **July**, 18–20.

84. Yu, X., Scholler, J. and Foged, N.T. (1996) Interaction between effects of parathyroid hormone and bisphosphonate on regulation of osteoclast activity by the osteoblast-like cell line UMR-106. *Bone* **19**, 339–345.

85. MacIntyre, I., Zaidi, M., Towhidul Alam, A.S.M. *et al.* (1991) Osteoclastic inhibition: an action of nitric oxide not mediated by cyclic GMP. *Proceedings of the National Academy of Sciences USA* **88**, 2936–2940.

86. Tobias, J.H. and Chambers, T.J. (1991) The effect of sex hormones on bone resorption by rat osteoclasts. *Acta Endocrinologica (Copenhagen)* **124**, 121–127.

87. Fuller, K., Owens, J. and Chambers, T.J. (1995) The effect of hepatocyte growth factor on the behaviour of osteoclasts. *Biochemical and Biophysical Research Communications* **212**, 334–340.

88. Holloway, W.R., Collier, F.M., Herbst, R.E. *et al.* (1996) Osteoblast-mediated effects of zinc on isolated rat osteoclasts: inhibition of bone resorption and enhancement of osteoclast number. *Bone* **19**, 137–142.

89. Murrills, R.J., Stein, L.S., Fey, C.P. *et al.* (1990) The effects of parathyroid hormone (PTH) and PTH-related peptide on osteoclast resorption of bone slices *in vitro*: an analysis of pit size and the resorption focus. *Endocrinology* **127**, 2648–2653.

90. Hakeda, Y., Hiura, K., Sato, T. *et al.* (1989) Existence of parathyroid hormone binding sites on murine hemopoietic blast cells. *Biochemical and Biophysical Research Communications* **163**, 1481–1486.

91. Oreffo, R.O., Bonewald, L., Kukita, A. *et al.* (1990) Inhibitory effects of the bone-derived growth factors, osteoinductive factor and transforming growth factor-beta on isolated osteoclasts. *Endocrinology* **126**, 3069–3075.

92. Arnett, T.R., Boyde, A., Jones, S.J. *et al.* (1994) Effects of medium acidification by alteration of carbon dioxide or bicarbonate concentrations on the resorptive activity of rat osteoclasts. *Journal of Bone and Mineral Research* **9**, 375–379.

93. Hall, T.J., Higgins, W., Tardif, C. *et al.* (1991) A comparison of the effects of inhibitors of carbonic anhydrase on osteoclastic bone resorption and purified carbonic anhydrase. *Calcified Tissue International* **49**, 328–332.

94. Debari, K., Sasaki, T., Udagawa, N. *et al.* (1995) An ultrastructural evaluation of the effects of cysteine-proteinase inhibitors on osteoclastic resorptive functions. *Calcified Tissue International* **56**, 566–570.

95. Delaisse, J.M., Boyde, A., Maconnachie, E. *et al.* (1987) The effects of inhibitors of cysteine proteinases and collagenase on the resorptive activity of isolated osteoclasts. *Bone* **8**, 305–313.

96. Yoneda, T., Lowe, C., Lee, C.-H. *et al.* (1993) Herbimycin A, a pp60^{c-src} tyrosine kinase inhibitor, inhibits osteoclastic bone resorption in vitro and hypercalcemia in vivo. *Journal of Clinical Investigation* **91**, 2791–2795.

97. Hall, T.J., Schaeublin, M. and Missbach, M. (1994) Evidence that c-src is involved in the process of osteoclastic bone resorption. *Biochemical and Biophysical Research Communications* **199**, 1237–1244.

98. McSheehy, P.M.J., Farina, C., Airaghi, R. *et al.* (1995) Pharmacological evaluation of the calcitonin analogue SB 205614 in models of osteoclastic bone resorption in vitro and in vivo: comparison with salmon calcitonin and elcatonin. *Bone* **16**, 435–444.

99. Domon, T. and Wakita, M. (1991) Electron microscopic and histochemical studies of the mononuclear osteoclast of the mouse. *American Journal of Anatomy* **192**, 35–44.

100. Goto, K. (1918) Mineral metabolism in experimental acidosis. *Journal of Biological Chemistry* **36**, 355–376.

101. Jaffe, H.L., Bodansky, A. and Chandler, J.P. (1932) Ammonium chloride decalcification as modified by calcium intake: the relation between generalized osteoporosis and ostitis fibrosa. *Journal of Experimental Medicine* **56**, 823–834.

102. Barzel, U.S. and Jowsey, J. (1969) The effects of chronic acid and alkali administration on bone turnover in adult rats. *Clinical Science* **36**, 517–524.

103. Spowage, M. and Arnett, T.R. (1995) Time dependence of proton-activated osteoclastic resorption. *Calcified Tissue International* **56**, 445.

104. Bushinsky, D.A. (1989) Net calcium efflux from live bone during chronic metabolic but not respiratory acidosis. *American Journal of Physiology,* **256**, F836–842.

105. Meghji, S., Henderson, B., Morrison, M.S. *et al.* (1996) Ca^{2+} release from cultured mouse calvaria is very sensitive to ambient pH. *Journal of Bone and Mineral Research* **11**, 1824 (Abstr. P37).

106. Goldhaber, P. and Rabadjija, L. (1987) H^{+} stimulation of cell-mediated bone resorption in tissue culture. *American Journal of Physiology* **253**, E90–E98.

107. Spowage, M. and Arnett, T.R. (1996) Activation of rat osteoclasts by pertussis toxin. *Bone* **17**, 577.

108. Matayoshi, A., Brown, C., DiPersio, J.F. *et al.* (1996) Human blood-mobilized hematopoietic precursors differentiate into osteoclasts in the absence of stromal cells. *Proceedings of the National Academy of Sciences USA* **93**, 10785–10790.

109. Teti, A., Grano, M., Colucci, S.,*et al.* (1991) Osteoblast–osteoclast relationships in bone resorption: osteoblasts enhance osteoclast activity in a serum-free co-culture system. *Biochemical and Biophysical Research Communications* **179**, 634–640.

110. Teti, A., Grano, M., Colucci, S. *et al.* (1992) Osteoblastic control of osteoclast bone resorption in a serum-free co-culture system. Lack of effect of parathyroid hormone. *Journal of Endocrinological Investigation* **15** (Suppl. 6), 63–68.

111. Britto, J.M., Fenton, A.J., Holloway, W.R. *et al.* (1994) Osteoblasts mediate thyroid hormone stimulation of osteoclastic bone resorption. *Endocrinology* **134**, 169–176.

112. Rao, L.G., Murray, T.M. and Heersche, J.N.M. (1983) Immunohistochemical demonstration of parathyroid hormone binding to specific cell types in fixed rat bone tissue. *Endocrinology* **113**, 805–810.

113. Teti, A., Rizzoli, R. and Zambonin-Zallone, A. (1991) Parathyroid hormone binding to cultured avian osteoclasts. *Biochemical and Biophysical Research Communications* **174**, 1217–1222.

114. Agarwala, N. and Gay, C.V. (1992) Specific binding of parathyroid hormone to living osteoclasts. *Journal of Bone and Mineral Research* **7**, 531–539.

115. Perry, H.M. 3rd, Skogen, W., Chappel, J. *et al.* (1989) Partial characterization of a parathyroid hormone-stimulated resorption factor(s) from osteoblast-like cells. *Endocrinology* **125**, 2075–2082.

116. Fuller, K., Gallagher, A.C. and Chambers, T.J. (1991) Osteoclast resorption stimulating activity is associated with the osteoblastic surface and/or the extracellular matrix. *Biochemical and Biophysical Research Communications* **181**, 67–73.

117. Greenfield, E.M., Alvarez, J.I., McLaurine, E.A. *et al.* (1992) Avian osteoblast conditioned media stimulate bone resorption by targeting multinucleating osteoclast precursors. *Calcified Tissue International* **51**, 317–323.

118. Collin, P., Guenther, H.L. and Fleisch, H. (1992) Constitutive expression of osteoclast-stimulating activity by normal clonal osteoblast-like cells: effects of parathyroid hormone and 1,25-dihydroxyvitamin D_3. *Endocrinology* **131**, 1181–1187.

119. Kuroki, Y., Shiozawa, S., Sugimoto, T. *et al.* (1994) Constitutive c-fos expression in osteoblastic MC3T3-E1 cells stimulates osteoclast maturation and osteoclastic bone resorption. *Clinical and Experimental Immunology* **95**, 536–539.

120. Katoh, M., Kitamura, K. and Kitagawa, H. (1995) Induction of bone resorbing activity by mouse stromal cell line, MC3T3-G2/PA6. *Bone* **16**, 97–102.

121. Vitte, C., Fleisch, H. and Guenther, H.L. (1996) Bisphosphonates induce osteoblasts to secrete an inhibitor of osteoclast-mediated resorption. *Endocrinology* **137**, 2324–2333.

122. Ishii, T., Saito, T., Morimoto, K. *et al.* (1993) Estrogen stimulates the elaboration of a cell/matrix surface-associated inhibitory factor of osteoclastic bone resorption from osteoblastic cells. *Biochemical and Biophysical Research Communications* **191**, 495–502.

123. Morris, C.A., Mitnick, M.E., Weir, E.C. *et al.* (1990) The parathyroid hormone-related protein stimulates human osteoblast-like cells to secrete a 9,000 Dalton bone resorbing protein. *Endocrinology* **126**, 1783–1785.

124. Greenfield, E.M., Gornik, S.A., Horowitz, M.C. *et al.* (1993) Regulation of cytokine expression in osteoblasts by parathyroid hormone: rapid stimulation of interleukin-6 and leukemia inhibitory factor mRNA. *Journal of Bone and Mineral Research* **8**, 1163–1171.

125. Greenfield, E.M., Shaw, S.M., Gornik, S.A. *et al.*, (1995) Adenyl cyclase and interleukin 6 are downstream effectors of parathyroid hormone resulting in stimulation of bone resorption. *Journal of Clinical Investigation* **96**, 1238–1244.

126. Greenfield, E.M., Horowitz, M.C. and Lavish, S.A. (1996) Stimulation by parathyroid hormone of interleukin-6 and leukemia inhibitory factor expression in osteoblasts is an immediate-early gene response induced by cAMP signal transduction. *Journal of Biological Chemistry* **271**, 10984–10989.

127. Athanasou, N.A. and Quinn, J.M. (1992) Bone resorption by macrophage polykaryons of a pilar tumor of scalp. *Cancer* **70**, 469–475.

128. Lakkakorpi, P.T. and Vaananen, H.K. (1991) Kinetics of the osteoclast cytoskeleton during the resorption cycle in vitro. *Journal of Bone and Mineral Research* **6**, 817–826.

129. Lakkakorpi, P.T. and Vaananen, H.K. (1996) Cytoskeletal changes in osteoclasts during the resorption cycle. *Microscopy Research and Technique* **33**, 171–181.

130. Laitala-Leinonen, T., Howell, M.L., Dean, G.E. *et al.* (1996) Resorption-cycle-dependent polarization of mRNAs for different subunits of V-ATPase in bone resorbing osteoclasts. *Molecular Biology of the Cell* **7**, 129–142.

131. Salo, J., Metsikko, K., Palokangas, H. *et al.* (1996) Bone-resorbing osteoclasts reveal a dynamic division of basal plasma membrane into two different domains. *Journal of Cell Science* **109**, 301–307.

CHAPTER FOUR

Bone organ cultures

Sajeda Meghji, Peter A. Hill and Malcolm Harris

4.1 INTRODUCTION

The previous chapters have highlighted the use of isolated osteoblasts and osteoclasts in skeletal research. While it is obviously important to investigate the functions of individual bone cell populations, the biology of bone is that of an organ system with distinct but interacting cells. Organ culture of bone provides a model that is, in certain respects, closer to the *in vivo* situation. Bone organ culture naturally encompasses the interactions that occur between the different cell types present in bone and bone marrow. Moreover, as bone is a tissue with a predominance of extracellular matrix, organ or explant culture techniques allow the study of the interaction of bone cells with the natural matrix of bone.

Organ culture techniques have been in use for more than 50 years to investigate the biology of bone and cartilage and the pioneering work of the late Dame Honor Fell is known to many worldwide [1]. This technique can be used to analyse bone both at the tissue and at the cellular level and to investigate the response of bone cells to specific stimuli. Organ culture can also be used to investigate the interactions which occur between bone cells during the tissue response to such stimuli. The principal advantage of organ culture technology is the ability to isolate local effects (mediators, mechanical stress, etc.) from systemic factors (steroids, other hormones, toxins, etc.). With explants of rodent or chick calvaria, or of long bones, the maintenance of anatomical order and natural cell-to-cell interactions guarantees that the response of bone cells to a particular stimulus more closely mimics the response that would have occurred if the bone was still within the living animal.

However, it must be emphasized that the artificial environment created in the organ/explant culture requires that the investigator exercises caution in the interpretation of results and in extrapolating the results to the *in vivo* situation. It must be noted that despite the decreased complexity, when compared with the whole animal, organ and explant cultures contain heterogeneous cell populations.

In this chapter the organ and explant culture systems that are currently in use for studying bone biology will be reviewed. The relative merits and demerits (limitations) of these various techniques have been evaluated both for the study of osteoclastic bone resorption and osteoblastic bone formation. One of the principal advantages of organ culture is that both of these interdependent cell populations coexist in their normal milieu.

Methods in Bone Biology. Edited by Timothy R. Arnett and Brian Henderson.
Published in 1997 by Chapman & Hall, London. ISBN 0 412 75770 2.

4.2 ORGAN CULTURE TECHNIQUES FOR STUDYING BONE RESORPTION

The process of osteoclastic bone resorption involves a series of steps which includes the formation of mononuclear osteoclast precursors from early haemopoietic progenitors within the bone marrow micro-environment, the differentiation of these precursors into multinucleated cells and finally the resorbing activity of these mature cells (Chapter 2). Over the years a variety of bone explant systems have been developed to study the effects of hormones and other factors on the various stages in the life cycle of the osteoclast, using organ culture techniques. To study the effects of bone-resorbing agents on osteoclast activity, two main bone explant systems are in current use: fetal rat long bones in stationary cultures [2] and calvarial bones from neonatal mice in both stationary culture [3] and roller tubes [4].

To examine bone resorption in organ culture, explants of bone are maintained in chemically defined media supplemented with low levels of heat-inactivated serum (usually 5%) for relatively short incubation times, usually 2–4 days. At the end of the incubation period, resorption is measured in a number of related ways (see below).

4.2.1 Fetal rat long bone cultures

The Raisz–Niemann model requires that pregnant mice (day 16 of gestation) or rats (day 18 of gestation) receive radioactive calcium ($^{45}Ca^{2+}$) to label fetal long bones [5]. After a 24-hour labelling period, the fetal long bones (radii/ulnae) are removed and placed in defined media on stainless steel grids. Following a period of biochemical equilibration of the organ cultures, test substances are added and the subsequent release of radiolabelled calcium is monitored as an indication of resorptive activity. Bones are usually paired, with one serving as a control relative to the test material added. These bones display a large surface area to volume ratio and pairs of exactly matching bones can be obtained only by careful microdissection. The use of radiolabelled fetal bones is a relatively expensive and hazardous technique since the pregnant mother has to be injected with quite large quantities of isotope in order to get enough $^{45}Ca^{2+}$ into the fetal skeletons. Consequently this method has been modified so that 'cold' Ca^{2+} released into medium is measured instead of pre-incorporated $^{45}Ca^{2+}$. Measurements are made using automated Ca^{2+} analysers based on titration or atomic absorption techniques. Some workers believe that the release of cold Ca^{2+} gives a more valid representation of bone resorption since it reflects changes in total mineral dissolution, whereas $^{45}Ca^{2+}$ release only demonstrates changes in resorption of relatively immature mineral [6]. The radiolabel analysis is usually compared with the histological profile of the organ culture to look for parallel alterations in osteoclast number or resorption. In this way, individual hormonal effects, immune modulation of bone resorption and cytokine regulation of resorptive activity have been investigated. The strengths of the system lie in its reproducibility, inherent balance toward resorptive activity, potential for retaining intact many of the original cell interactions and its simplicity of design. Unfortunately direct effects cannot be distinguished from indirect actions on a target cell population. This was apparent when it was reported that parathyroid hormone (PTH) stimulated bone resorption by a direct action on osteoclasts. The subsequently accepted view, based on other studies, is that PTH may exert its stimulatory action through an effect on intermediary cells, possibly osteoblasts, which in turn influence the osteoclast [7, 8]. The organ culture system does not allow one to discriminate between these options, nor necessarily to extrapolate from results obtained with fetal or neonatal animals to predicted responses that might occur in older organisms. Despite these limitations, the bone organ culture represents an exceedingly valuable system for determining the mode of action of bone resorptive agents [9–13].

4.2.2 Calvarial bone resorption assay

Another resorption assay in routine use is the method originally developed by Reynolds and Dingle and illustrated in Fig. 4.1. Again this may be used as the original radiolabelled technique or the more practical cold calcium release.

(a) Radiolabelled calcium assay
For this method, the skeletons of 1-day-old mice are labelled 6 days before experimentation by one subcutaneous injection of $^{45}CaCl_2$. Half-calvariae explants consisting of frontal and parietal bones, divided into two pieces along the median sagittal suture, are prepared from the pre-labelled mice by careful microdissection, so as not to damage the periosteum. The dissection is carried out in culture medium with a lowered level of bicarbonate (50 mg/100 ml) so as to equilibrate with air. The separated half-calvariae are then explanted into dishes of medium. A paired system is usually employed in which one explant is cultured in control medium and the other one of the pair is explanted into medium containing the test substance. The essential point is that the explants rest on stainless steel rafts (Minimesh FDP quality, Expanded Metal Co., West Hartlepool, UK) in the plastic dishes so that the explant is at the interface of the gas phase and the liquid medium, incubated at 37°C in a humidified atmosphere.

Nearly all experimental work in recent years has employed chemically defined media, often with a small supplement of serum or 0.1% w/v bovine serum albumin (see below). Chemically defined media are necessary if biochemical methods are to be used for quantitative measurements of metabolic products released from the explants into the medium, since large amounts of serum often interfere with biochemical assays. Biggers, Gwatkin and Heyner [14] developed one of the most successful chemically defined media for the growth of bone and cartilage and it is commonly referred to as BGJ. Other media in common use are CMRL 1066 and the Fitton–Jackson modification of BGJ. For some studies the synthetic medium alone is quite satisfactory, but the addition of a small quantity of serum is often necessary in order to observe the maximum effect of hormones [3]. We frequently use 5% heat inactivated rabbit or calf serum. Quantitation of the change in bone resorption after a certain treatment is done by counting the $^{45}Ca^{2+}$ in the medium (0.5 ml aliquots), from both the treated and untreated paired explants, by liquid scintillation counting. The bones can be counted at the conclusion of the experiment by dissolving them in either formic or hydrochloric acid. For each explant, the percentage release of total bone isotope into the medium is calculated, and the difference in the percentages between the control and the treated explants is the value for the change in bone resorption. The time course of action of a test substance can be investigated by taking small aliquots of the media throughout the *in vitro* period [3]. For some experiments, and for the *in vivo/in vitro* method described

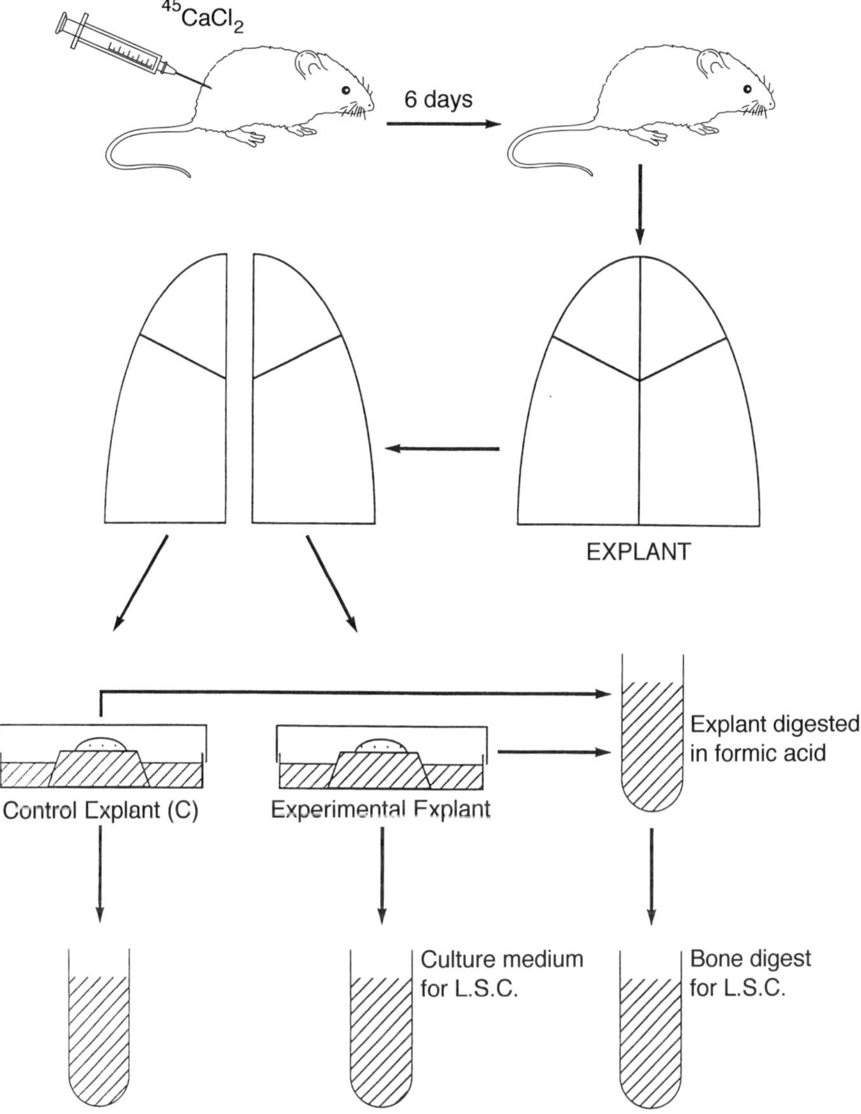

Figure 4.1 Calvarial bone resorption assay originally developed by Reynolds and Dingle [3]; liquid scintillation counting (LSC).

below, it is necessary to have an absolute measurement of the amount of resorption in the bone explants from each mouse that occurs during the *in vitro* period.

The $^{45}Ca^{2+}$ released from a living explant is mainly liberated by two mechanisms: an exchange of the isotope with non-radioactive calcium in the medium and net dissolution of bone mineral by cell-mediated resorption [15]. To arrive at a value for cell-mediated resorption (CMR), the amount of isotope released into the medium by a dead explant (thrice frozen and thawed) is subtracted from that released by its living half calvaria. For consistent results it is necessary to reduce the amount of isotope released by exchange to a minimum, otherwise CMR will be small compared with the total isotope released into the medium. It is for this reason that the bone explants should be pre-labelled 6 days before experimentation [16].

Using this methodology, Reynolds *et al.* [17] were able to demonstrate that 1,25 dihydroxyvitamin D_3 (1,25 $(OH)_2D_3$) was the most potent of the vitamin D metabolites in inducing resorption. Similarly interleukin 1 was found to be active in this assay at concentrations as low as 50 pM [18, 19]. Gowen *et al.* [20] have used a similar assay system to demonstrate that interferon gamma can prevent cytokine-induced resorption but does not inhibit PTH- or 1,25 dihydroxyvitamin D_3-stimulated resorption. This selectivity may be important in the local control of remodelling. Furthermore, although tumour necrosis factor α and TNF-β stimulate bone resorption [13] and prostaglandin synthesis by osteoblasts [21], it has been shown that prostaglandin synthesis is not a prerequisite for the stimulatory action of TNFs on bone resorption in explants of neonatal murine calvariae [22].

(b) Cold calcium assays

As mentioned earlier, the use of radiolabelled fetal bones is a relatively expensive and hazardous technique, and further complicated because, at least in the UK, it requires a Home Office animal licence. Zanelli *et al.* [23] developed a method that involved the measurement of released stable calcium colorimetrically. Using this method, Harris *et al.* [24] were the first to show that the osteolytic activity produced by cysts of the jaw was due to prostaglandins. This seemed an excellent method, particularly as the release of cold Ca^{2+} would give a more valid representation of bone resorption since it reflects changes in total mineral dissolution, whereas $^{45}Ca^{2+}$ release may only demonstrate changes in resorption of relatively immature mineral. This method was further adapted by Meghji *et al.* [25], who instead of using a paired system (i.e. one half calvariae used for control and one for test) pooled the halved calvaria and used five explants per group. The advantage of this randomized system is that it enables extensive experiments to be carried out while providing sufficient replicates for statistical significance testing. This method resulted in the first demonstration that 5-lipoxygenase products of arachidonic acid were potent bone-resorbing agents [25] and that molecular chaperones have a role in bone remodelling [26]. The response of calvaria to an osteolytic agent can be seen in Fig. 4.2.

(c) Modification of the original techniques

When using these bone organ cultures as a bioassay for assessing different stimulators of bone resorption, one major problem is the high rate of spontaneous mineral mobilization that can sometimes be observed in control cultures. To some extent this may reflect different rates of bone resorption in the animals at the time of dissection, or it may be caused by excessive prostaglandin release during the initial 24-hour culture period. The latter would seem to be important since prostaglandin levels do not correlate with calcium liberation from neonatal mouse calvaria during the first 24 hours of culture but are merely an artefact resulting from the surgery of preparing the calvaria [27]. Furthermore, Lerner [28] has shown that the responsiveness of the assay can be improved by pre-incubating the explants with the cyclo-oxygenase inhibitor indomethacin for 24 hours before testing resorbing agents. However, this is not acceptable, as Marshall *et al.* [29] using calvaria from 4-day-old mice showed this 24-hour pre-addition of indomethacin caused a large reduction in the number of tartrate-resistant acid phosphate (TRAP) positive

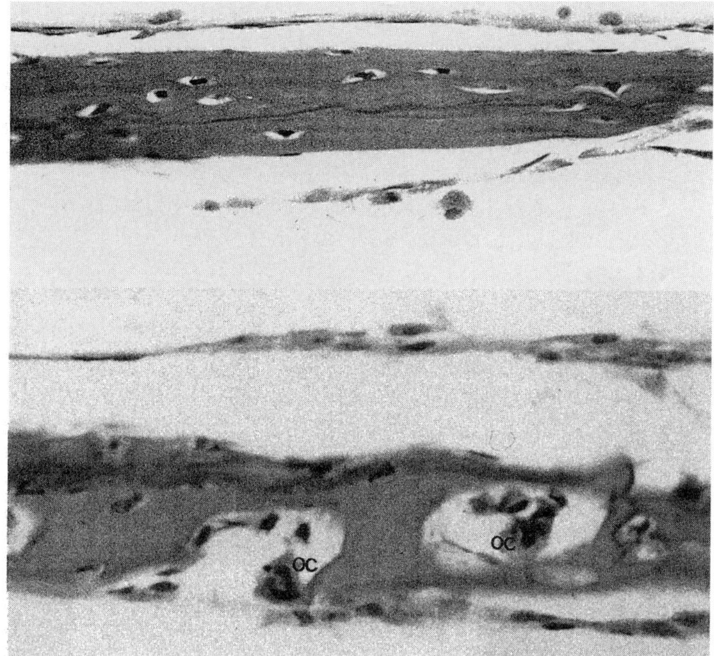

Figure 4.2 Bone resorption in organ culture. Fetal mouse calvaria were cultured (after the 24-hour pre-incubation) for 48 hours in BGJ medium supplemented with heat-inactivated fetal bovine serum with and without a bone-resorbing agent. Large multinucleated osteoclasts (OC) in lacunae of resorbed bone can be seen on the test bone.

osteoclasts relative to the control bones. This reduction did not occur if prostaglandin E_2 (PGE_2) was added with the indomethacin. This reduction of osteoclasts is possibly due to the inhibition of the differentiation of the osteoclast precursors. Thus it would appear that the best way to minimize the high background resorption is by careful dissection and minimum handling of the bones.

Our recent studies have shown that osteoclast-mediated calcium release from cultured mouse calvaria is extremely sensitive to extracellular pH. The effect closely parallels the action of extracellular protons on osteoclasts (Chapter 2), i.e. resorption is 'switched off' above pH 7.2 and is greatest at pH 6.8 or below. Peak proton-stimulated calcium release from cultured calvaria is equivalent to the maximal resorptive effects elicited by 1,25 dihydroxyvitamin D_3, parathyroid hormone or prostaglandin PGE_2 and is completely blocked by calcitonin [78]. This important variable is thus neglected at the investigator's peril.

Ljunggren *et al.* [30] introduced a modified bone culture system, based on the incubation of small, equally sized bone fragments from the neonatal mouse calvarial bones. When they divided the parietal bone into four fragments they noted that the rate of bone resorption in the anterior fragments was less than that in the posterior parts. They found that four parietal bone fragments only showed identical basal and PTH-stimulated release of $^{45}Ca^{2+}$ after removal of the anterior aspect of the parietal bone (Fig. 4.3). The advantage of this dissection technique is that it enables extensive experiments to be undertaken such as the assessment of the dose–response relationships of osteolytic agents. They also demonstrated that

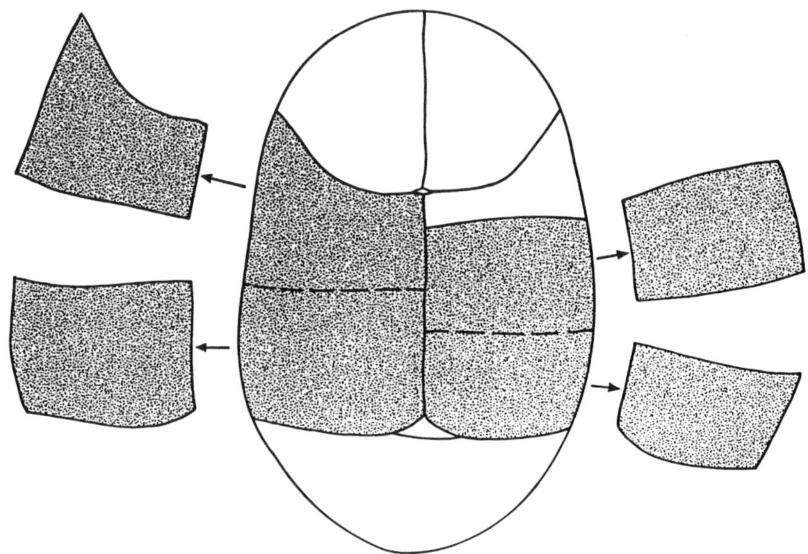

Figure 4.3 Dissection of a calvarial bone into four fragments; Ljunggren *et al.* [30].

the rate and magnitude of resorption were the same in bones cultured on grids or submerged, as was the release of the enzymes β-glucoronidase and lactate dehydrogenase. However, the rate of DNA synthesis was slower in bones that were cultured submerged. This indicates that the rate of cell division is higher in the cells close to the interface between the culture medium and gas phase, but the cell viability is probably the same in the bones cultured on grids or submerged. These observations are in agreement with previous reports in neonatal mouse calvariae showing that the rate of bone resorption stimulated by PTH, PGE_2, dibutyryl cyclic adenosine monophosphate (cAMP) and forskolin is unaffected by hydroxyurea at a concentration causing approximately 90% inhibition of [^{3}H]-thymidine uptake [31]. It also supports the findings of Lorenzo *et al.* [32] and Krieger *et al.* [33] who independently demonstrated that the stimulatory effect of PTH on bone resorption and osteoclast formation in bone organ culture is independent of cell replication.

In most experiments, the magnitude of bone resorption is quantified by analysing the mobilization of mineral from bones, either by measuring the release of $^{45}Ca^{2+}$ from pre-labelled bones or by determining the release of cold calcium and inorganic phosphate. However, in some experiments the rate of matrix degradation is of interest. Bone matrix degradation is generally estimated by measuring the release of $^3H^+$ from the bones pre-labelled with [^{3}H]-proline [34]. The release of $^3H^+$ parallels the release of ^{3}H-hydroxyproline and is thus a reliable indicator of collagen breakdown [35]. The greatest problem in using labelled hydroxyproline to assess matrix breakdown is that of incorporating enough label into the bone to measure its release accurately. When fetal animals are labelled via the mother, several millicuries of proline must be injected to attain sufficient labelling. Although injection of the new-born animal with a small amount of isotope may seem more attractive, much of this label may be in the cell rather than matrix protein [36]. Colorimetric measurement of stable hydroxyproline [37] may be a simpler and more reliable option.

Another method of measuring bone matrix breakdown is analysing the release of the carboxyterminal telopeptide of type I collagen (ICTP). Ljunggren and Ljunghall [38] adapted the radioimmunoassay method originally developed by Melkko *et al.* [39] to assess the extent of bone resorption based upon the release of ICTP from bone. More recently Foged *et al.* [40] have developed an ELISA for the α 1 chain of type I collagen and have demonstrated that a correlation exists between the extent of osteoclast pit formation and the release of antigenic collagen fragments. It would seem that the latter method may supersede the radioactive methods for measuring bone resorption in organ cultures. The measurement of collagen breakdown products from bone explants is also discussed in Chapter 9.

4.2.3 Mouse vertebral bone assay

Mouse vertebral bones have also been used in resorption assays [41] on the basis that they may be a better model for studying osteoporosis-induced bone changes. There is little evidence for large differences in sensitivity between various mammalian culture systems used despite differences in age, species, tissue site and developmental pattern of the bones [42]. As stated above, the only instance where major discrepancies have occurred concerns experiments in which indomethacin is used to investigate the possible role of prostaglandin synthesis in a resorptive response. Sometimes contrasting results are found depending on whether calvariae or limb bones are used. For example, epidermal growth factor and transforming growth factor α (TGFα) show osteolytic activity in the calvarial system [6, 43] but not in the fetal long bone assay [44, 45]. Whether these results show fundamental differences in the roles of prostaglandins in the two types of bone cultures is still debatable.

One of the most important experimental advantages of the use of organ culture of bone and cartilage is the ability to isolate the tissue from systemic effects, and to be able to test the actions of substances under controlled conditions. However, this can also be one of the disadvantages of the method, unless care is taken to integrate the organ culture results with *in vivo* findings.

For this reason an *in vivo/in vitro* method was developed by Reynolds [16] for studying bone resorption, in order to bridge the gap in interpretation. The essential feature of the *in vivo/in vitro* methods is to use the *in vitro* culture period as an assay to gain information on the prevailing bone resorption in the mouse at the time of explantation. A direct test with organ cultures can tell us if a test substance is able to influence bone resorption; the *in vivo/in vitro* method can give us data on whether the substance acts *in vivo*.

4.2.4 Bone resorption studies with the *in vivo/in vitro* method

For the *in vivo/in vitro* technique, 6-day-old mice pre-labelled with ^{45}CaCl$_2$ (0.18 mBq) are prepared as described above. At this time they are weight-paired, one mouse being injected with a control solution and the other paired mouse receiving one dose of test substance. A period of 18 hours to 4 days is allowed to elapse before the mice are killed to prepare explants [16]. Mice over about 12 days of age are unsuitable for culture. At death, pairs of explants of half calvariae are prepared from each mouse; one bone of each pair is killed by thrice freezing and thawing. Thus for each mouse the amount of cell-mediated resorption can be calculated from the difference in isotope release, living minus dead half; a convenient *in vitro* test period is 48 hours. Changes in the resorption can then be calculated as the difference in the values obtained between mice treated with test substance and weight-paired mice that were not treated.

4.3 ORGAN CULTURE TECHNIQUES FOR STUDYING OSTEOCLAST FORMATION

An important event in bone resorption involves the formation and recruitment of osteoclasts to future resorption sites. Osteoclasts are generated from progenitor cells of the mononuclear phagocyte system [46]. At an early stage of embryonic life, the mononucleated precursors for the osteoclast are disseminated via the bloodstream and deposited in the mesenchyme surrounding the bone rudiments, where they subsequently proliferate and differentiate into mature osteoclasts [47, 48].

Bone explant techniques have enhanced our understanding of osteoclast ontogeny. Both the calvarial system of Ko and Bernard [49] and the mouse metatarsal long bone model designed by Burger *et al.* [48, 50] (see below) have enabled probing of the cell lineage and cell interaction events occurring during osteoclast development.

The interest in using metatarsal bones of embryonic mice for investigating early events of the bone resorption processes was first emphasized by Burger *et al.* [50]. In long bone development the primitive marrow cavity is formed by resorption of the calcified cartilage matrix in the centre of the long bone model. At this age (16 days post-gestation), the metatarsal long bones have no mature osteoclasts, but only preosteoclasts that are differentiating into TRAP-positive cells. It was shown that enzymatic removal of the perichondrium–periosteum from the 17-day-old fetal metatarsal bones prevented invasion and resorption during subsequent organ culture. This is because the preosteoclasts are just under the perichondrium–periosteum and have been damaged or removed. This suggested that the osteoclast progenitors are transported via the bloodstream and seed in the periosteum of long bones at an early stage of embryonic development, long before the active formation of the marrow cavity [48–51]. This was confirmed when osteoclast-free ('stripped') metacarpals were cultured with a source of early haemopoietic progenitors such as fetal liver or spleen. These co-culture experiments resulted in the formation of osteoclasts which had the ability to excavate the primitive marrow cavity [47, 50, 51]. The differentiation kinetics of osteoclasts in the periosteum of embryonic bones has been established in a series of experiments by Scheven *et al.* [52]. Their findings are as follows: at 15 days post-gestation, mitotically active, blood-borne progenitors differentiate in the periosteum into post-mitotic osteoclast precursors; at 17 days post-gestation these acquire TRAP+ activity; at 18 days post-gestation, multinucleated osteoclasts are formed by fusion [52]. These cells invade the calcified cartilage zone, resulting in the formation of a primitive marrow cavity (Fig. 4.4).

This metatarsal experimental model may therefore provide a suitable model for the study of factors influencing the proliferation of osteoclast progenitors (i.e. long-term effects) as well as factors regulating fusion, mobility and activity of post-mitotic osteoclasts (i.e. short-term effects) [53]. Furthermore, metatarsals of 17-day mouse embryos are at an interesting developmental stage for investigating the molecular mechanisms involved in the migration of preosteoclasts to specific sites of the bone. The core of the 17-day embryo diaphysis consists of intact calcified

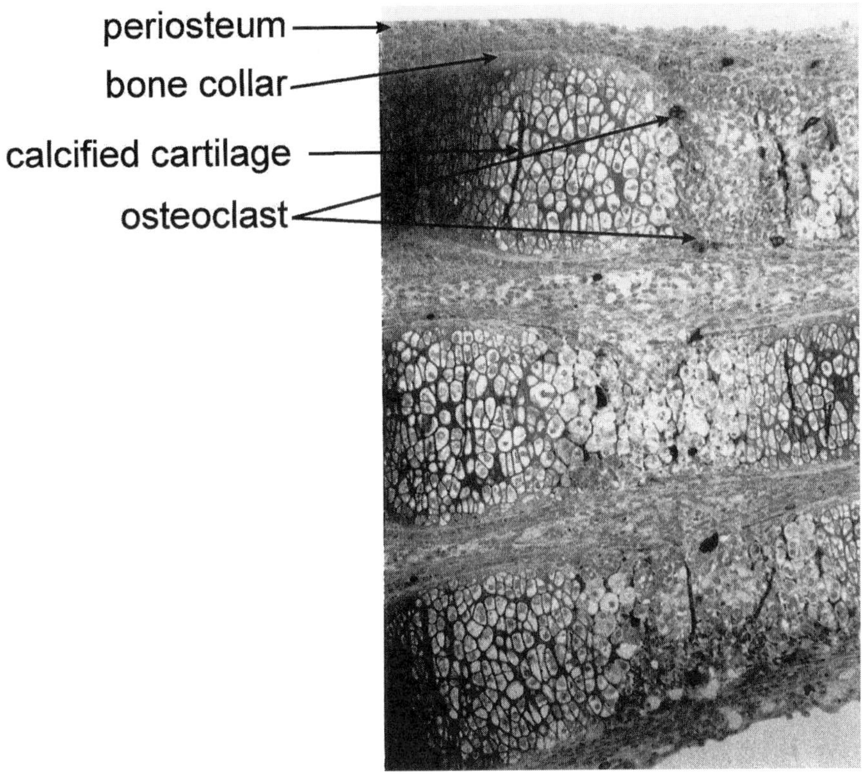

Figure 4.4   Embryonic metacarpal at 18 days post-gestation. Multinucleated osteoclasts invade the calcified cartilage zone, resulting in the formation of a primitive marrow cavity.

cartilage that must be replaced by a marrow cavity. They have only preosteoclasts located in the periosteum and separated from the calcified cartilage by an osteoid layer 15–25 μm thick and a very thin and discontinuous layer of mineral. Upon culture, these metatarsals develop as they do *in vivo*. The maturing preosteoclasts move across the osteoid and invade the calcified cartilage, resorbing it into a 'marrow' cavity, while the osteoid further develops into a bone collar. Both histomorphometric and radioisotopic methods can be used with the fetal metatarsal bone cultures to provide reliable information on the effects of factors influencing osteoclast formation and migration. This experimental system has recently shown that the matrix metalloproteinases are obligatory for the migration of preosteoclasts across the osteoid seam and for the resorption of the calcified cartilage, whereas the plasminogen activator/plasmin cascade appears to play no part in the process [54, 55].

4.4 DIFFERENCES IN RESPONSE TO BONE-RESORBING AGENTS IN THE VARIOUS ORGAN CULTURE ASSAYS

Although these various organ culture techniques have provided invaluable information about factors modulating bone resorption, there has not been uniformity in the findings and considerable discrepancies have been reported for the effects of the same factors on resorption. These differences may be attributable to the stage of development of the particular bone. In contrast to 17-day-old fetal metatarsal long bone rudiments, calvarial bones (as well as radii/ulnar long bone rudiments from the same age fetuses) have fully developed multinucleated osteoclasts within the mineralized matrix and bone resorption depends primarily on the activity of these cells. This difference in the stage of osteoclast development between these two distinct long bone rudiments accounts for the diversity in response to a variety of resorbing agents [56, 57].

Similarly, differences have been shown between the calvarial and long bone system. Resorption in the calvarial system appears to be dependent on the cyclo-oxygenase pathway [6, 45, 58, 59]. However, the only factors known to induce opposite effects in the two systems are TGFβ1 and TGFβ2 [6, 60].

The manner in which the explants are prepared and cultured also appears to be important. Several effects of different experimental conditions on osteoclast-mediated resorption have been previously described. For example, composition of the culture medium, atmospheric oxygen tension and size of the bones were reported to determine tissue vitality and degree of resorption of long bones in culture [61]. For these reasons, several experimental approaches for the culture of embryonic bones have been developed, including stationary cultures performed on the surface of liquid or semi-solid medium, cultures in a relatively large volume of medium that was gassed continuously and cultures in roller tubes with overlying liquid medium. The standardization of experimental culture conditions in which fetal bone explants are maintained is paramount if $^{45}Ca^{2+}$ release is to be used as a reliable marker of osteoclastic bone resorption.

To create acceptable culture conditions for bone tissue, several experimental approaches have been developed in the past. Goldhaber [4] previously reported that it is possible to regulate the amount of bone resorption by varying the oxygen concentration in the atmosphere. Satisfactory results were obtained with the use of stationary cultures performed on the surface of the liquid or with a semi-solid medium [2, 62], with cultures in a relatively large volume of medium that was gassed continuously [61] or with cultures carried out in roller tubes with overlying liquid medium [4]. Furthermore, based on histological findings in long-term cultures, Fell and Mellanby [1] reported that radii/ulnae of 17-day-old mouse embryos were more suitable for culture than humeri/femora of the same fetuses, since the latter were found to be too bulky to permit good diffusion and tissue survival. The importance of oxygen diffusion into bone explants in culture is also the reason that 19-day-old fetal rat radii/ulnae, which are usually cultured after removal of the cartilage ends, thus providing a smaller bone explant, show increased responsiveness to hormones [2, 61]. A study using vitamin D metabolites by Puzas and Brand [63] showed that increasing the volume of culture medium to which a bone explant is exposed (whilst maintaining a constant metabolite concentration) has the effect of increasing the amount of metabolite in the cell compartment. Moreover, daily changes of the ligand containing fresh metabolite increased the amount of ligand in the cell compartment as well.

Van Beek *et al.* [64] found that when intact fetal radii/ulnae were cultured under the surface of the medium, rather than at the interphase of medium and air, there was a significantly greater amount of $^{45}Ca^{2+}$ released, which cannot be attributable to either osteoclast-mediated resorption or physicochemical exchange. Also, histologically TRAP+ cells were almost absent after culture and the bone marrow cells in the centre of the bone appeared necrotic, possible due to lack of oxygen [64].

4.5 ORGAN CULTURE TECHNIQUES TO STUDY BONE FORMATION

Similar culture systems can be used to study bone formation. The majority of studies have used organic matrix synthesis, rather than *de novo* mineralization, as an index of bone formation. To investigate organic matrix synthesis, the cultured explants are not pre-labelled but are allowed to incorporate isotopes that label the matrix from the medium. Most work has focused on the incorporation of [^{3}H] proline into collagenase-digestible or limited pepsin-digestible protein [65] and it has been found that agents that induce resorption – for example, PTH, tumour necrosis factor, vitamin D, bacterial components, prostaglandins and leukotrienes [65–69] – can also inhibit collagen synthesis.

The simultaneous measurement of bone resorption and collagen synthesis in neonatal mouse calvariae is possible [70]. Bone resorption is estimated by measuring the release of previously incorporated ^{45}Ca^{2+} as before, and bone formation is determined by the incorporation of [^{3}H]-proline into collagenase-digestible and non-collagenous protein by pulsing the explants with radioisotope for the last 2 hours of culture [64]. It is also possible to assess DNA synthesis by pulsing the explants with [^{3}H]-thymidine [71].

Although these radioactive methods provide a reliable estimate of organic matrix synthesis, no information is obtained about the mineralization process and the total bone formation. Saito *et al.* [72] have introduced an organ culture system using chick embryonic femurs, whereby both bone resorption and mineralization can be measured by labelling the explants with ^{45}Ca^{2+} *in vitro*.

4.6 MORPHOLOGICAL ANALYSIS OF BONE

As an alternative to radioisotopes, Schwartz *et al.* [73] used morphological methods to demonstrate bone growth and resorption in cultured fetal murine radii and ulnae. They cultured bones for 6 days in BGJ medium supplemented with 20% fetal calf serum and 150 μg vitamin C/ml and estimated bone growth and mineralization by measuring the total length of the bone and the diaphysis by light and transmission electron microscopy. They demonstrated that there is a continuous measurable increase in the total length of the fetal long bones over the 6 days of culture. Transmission electron microscopy studies showed differentiation of mesenchymal cells to osteoblasts, formation of new bone matrix and bone mineralization similar to that found in developmentally matched controls. Osteoclasts were also seen in the diaphysis at 2 and 4 days of culture [73]. This model appears to be a reliable technique for the measurement of bone formation and resorption.

Gronowicz *et al.* [74] developed a fetal rat parietal bone organ culture system that mineralizes *in vitro*, using a defined serum-free medium. When the 20-day-old fetal rat parietal bones were cultured for 5 days with a physiological concentration of phosphate (3 mM), there was a doubling of the calcium content of the bones which was comparable to that occurring *in vivo* after 2 days. There was also a marked increase in the mineralized bone area which mirrored the increase in the calcium content of the bones. Fluorescent calcein labelling showed that calcification occurred in an ordered pattern similar to that which occurred *in vivo* over 2 days. However, when the parietal bones were cultured with either β-glycerol phosphate or a high non-physiological concentration of inorganic phosphate, an ectopic pattern of calcification occurred which differed from mineralization *in vivo* or in bones cultured with a physiological concentration of phosphate. Kaji *et al.* [75] developed culture conditions permitting osteogenesis to proceed in 9-day-old embryonic chick femurs. In contrast to the rat parietal bone explants, when the femurs were cultured with BGJb-HW2 medium supplemented with 5 mM β-glycerophosphate, histological examination showed the formation of a thick and active periosteum, numerous osteoblastic cells, a sufficient amount of osteoid tissue and well-developed calcified trabeculae without any pathological changes. Altering the calcium and phosphate concentrations in the BGJb-HW2 medium or the 10% addition of chick embryo extract (CEE) did not induce ossification. Furthermore, combinations of the 10% CEE with a high Ca × Pi product or with 5 mM β-glycerophosphate often caused pathological abnormalities in the periosteum.

It would seem that reliable estimates of bone formation in organ culture can be obtained by combining data from both morphological analysis and radioactive estimations of organic and inorganic matrix synthesis.

4.7 PROTOCOLS

For a number of the protocols described there is a requirement in the UK for Home Office approval. Readers should check with their animal-care authorities in their own countries of origin before using these methods.

4.7.1 Neonatal mouse calvarial assay for assessing bone resorption

(a) Radioactive method

1. Neonatal mice are injected subcutaneously with 0.04 mBq ^{45}CaCl$_2$ (Amersham International) 1 or 2 days after birth.
2. Six days later the mice are killed and the calvarial bones are removed by careful microdissection so as not to damage either the periosteum or endosteum. The dissection is carried out in culture medium with a lowered level of bicarbonate (50 mg/100 ml) so as to equilibrate with air.
3. The calvarial explants (consisting of frontal and parietal bones) are divided into equal halves along the median sagittal suture.
4. The separated half-calvariae are explanted onto the surfaces of stainless steel rafts in 30 mm Petri dishes containing 3 ml of serum-free BGJ medium (Fitton–Jackson modification) supplemented with 1 mg bovine serum albumin/ml.
5. After a 24-hour preincubation period, the bones are washed three times in ice-cold Tyrode's solution and further washed for three hours in basic medium.
6. The bone explants are then transferred to separate 30 mm culture dishes containing stainless steel rafts. The explant rests on the raft so that it is at the interface of the gas phase (95% air / 5% CO$_2$) and the liquid medium. A paired system is usually used in which one explant is cultured in control medium and the other of the pair is explanted into medium containing the test substance.
7. Bones are usually cultured for 48 or 120 hours (with a change of medium after 48 hours). At the end of the culture, the bones are demineralized in either 200 μl of formic acid or hydrochloric acid (6 mol/l) at 60°C for 1 hour, and an

aliquot is analysed for $^{45}Ca^{2+}$ using standard techniques for liquid scintillation counting. To arrive at a value for cell-mediated resorption the amount of isotope released into the medium by a dead explant (thrice frozen, at $-80°C$, and thawed) is subtracted from that released by its living half calvaria.

9. Mobilization of radioactive calcium is expressed as the percentage release to the medium of the initial radioactivity in the bones (calculated as the radioactivity in the medium and bones after culture).

10. In short-term kinetic studies, the magnitude of mineral mobilization is determined by withdrawal of small amounts of culture medium (50–300 µl) at stated time intervals, i.e. 6, 12, 24 and 36 hours, and $^{45}Ca^{2+}$ release is calculated as a percentage of initial radioactivity. These experiments are performed in a paired manner, i.e. one half of the individual calvarium is used as a control bone and the other half as an experimental bone. The animals used for these experiments are pre-labelled with 0.4 mBq $^{45}CaCl_2$.

11. Degradation of bone matrix proteins can be assessed by quantifying the release of $^3H^+$ from bones that have been pre-labelled *in vivo* for 96 hours by an injection of 0.37 mBq [^3H]-proline. Calvarial halves are incubated in BJG medium as before, and after culture the amount of ^3H in the culture media and calvarial bones is analysed and the percentage release of ^3H is calculated as for $^{45}Ca^{2+}$ release.

12. Measurement of lysosomal enzyme release that parallels bone resorption. The lysosomal enzyme, β-glucuronidase, which is released into the medium can be determined fluorimetrically using 4-methylumbelliferyl-β-D-glucuronide as the substrate. *N*-Acetyl-β-glucosaminidase can be analysed using *p*-nitrophenyl-*N*-acetyl-β-D-glucosaminide as the substrate.

13. Quantification of the morphological changes in bone sections of calvarial bones from *in vitro* organ cultures can be examined histologically. The bone matrix area between the lambdoid and coronal sutures is measured and the number of osteoclasts/mm^2 of bone is counted using an image analysis system. The extent of resorption along the inner and outer bone surfaces is also measured. The histological technique has been described previously.

(b) Non-radioactive method

Bone resorption can be measured by the release of calcium (measured colorimetrically [22]) from explants of neonatal mouse calvaria in culture.

1. Five-day-old mice are killed by cervical dislocation. The skin is removed to expose the cranial bones. The dissection is then continued using the lambdoidal suture as a guide line, taking care not to damage the periosteum. The frontoparietal bones are trimmed of any adhering connective tissue and interparietal bone.

2. Dissected calvaria are pooled in Hanks balanced salt solution (HBSS) and washed free of blood and adherent brain tissue. The calvaria can then be divided along the sagittal suture and placed in a dish with fresh HBSS.

3. Each half calvarium is cultured separately on a stainless steel grid (1 cm^2 Minimesh FDP quality, Expanded Metal Co., West Hartlepool, UK) in a 30 mm plastic Petri dish with 1.5 of Biggers, Gwatkin and Heyner medium (BGJ) (Flow Laboratories, Irvine, Scotland) 5% heat inactivated fetal calf serum. The bones are incubated for 24 hours at $37°C$ in a humidified atmosphere of 5% CO_2, 95%

air. The 24-hour pre-incubation period prior to addition of experimental factors is included for two reasons. Firstly, it enables the bones to adapt to their new environment and to permit calcium exchange to reach an equilibrium. Secondly, it has been reported that the release of prostaglandins (PG) from freshly explanted bone cultures is relatively high [76]. These authors recommend that the assay for bone-resorbing substances, which may themselves act by modifying PG synthesis in the bone, should be postponed until PG levels have normalized.

4. At the end of the 24-hour pre-incubation, the medium is removed and replaced with fresh medium containing test substances. The resorption in these cultures is compared with that in control cultures containing BGJ alone. The cultures are incubated for a further 48 hours and resorption is measured as the release of calcium into the culture medium over this period. Each experimental group contains five separate cultures.

5. At the end of the incubation period, the culture medium is removed from each dish with a Pasteur pipette into autoanalyser cups (conical-bottomed, 2 ml, Chem Lab Instruments, Essex, UK). Calcium concentrations are measured colorimetrically on an autoanalyser (Chem Lab Instruments, Essex), by using the metal complexing dye cresolphthalein complexone (CPC) [77]. Calcium is separated from proteins by continuous flow dialysis under acidic conditions. The concentration of dialysed calcium is then determined colorimetrically by complexing with CPC. One aliquot of 100 µl is removed from each sample through the stainless steel sampling probe followed by a 20-second wash with distilled water after each sample. The samples are mixed with 1 M HCl containing 8-hydroxy-quinoline (8HQ) at 2.5 g/l (which eliminated interference by magnesium), and dialysed against a solution of similar composition containing CPC at 0.7 g/l. The dialysate is then mixed with 2-amino-2-methyl-propano-1-ol (AMP) (90 g/l). The absorbance of the resultant purple-coloured solution is measured in a 15 mm flow cell at 570 nm, and plotted on a chart recorder at 0.5 cm/minute. Calcium concentrations are calculated from the absorbance peak heights measured against the standard curve. It is possible to capture the data on a computer, which can directly calculate calcium concentrations from the standard dose response.

4.7.2 Fetal long bone assay for assessing bone resorption

The fetal long bone assay (usually tibiae and ulnae from either rats or mice) for assessing bone resorption was originally introduced by Raisz [2]. As with calvariae, these bones display a large surface area to volume ratio and pairs of exactly matching bones can be obtained by microdissection.

1. Pregnant rats at the 18th day of gestation or pregnant mice at the 16th day of gestation (the day of conception and vaginal plug discovery are taken as day 0) are injected with 3.7 mBq of $^{45}CaCl_2$. One day later the animals are sacrificed and the fetuses are aseptically removed.

2. The mineralized shafts of the fetal long bones (tibiae and ulnae) are dissected free of surrounding tissue and cartilage. The explants are cultured on stainless steel rafts as for the calvarial explants.

3. As for steps 4 to 9 in the neonatal calvarial assay (section 4.7.1).

4.7.3 Cultures of 17-day-old fetal mouse metatarsals/metacarpals to assess osteoclast formation and migration

1. Timed-pregnant mice are injected subcutaneously at day 16 of gestation with 3.7 mBq of $^{45}CaCl_2$. The mothers are sacrificed one day later (day 17) and the embryos are removed aseptically.
2. The middle three metatarsals of the hindlimb are dissected free without removal of the cartilaginous epiphysis and care is taken not to damage the periosteum–perichondrium. Care should be taken to standardize the age of the rudiments by selecting litters in which the first phalanges of the feet have not yet started to calcify.
3. The three metatarsals are usually cultured together as a triad, one limb providing the controls and the other the tests, to allow paired comparisons. The cultures are done in liquid medium (see below).
4. The three metatarsals are placed on a piece of lens paper, itself deposited upon a stainless steel grid, which is in turn suspended above the centre well of the 35 mm tissue culture dish. The culture medium (1.5 ml) is placed in the well so that the explants are just at the surface level of the medium. Sterile water (2 ml) is placed in the outer well of the dish to ensure optimum humidification during the culture. The cultures are done in a BGJ medium containing $NaHCO_3$ (2.2 g/l), 10% (v/v) heat inactivated fetal calf serum, glutamine (200 μg/ml) and ascorbate (50 μg/ml).

(a) Monitoring resorption
1. Demineralization of the explants during culture is monitored by determining the amount of ^{45}Ca released into the medium, harvested at successive days by liquid scintillation counting.
2. At the end of the culture, the explants are dissolved in formic acid (three metatarsals to 200 μm) to measure the residual ^{45}Ca left in the bones, to be able to evaluate the total amount of isotope present in the metatarsals at the time of their explantation.

Note that the ^{45}Ca released during the first day of culture does not result from osteoclastic resorption but from a simple physicochemical exchange at the surface of the mineral. This value is subtracted from the total amount of ^{45}Ca initially present in the explants to obtain a value representative of cell-mediated resorption. The release of ^{45}Ca resulting from this resorption is thus taken into account from days 2 to 7 of the cultures. Autoradiographs of paraffin sections show that the ^{45}Ca injections result in labelling mainly the calcified core of the metatarsals and the innerside of the bone collar of the radii. Thus the release of ^{45}Ca from metatarsals is mainly due to resorption of calcified core, and that from the radii is mainly due to resorption of bone collar at its endosteal surface.

(b) Evaluation of osteoclast migration/invasion in the explanted metatarsals: preparation of tissue sections
Paraffin sections for histomorphometry are prepared according to the method of Scheven *et al.* [51].

1. Metatarsal bones are washed in cold Hanks balanced salt solution and fixed in 4% neutral buffered formalin for 3 hours at 4°C.

2. The bones are decalcified in 5% formic acid and 5% formalin for 3 hours at 4°C.
3. The bones are washed in distilled water and stained for TRAP by the *en bloc* procedure using naphthol AS-BI phosphate as substrate and hexazonium pararosanilin as coupler in Michaelis veronal acetate buffer for 1 hour at 37°C. As the inhibitor, L(+)-tartaric acid is used and is dissolved in 0.1 M acetate buffer (pH 5.0) and added to the substrate solution to a final concentration of 10 mM.
4. The bones are washed in distilled water and dehydrated in a graded series of ethanol and embedded in paraffin.
5. Serial sections of 5 μm are cut on a Reichert-Jung microtome and counterstained with Ehrlich's haematoxylin.
6. Plastic sections for a detailed characterization of the migration pathway of the osteoclasts can also be prepared. Metatarsals are fixed in 4% neutral buffered formalin for 18 hours at 4°C, and embedded in glycolmethacrylate according to the instructions of the Historesin Embedding Kit (Reichert-Jung, Heidelberg).
7. Sections of 3 μm are cut with glass knives on a Reichert-Jung microtome. Depending on the experiments, the sections are stained for TRAP, with methylene blue, with Masson-Goldner's or with von Kossa's stain.

REFERENCES

1. Fell, H.B. and Mellanby, E. (1952) The effect of hypervitaminosis A on embryonic limb-bones cultivated in vitro. *Journal of Physiology* **116**, 320–349.
2. Raisz, L. (1965) Bone resorption in tissue culture. Factors influencing the response to parathyroid hormone. *Journal of Clinical Investigation* **44**, 103–116.
3. Reynolds, J.J. and Dingle, J.T. (1970) A sensitive in vitro method for studying the induction and inhibition of bone resorption. *Calcified Tissue International* **4**, 339–349.
4. Goldhaber, P. (1961) Oxygen dependent bone resorption in tissue culture, in *The Parathyroids*, (eds R.O. Greep and R.V. Talamage), Charles C. Thomas, Springfield, IL, pp. 243–255.
5. Raisz, L. and Niemann, I. (1967) Early effects of parathyroid hormone and thyrocalcitonin in bone organ culture. *Nature* **214**, 486–488.
6. Tashjian, A.H., Voekel E.F., Lazzaro, M. *et al.* (1985) Alpha and beta transforming growth factors stimulate prostaglandin production and bone resorption in cultured murine calvaria. *Proceedings of the National Academy of Sciences USA* **82**, 4535–4538.
7. McSheehy, P.M.J. and Chambers, T.J. (1986) Osteoblastic cells mediate osteoclastic responsiveness to parathyroid hormone. *Endocrinology* **118**, 824–828.
8. Meghji, S., Sandy, J.R., Scutt, A. *et al.* (1988) Heterogeneity of bone resorbing factors produced by murine osteoblasts in vitro: modulation by parathyroid hormone and mononuclear cell products. *Archives of Oral Biology* **33**, 773–778.
9. Hill, P.A., Reynolds, J.J. and Meikle, M.C. (1995) Osteoblasts mediate insulin-like growth factor-I and -II stimulation of osteoclast formation and function. *Endocrinology* **136**, 124–131.
10. Akatsu, T., Takahashi, N., Udagawa, N. *et al.* (1991) Role of prostaglandins in interleukin-1 induced bone resorption in mice in vitro. *Journal of Bone and Mineral Research* **6** (2) 183–190.
11. Corboz, V.A., Cecchini, M.G., Felix, R. *et al.* (1992) Effect of macrophage colony stimulating factor on in vitro osteoclast generation and bone resorption. *Endocrinology* **130** (1), 437–442.

12. Girasole, G., Passeri, G., Jilka, R.L. and Manolagas, S.C. (1994) Interleukin-11: a new cytokine critical for osteoclast development. *Journal of Clinical Investigation* **93**: 1516–1524.

13. Bertolini, D.R., Nedwin, G.E., Bringman, T.S. *et al.* (1986) Stimulation of bone resorption and inhibition of bone formation in vitro by human tumour necrosis factor. *Nature* **319**, 516–518.

14. Biggers, J.D., Gwatkin, R.B.L. and Heyner, S. (1961) Growth of embryonic avian and mammalian tibiae on a relatively simple chemically defined medium. *Experimental Cell Research* **25**, 41–58.

15. Reynolds, J.J. (1972) Skeletal tissue in culture, in *The Biochemistry and Physiology of Bone*, 2nd edn, Vol. 1, (ed. G.H. Bourne), Academic Press, New York and London, pp. 69–126.

16. Reynolds, J.J. (1976) Organ cultures of bone: studies on the physiology and pathology of resorption, in *Organ Culture in Biomedical Research* (eds M. Balls and M. Monnickendam), Cambridge University Press, Cambridge, pp. 355–366.

17. Reynolds, J.J., Holick, M.F. and DeLuca, H.F. (1973) The role of vitamin D metabolites in bone resorption. *Calcified Tissue International* **12**, 295–301.

18. Heath, J.K., Saklatvala, J., Meikle, M.C. *et al.* (1985) Pig interleukin-1 (catabolin) is a potent stimulator of bone resorption in vitro. *Calcified Tissue International* **37**, 95–97.

19. Gowen, M. and Mundy, G.R. (1986) Actions of recombinant interleukin-1, interleukin-2, and interferon-gamma on bone resorption in vitro. *Journal of Immunology* **136**, 2478–2482.

20. Gowen, M., Nedwin, G.E. and Mundy, G.R. (1986) Preferential inhibition of cytokine-stimulated bone resorption by recombinant interferon gamma. *Journal of Bone and Mineral Research* **1**, 469–474.

21. Tatakis, D.N. and Dziak, R. (1989) Recombinant human lymphotoxin effects on osteoblastic cells. *Biochemical and Biophysical Research Communications* **162**, 435–440.

22. Lerner, U.H. and Ohlin, A. (1993) Tumour necrosis factors alpha and beta stimulate bone resorption in cultured mouse calvariae by a prostaglandin-independent mechanism. *Journal of Bone and Mineral Research* **8** (2), 147–155.

23. Zanelli, J.M., Lea, D.J. and Nisbet, J.A. (1969) A bioassay method in vitro for parathyroid hormone. *Journal of Endocrinology* **43**, 33–46.

24. Harris, M., Jenkins, M.V., Bennett, A. and Wills, M.R. (1973) Prostaglandin production and bone resorption by dental cysts. *Nature* **245**, 213–215.

25. Meghji, S., Sandy, J.R., Scutt, A. *et al.* (1988) Stimulation of bone resorption by lipoxygenase metabolites of arachidonic acid. *Prostaglandins* **36**, 139–149.

26. Kirby, A.C., Meghji, S., Nair, S.P. *et al.* (1995) The potent bone resorbing mediator of *Actinobacillus actinomycetemcomitans* is homologous to the molecular chaperone GroEL. *Journal of Clinical Investigation* **96**, 1185–1194.

27. Leis, H.J., Malle, E., Hoffmann, O. *et al.* (1990) Experimental considerations on the measurement of prostaglandins during long-term incubations of neonatal mouse calvaria. *Biochemical and Biophysical Research Communications* **169** (2), 545–550.

28. Lerner, U.H. (1987) Modifications of the mouse calvarial technique improve the responsiveness to stimulators of bone resorption. *Journal of Bone and Mineral Research* **2**, 375, 383.

29. Marshall, M.J., Holt, I. and Davie, M.W.J. (1995) The number of tartrate-resistant acid phosphatase-positive osteoclasts on neonatal mouse parietal bones is decreased when prostaglandin synthesis is inhibited and increased in response to prostaglandin E_2, parathyroid hormone and 1.25 dihydroxyvitamin D_3. *Calcified Tissue International* **56**, 240–245.

30. Ljunggren, O., Ransjo, M. and Lerner, U.H. (1991) In vitro studies on bone resorption in neonatal mouse calvariae using a modified dissection technique giving four samples of bone from each calvaria. *Journal of Bone and Mineral Research* **6**, 543–550.

31. Lerner, U.H. and Hanstrom, L. (1989) Stimulation and inhibition of mitotic activity in cultured mouse calvarial bones by cyclic AMP analogues and phosphodiesterase inhibitors is unrelated to the delayed resorptive effects of cyclic AMP. *Experimental Clinical Endocrinology* **8**, 89–95.

32. Lorenzo, J.A., Raisz, L.G. and Hock, J.M. (1983) DNA synthesis is not necessary for osteoclastic responses to parathyroid hormone in cultured foetal rat long bones. *Journal of Clinical Investigation* **72**, 1924–1929.

33. Krieger, N.S., Feldman, R.S. and Tahjian, A.H. Jr (1982) Parathyroid hormone and calcitonin interactions in bone: irradiation-induced inhibition of escape in vitro. *Calcified Tissue International* **34**, 197–203.

34. Hill, P.A., Docherty, A.J.P., Bottomley, K.M.K. *et al.* (1995) Inhibition of bone resorption in vitro by selective inhibitors of gelatinase and collagenase. *Biochemical Journal* **308**, 167–175.

35. Lerner, U.H., Hanstrom, L. and Sjostrom, S. (1990) Stimulation of bone resorption and cell proliferation in vitro by human gingival fibroblasts from patients with periodontal disease. *Bone and Mineral* **10**, 225–242.

36. Brand, J.S. and Raisz, L.G. (1972) Effects of thyrocalcitonin and phosphate ion on the parathyroid hormone stimulated resorption of bone. *Endocrinology* **90**, 479–487.

37. Bingham, P.J. and Raisz, L.G. (1974) Bone growth in organ culture: effects of phosphate and other nutrients on bone and cartilage. *Calcified Tissue Research* **14**, 31–48.

38. Ljunggren, O. and Ljunghall, S. (1992) Carboxyterminal telopeptide of type I collagen, ICTP, as a marker of matrix degradation in neonatal mouse calvarial bones, in vitro. *Bioscience Reports* **12** (5), 407–411.

39. Melkko, J., Niemi, S., Risteli, L. *et al.* (1990) Radioimmunoassay of the carboxyterminal propeptide of human type I procollagen. *Clinical Chemistry* **36** (7), 1328–1332.

40. Foged, N.T., Lou, H. and Delaisse, J.M. (1995) Characterization of the collagen degradation in continous cultures of osteoclasts. *Journal of Bone and Mineral Research* **10** (S1), S294.

41. Stewart, P.J. and Stern, P.H. (1987) Vertebral bone resorption in vitro: effects of parathyroid hormone, calcitonin, 1,25 dihydroxyvitamin D_3, epidermal growth factor, prostaglandin E_2 and oestrogen. *Calcified Tissue International* **40**, 21–26.

42. Stern, P.H. and Kreiger, N.S. (1983) Comparison of foetal rat limb bone and neonatal mouse calvariae: effects of parathyroid hormone and 1,25 dihydroxyvitamin D_3. *Calcified Tissue International* **35**, 172–176.

43. Tashjian, A.H. and Levine, L. (1978) Epidermal growth factor stimulates prostaglandin production in cultured murine calvaria. *Biochemical and Biophysical Research Communications* **85**, 966–975.

44. Stern, P.H., Kreiger, N.S., Nissenson, R.A. *et al.* (1985) Human transforming growth factor-alpha stimulates bone resorption in vitro. *Journal of Clinical Investigation* **76**, 2016–2019.

45. Raisz, L.G., Simmons, H.A., Sandberg, A.L. *et al.* (1980) Direct stimulation of bone resorption by epidermal growth factor. *Endocrinology* **107**, 207–203.

46. Suda, T., Takahashi, N. and Martin, T.J. (1992) Modulation of osteoclast differentiation. *Endocrine Review* **13**, 66–80.

47. Burger, E., Van der Meer, J. and Nijeide, P. (1984) Osteoclast formation from mononuclear phagocytes: role of bone forming cells. *Journal of Cell Biology* **99**, 1901–1906.

48. Burger, E.H., Thesingh, C.W., Van der Meer, J.W.M. and Nijweide, P.J. (1984) Development of osteoclasts in the mouse-localization of osteoclast precursors and role of bone forming cells, in *Endocrine Control of Bone and Calcium Metabolism* (eds D.V. Cohn, T. Fujita, J.T. Potts and R.V. Talmage), Elsevier, Amsterdam, pp. 125–130.

49. Ko, J. and Bernard, G. (1981) Osteoclast formation in vitro from bone marrow mononuclear cells in osteoclast free-bone. *American Journal of Anatomy* **106**, 415–425.

50. Burger, E.H., van der Meer, J.W.H.., van de Gevel, J.S. *et al.* (1982) In vitro formation of osteoclasts from long-term cultures of bone marrow phagocytes. *Journal of Experimental Research* **156**, 1604–1614.

51. Thesingh, C.W. and Burger, E.H. (1983) The role of mesenchyme in embryonic long bones as early deposition site for osteoclast progenitor cells. *Developmental Biology* **95**, 429–438.

52. Scheven, B.B., Kawilarang-de Haas, E.W., Wassenaar, A.M. *et al.* (1986) Differentiation kinetics of osteoclast in the periosteum of embryonic bones in vivo and in vitro. *Anatomical Record* **214**, 418–423.

53. Van der Pluijm, G., Most, W., Van der Wee-Pals, L. *et al.* (1991) Two distinct effects of recombinant human tumour necrosis factor-alpha on osteoclast development and subsequent resorption of mineralised matrix. *Endocrinology* **129**, 1596–1604.

54. Leloup, G., Delaisse, J.M. and Vaes, G. (1994) Relationship of the plasminogen activator/plasmin cascade to osteoclast invasion and mineral resorption in explanted foetal metatarsal bones. *Journal of Bone and Mineral Research* **9**, 891–902.

55. Blavier, L. and Delaisse, J.M. (1995) Matrix metalloproteinases are obligatory for the migration of preosteoclasts to the developing marrow cavity of primitive long bones. *Journal of Cell Science* **108** 3649–3659.

56. Van der Pluijm, G., Mouthaan, H., Baas, C. *et al.* (1994) Integrins and osteoclastic resorption in three bone organ cultures: differential senstivity to synthetic Arg-Gly-Asp peptides during osteoclast formation. *Journal of Bone and Mineral Research* **9**, 1021–1028.

57. Most, W., Schot, L., Ederveen, A. *et al.* (1995) In vitro and ex vivo evidence that estrogens suppress increased bone resorption induced by ovariectomy or PTH stimulation through an effect on osteoclastogenesis. *Journal of Bone and Mineral Research* **10**, 1523–1530.

58. Tashjian, A.H. and Levine, L. (1978) Epidermal growth factor stimulates prostaglandin production and bone resorption in cultured mouse calvaria. *Biochemical and Biophysical Research Communications* **85**, 966–975.

59. Ibbotson, K.J., Twardzik, D.R., D'Souza, S.M. *et al.* (1985) Stimulation of bone resorption in vitro by synthetic transforming growth factor alpha. *Science* **228** (4702), 1007–1009.

60. Pfeilschifter, J., Seyedin, S.M. and Mundy, G.R. (1988) Transforming growth factor beta inhibits bone resorption in foetal rat long bone cultures. *Journal of Clinical Investigation* **82**, 680–685.

61. Stern, P. and Raisz, L.G. (1979) Organ culture of bone, in *Skeletal Research: an Experimental Approach*, (eds D.J. Simmons and A.S. Kunin), Academic Press, New York, pp. 21–59.

62. Boonekamp, P.M., van der Wee Pals, L.J.A., van Wijk-van Lennep, M.M.L. *et al.* (1986) Two modes of action of bisphosphonates on osteoclastic resorption of mineralised matrix. *Bone and Mineral* **1**, 27–39.

63. Puzas, L.E. and Brand, J.S. (1985) In vitro use of vitamin D3 metabolites: culture conditions determine cell uptake. *Calcified Tissue International* **37** (5), 474–477.

64. Van Beek, E., Ruit, M.O., van der Wee Pals, L.J.A. *et al.* (1995) Effects of experimental conditions on the release of ^{45}Calcium from prelabeled foetal mouse long bones. *Bone* **17** (1), 63–69.

65. Meghji, S., Sandy, J.R., Harvey, W. *et al.* (1992) Stimulation of bone collagen and non-collagenous protein synthesis by products of 5- and 12-lipoxygenase: determination by use of a simple quantitative assay. *Bone and Mineral* **18**, 119–132.

66. Lowik, C.W.G.M., van der Pluijm, G., Bloys, H. *et al.* (1989) Parathyroid hormone (PTH) and PTH-like protein (PLP) stimulate IL-6 production by osteogeneic cells: a possible role of IL-6 in osteoclastogenesis. *Biochemical and Biophysical Research Communications* **162**, 1546–1552.

67.   Kream, B.E., Rowe, D.W., Gworek, S.C. and Raisz, L.G. (1980) Parathyroid hormone alters collagen synthesis and procollagen mRNA levels in foetal rat calvaria. *Proceedings of the National Academy of Sciences USA* **77**, 5654–5658.

68.   Meghji, S., Henderson, B., Nair, S. and Wilson, M. (1992) Inhibition of bone DNA and collagen production by surface associated material from bacteria implicated in the pathology of periodontal disease. *Journal of Periodontology* **63**, 736–742.

69.   Chyun, Y.S. and Raisz, L.G. (1984) Stimulation of bone formation by prostaglandin E_1 and $F_1\alpha$. *Prostaglandins* **27**, 96–103.

70.   Hefley, T.J., Kreiger, N.S. and Stern, P.H. (1986) Simultaneous measurement of bone resorption and collagen synthesis in neonatal mouse calvaria. *Analytical Biochemistry* **153**, 166–171.

71.   Meghji, S., Wilson, M., Henderson, B. and Kinnane, D. (1992) Anti-proliferative and cytotoxic activity of surface-associated material from periodontopathogenic bacteria. *Archives of Oral Biology* **37**, 637–644.

72.   Saito, M., Kawashima, K. and Endo, H. (1987) The establishment of a new biological assay system for simultaneous measurement of bone resorption and bone mineralization in organ cultures of chick embryonic femur. *Journal of Pharmacobiological Dynamics* **10** 487–493.

73.   Schwartz, Z., Ornoy, A. and Soskolne, W.A. (1985) An in vitro assay of bone development using foetal long bones of mice: morphological studies. *Acta Anatomica* **124**, 197–205.

74.   Gronowicz, G., Woodiel, F.N., McCarthy, M.B. and Raisz, L.G. (1989) In vitro mineralization of parietal bones in defined serum-free medium; effect of beta-glcerol phosphate. *Journal of Bone and Mineral Research* **4**, 313–324.

75.   Kaji, T., Kawatani, R., Hoshino, T. *et al.* (1990) A suitable culture medium for ossification of embryonic chick femur in organ culture. *Bone and Mineral* **9**, 89–100.

76.   Katz, J.M., Wilson, S.J.M. and Gray, D.H. (1981) Bone resorption and prostaglandins production by mouse calvaria in vitro: response to exogenous prostaglandins and their precursor fatty acids. *Prostaglandins* **22**, 537–551.

77.   Gitelman, H.J. (1967) An improved automated procedure for the determination of calcium in biological specimens. *Analytical Biochemistry* **18**, 520–531.

78.   Meghji, S., Henderson, B., Morrison, M.S. *et al* (1996) Ca^{2+} release from cultured mouse calvaria is very sensitive to ambient pH. *Journal of Bone and Mineral Research* **11**, 1824 (Abstr. P37)

CHAPTER FIVE

Methods for studying cell death in bone

Brendan F. Boyce, David E. Hughes and Kenneth R. Wright

5.1 INTRODUCTION

Apoptosis is a common form of cell death that controls cell numbers in most tissues during their development as well as in a wide variety of normal and pathological settings. It has been defined by a series of characteristic morphological changes that take place in nuclear chromatin and cytoplasm of cells dying by a mechanism that differs from ischaemic necrosis in that it typically affects single cells, rather than groups of cells [1, 2]. These changes appear to be regulated in affected cells; hence the alternative terms: programmed, biological, physiological or controlled cell death.

Although the term apoptosis was coined only 25 years ago, this form of cell death has been recognized by pathologists for a long time and has been described by a variety of terms, including zeiosis, popcorn-type cytolysis, necrobiosis, pyknosis and karyorrhexis [2]. However, its major biological significance has only been recognized widely by pathologists and cell and molecular biologists since the late 1980s, as appreciation of its role in the pathogenesis of cancer and aberrant immune responses has grown along with an increased understanding of its importance in fetal development and normal immune responses. During the last 8 years numerous genes involved in cell survival and death have been identified and researchers have become aware that cell death, like cell division, is an important determinant of cell behaviour.

During embryonic development, apoptosis controls cell numbers in mesenchymal, neural and epithelial cells and facilitates the deletion of tissue during organ folding and rotation at precise stages in fetal growth. It is responsible for the removal of the web of soft tissue between developing fingers, a process that involves BMP signalling [3], and is the major mechanism for deletion of B and T cells during negative selection of potentially auto-reactive lymphocytes in the immune response and of helper T cells in HIV-infected patients. Apoptosis is common in some malignant tumours, and the relative rates of apoptosis and mitosis may account for the slow growth of basal cell carcinomas of the skin (high mitosis and apoptosis rates) and the aggressive growth of malignant melanomas (high mitosis and low apoptosis rate) [4]. Recent studies have indicated that induction of apoptosis in tumour cells

Methods in Bone Biology. Edited by Timothy R. Arnett and Brian Henderson.
Published in 1997 by Chapman & Hall, London. ISBN 0 412 75770 2.

is a major mechanism of action of irradiation [5], of many chemotherapeutic agents [6] and of hormonal therapy in hormone-responsive tumours. Conversely, trophic hormone withdrawal causes normal cells to undergo apoptosis in target organs, such as the prostate, lactating breast and adrenal glands. Hormone withdrawal is associated with decreased expression of the cell survival gene, *bcl-2* , and in some cases with enhanced expression of *p53* [7].

Apoptosis is involved in the normal host response to many viral infections and in most instances virus-infected cells eliminate themselves by this mechanism to stop viral replication and thus minimize damage to uninfected neighbours [8]. To counteract this response and ensure their survival, many viruses have developed apoptosis-inhibiting genes to prevent the death of cells they have infected. Many of these genes have been conserved among species. For example, the adenovirus gene, *E1B 19kD* [9] and the Epstein–Barr virus gene, *BHRF1* [10], share sequence similarity with the mammalian *bcl-2* gene, while the cowpox virus *crmA* gene encodes a serpin-like protease inhibitor which inhibits IL-1 converting enzyme (ICE) [11]. ICE is a member of a growing family of apoptosis-promoting cysteine proteases (now called caspases [12]) which are involved in many types of apoptotic deaths.

Apoptosis can be induced or prevented by many factors, including oncogenes, tumour suppressor genes, growth factors, cytokines, hormones and integrin-mediated signal transduction (Table 5.1) (see recent reviews [7, 13, 14, 15]), but the observed effect of individual agents can be survival or death depending upon the cell type. For example, oestrogen replacement therapy after oophorectomy causes proliferation of endometrial cells lining the uterine cavity, but leads to apoptosis of osteoclasts (see later), while, conversely, tumour necrosis factor α causes apoptosis of a variety of cell types but prevents apoptosis of osteoclasts.

This chapter will review the morphological features of apoptosis, its occurrence and regulation in bone cells and the methods that are presently being used to study it.

Table 5.1   Inducers and inhibitors of apoptosis

Inducers	*Inhibitors*
Cell-damage related	Trophic hormones
Irradiation (p53-mediated)	Cytokines – most interleukins, TNF
Free radicals of oxygen and nitrogen, hydrogen peroxide	Sex steroids
Cytotoxic T cells (Fas/TNF receptor and granzme b)	Extracellular matrix
Oncogenes: myc, Max, Mad, fos	Growth factors – CSFs, IGF-1, NGF
Tumour suppressors: p53	Bcl-2, Bcl-x$_L$
Physiological mechanisms	Cysteine protease inhibitors (caspases)
Hormones: glucocorticoids, sex hormones	Calpain inhibitors
Growth factor/hormone withdrawal (*bcl-2*-mediated)	Ca^{2+} channel blockers
Loss of matrix attachment	Viral proteins:
Growth factors/cytokines:	adenovirus E1B 19kD
TGF-β, TNF, (Fas ligand)	baculovirus p35 and 1AP
Chemotherapy: most cytotoxic drugs	cowpoxvirus CrmA (serpin)
	Epstein–Barr varus
	BHRF1 and LMP-1

5.2 MORPHOLOGICAL FEATURES OF APOPTOSIS

Apoptotic cell death has been defined on the basis of characteristic light and electron microscopic nuclear and cytoplasmic changes which were identified by Wyllie and Kerr [1] in sections of normal and diseased tissues and in cultured cells, particularly thymocytes which had been induced to die in response to glucocorticosteroids [2]. These microscopic changes have been observed consistently in almost all apoptotic cell deaths, independently of the stimuli or circumstances in which they occur. Although a variety of other techniques has been developed to facilitate recognition of the occurrence or quantification of apoptosis, we believe that morphology remains the gold standard for identification of apoptotic cells. However, refinement and improvement of these techniques and the development of others should greatly assist progress of the study of the factors which promote and regulate apoptosis.

The earliest morphological feature of apoptosis is clumping of chromatin into dense aggregates around the nuclear membrane (Fig. 5.1), in contrast with the even dispersal seen typically in viable cells, followed by condensation of chromatin into dense balls (Fig. 5.2). Disintegration of the nuclear membrane ensues and condensed aggregates of chromatin can be seen readily within the cytoplasm (Fig. 5.2). These nuclear changes are accompanied by fluid movement out of the cell and transglutaminase-mediated cytoplasmic protein cross-linking [16] which result in cytoplasmic condensation and cell shrinkage (Fig. 5.2). Expression of tissue transglutaminase has been used to identify apoptotic cells immunohistochemically [16], but because some viable cells express this enzyme constitutively, a positive staining reaction can be interpreted as apoptosis only if the typical morphological features arc also present.

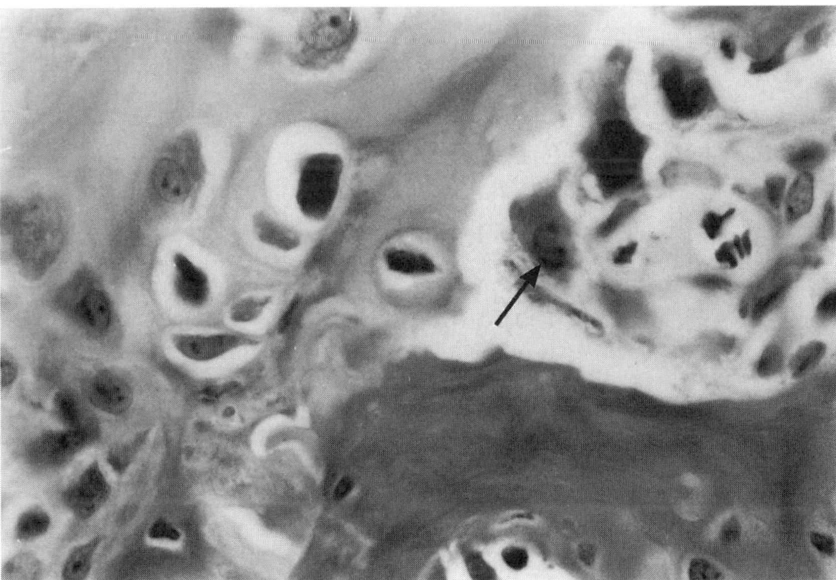

Figure 5.1 Early features of apoptosis. Nuclear chromatin has relocated to clumps around the nuclear membrane (arrow) of an osteoclast in the early stages of apoptosis. (Haematoxylin and eosin, magnification ×845.)

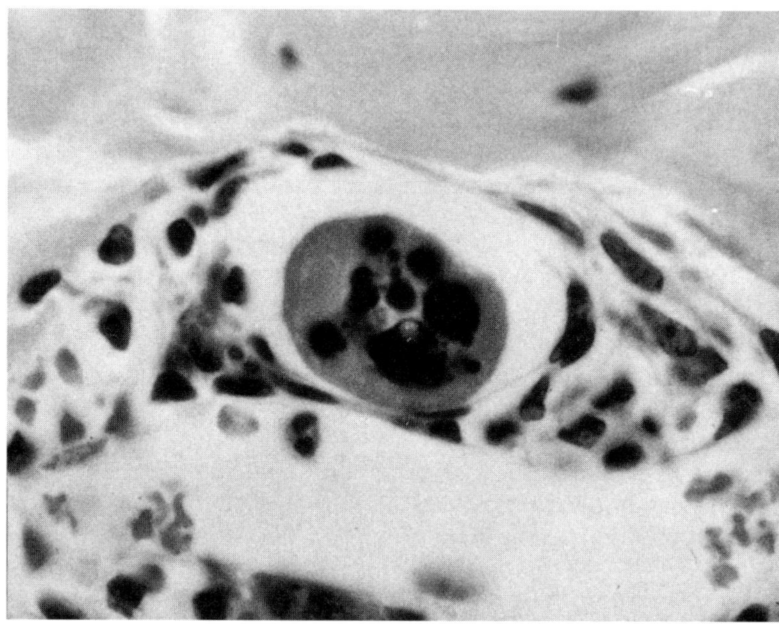

Figure 5.2   Apoptotic osteoclast. Nuclear condensation and fragmentation, cytoplasmic condensation and retraction of the cell from the bone surface are all typical features of apoptosis in this osteoclast in a bone section from a TRAP-tag transgenic mouse in which the TRAP promoter has targeted T-antigen to the osteoclast. (Haematoxylin and eosin, magnification ×845.)

During contraction, apoptotic cells lose attachment to neighbouring cells or matrix (Fig. 5.2) and can often be seen within a clear space in tissue sections. This shrinkage is associated with the appearance of numerous cell surface convolutions and is followed by disintegration of the cell into multiple membrane-bound, condensed apoptotic bodies (Fig. 5.3). Typically these bodies are phagocytosed rapidly by neighbouring cells (Fig. 5.4) which are able to acquire a phagocytic function for this purpose or are phagocytosed by professional phagocytes. Phagocytes do not become activated in these circumstances and consequently other immune cells are not recruited to the site. The whole process lasts a few minutes (e.g. cytotoxic T cell-induced apoptosis) to a few hours, depending on the stimulus and, although there is no inflammatory reaction, inflammatory agents, cytokines and growth factors are involved, some having stimulatory and others inhibitory actions. In this respect, apoptosis differs significantly from necrosis, in which activation and attraction of inflammatory cells are key features. Furthermore, necrosis is characterized by influx of fluid into dying cells, which consequently swell up rather than contract, and by retention of the nuclear membrane and diffuse dispersal of genomic DNA until close to the time when the cell bursts and disintegrates.

Apoptotic bodies that contain nuclear fragments can be visualized readily in tissue sections. However, many of them do not contain nuclear fragments. Thus, they often go undetected under light microscopy, particularly because most cells do not have distinctive cytoplasmic markers that would permit identification of small apoptotic bodies, either within or between viable cells, using conventional or special stains. Phagocytosed apoptotic bodies can be seen by electron microscopy,

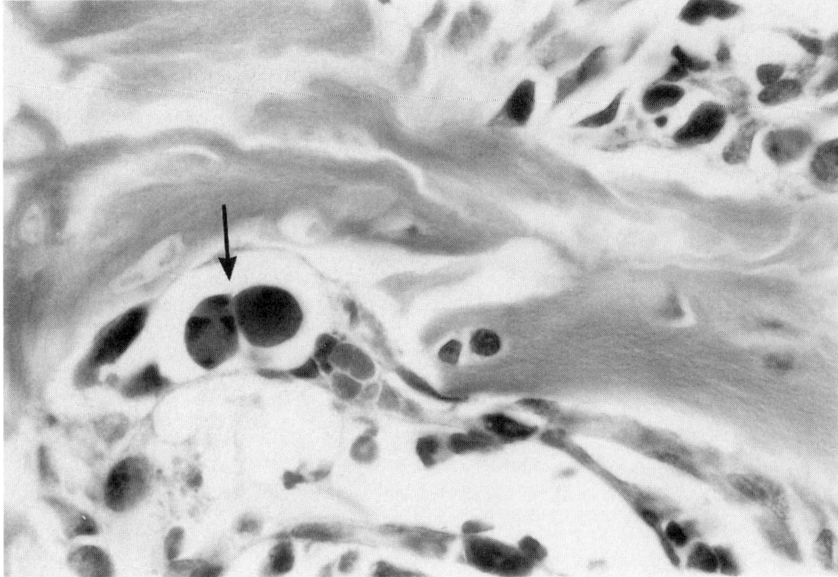

Figure 5.3 Apoptotic bodies of a dead osteoclast. Two discrete fragments (arrow) of an apoptotic osteo-clast are present within a clear space adjacent to the bone surface. Condensed fragments of nuclear chromatin are present within the apoptotic body on the left, but none is seen in the TRAP-positive cytoplasm of the one on the left. (TRAP and haematoxylin stain, magnification ×845.)

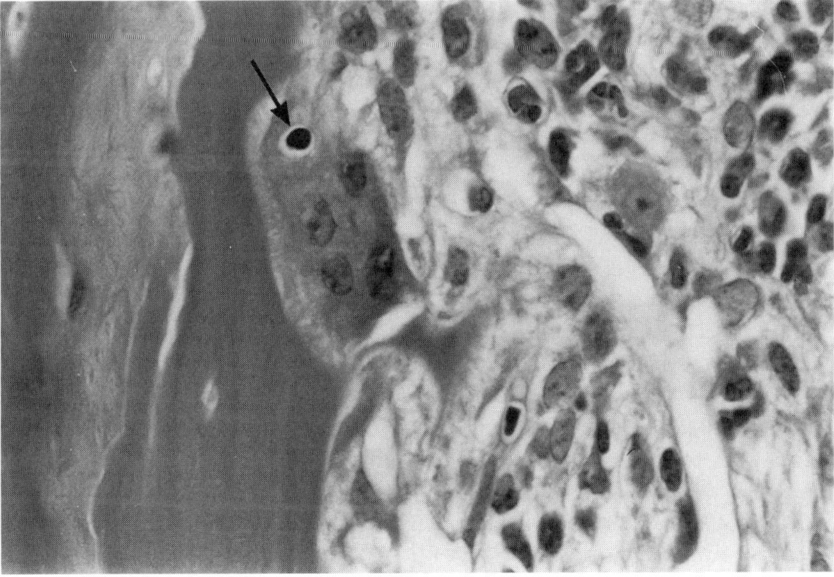

Figure 5.4 Apoptotic osteocyte phagocytosed by an osteoclast. The densely haematoxyphilic nucleus (arrow) of an apoptotic osteocyte is present within the cytoplasm of an osteoclast that is actively resorbing the bone matrix in this section of mouse calvaria and contrasts with the lightly stained chro-matin and nucleoli of viable nuclei. (Haematoxylin and eosin, magnification ×845.)

which is also used routinely to confirm the morphological features of apoptosis [2]. However, as will be shown later, this technique is likely to permit identification of such bodies only if they are present in relatively high numbers, because the samples that are used for analysis typically are small and do not contain large numbers of cells.

In contrast with many other cell types, intact apoptotic osteoclasts and fragmented apoptotic bodies of disintegrated osteoclasts can be visualized readily in tissue sections if they are stained for tartrate-resistant acid phosphatase (TRAP) activity and viewed under light microscopy (Fig. 5.3). This is because not only are osteoclasts larger than most cells, but also they retain strong cytoplasmic TRAP activity during apoptosis (see later).

The striking nuclear changes of apoptosis result from the effects of the activity of endogenous endonucleases which are activated during the early stages of apoptosis [1, 2]. These split genomic DNA at nucleosomes, using a Ca^{2+}/Mg^{2+}-dependent mechanism [17], into nucleosomal fragments of varying sizes. These give rise to the characteristic 'ladders' that are seen on gel electrophoresis [2] – a standard technique for confirmation of apoptosis, particularly in cultured cells that will be described later.

5.3 APOPTOSIS IN BONE CELLS

5.3.1 Osteoclasts

The first reports of osteoclast apoptosis came in 1993 at the Annual Meeting of the American Society for Bone and Mineral Research [18, 19]. Fuller *et al.* [20] reported that osteoclasts undergo apoptosis *in vitro* upon withdrawal of M-CSF – a finding similar to that reported previously when growth factors were withdrawn from haematopoietic precursors and neutrophils [21] and indicating the requirement for M-CSF not only for osteoclast formation but also for osteoclast survival. We observed that 5% of osteoclasts (versus 0.3% in controls) were apoptotic in transgenic mice in which the TRAP promoter was used to target the transforming simian virus 40 (SV40) large T antigen to osteoclasts [22]. This finding was unexpected because previous studies had indicated that T antigen binds to and inactivates p53 protein [23] and since p53 promotes apoptosis in some circumstances [5, 24] we anticipated that this binding would result in reduced, rather than increased, apoptosis. The mechanism of induction of apoptosis in these circumstances remains unexplained.

Osteoclast apoptosis actually had been observed before 1993 by several investigators in a variety of circumstances. For example, it occurs in response to reintroduction of calcium into the diet of calcium-deficient rats [25], and following the administration of oestrogen to ovariectomized mice [26] or of bisphosphonates to rodents [27, 28, 29]. One of us (BFB) had also seen, but not commented upon, apoptotic osteoclasts in the calvariae of mice following treatment with high doses of interleukin 1 [30] (see below). Although the investigators in all of these studies recognized disintegrating osteoclasts in their experiments, none of us appreciated at the time that we were observing osteoclasts undergoing apoptosis.

Because osteoclasts are multinucleated and larger than most other cells, when they undergo apoptosis they can be recognized more readily in bone sections than apoptotic mononuclear cells. All of their nuclei appear to become apoptotic synchronously (Fig. 5.2), thus enhancing the chances of detecting their condensed nuclear fragments in a single 3–5 μm tissue section during the later stages of apoptosis. The presence of cytoplasmic TRAP not only assists in the identification of viable osteoclasts; it also further enhances detection of those undergoing apoptosis.

The intensity of TRAP staining in apoptotic osteoclasts typically is higher than in viable cells, presumably as a result of its continued production or failed secretion during apoptosis coupled with the loss of cell fluid that accompanies this form of death and accounts in part for the cell shrinkage. This increased staining intensity enhances identification, not only of large apoptotic intact osteoclasts, but also of small osteoclast apoptotic bodies. These can be seen readily in decalcified TRAP-stained sections taken through long bone metaphyses of normal mice and of mice treated with agents, such as bisphosphonates, that promote osteoclast apoptosis as part of their anti-osteoclastic resorptive action [31, 32].

In sections of normal mouse metaphyses, small TRAP-positive apoptotic bodies are relatively common. Most of these are less than 7 μm in diameter and few of them contain condensed nuclear fragments. Thus, it is not possible to determine whether most of these are the remains of fragmented multinucleated osteoclasts, mononuclear osteoclasts or osteoclast precursors. We have seen them much less commonly in decalcified sections of human bone from sites of both normal and increased bone turnover (unpublished observations). The reasons for this difference between the behaviour of human and murine osteoclasts are not clear at present, but may be species-related or may reflect differences in turnover rates of osteoclasts and their precursors in growth plates of growing rodents versus remodelling sites in adult human trabecular bone.

To determine whether osteoclasts undergo apoptosis in bone remodelling units, we reviewed sections from experiments in which interleukin 1 (IL-1) had been injected once a day for 3 days into the subcutaneous tissues overlying the calvarial bones of young mice [30] and found that, 5 days after the last IL-1 injection, 13% of the osteoclasts were apoptotic [33]. Osteoclasts at the advancing edge of resorption lacunae were viable, and most of the apoptotic osteoclasts were at the ensuing reversal site. We proposed subsequently that the youngest and presumably most active osteoclasts maintain forward progression of bone remodelling units through recruitment of osteoclast precursors to the advancing front, while older osteoclasts remain behind to complete lateral and downward progression before undergoing apoptosis [34]. Prolongation of the life span of these older osteoclasts is likely to result in an increase in the depth of bone remodelling units and thus could account for the penetration of trabecular plates and the loss of connectivity that characterizes post-menopausal osteoporosis (see below).

Several papers confirming that osteoclasts die by apoptosis have been published recently. For example, based on *in vitro* and *in vivo* observations in mice, we have suggested that a major part of the anti-resorptive action of bisphosphonates is induction of osteoclast apoptosis [35], an observation that has been confirmed *in vitro* by Selender *et al.* [32]. Kameda *et al.* [36] reported that neonatal rabbit osteoclasts undergo apoptosis in culture and that this involves protein kinase C signalling. Furthermore, they found that vitamin K_2, but not vitamin K_1, inhibits bone resorption by inducing osteoclast apoptosis [37]. Lutton *et al.* [38] have used time-lapse video microscopy to observe rat osteoclasts undergoing apoptosis *in vitro* in response to cyclosporine A, tamoxifen, corticosterone and dexamethasone. In these experiments, cell shrinkage was apparent within 25 minutes of exposure of the cells to these agents and marked changes in the morphology of osteoclasts became apparent within 2 to 4 hours. The apoptosis-inducing effects of tamoxifen have been confirmed by Arnett *et al.* [39] and Hughes *et al.* [40].

We have found that the sex steroids, oestrogen [40] and testosterone [41], induce osteoclast apoptosis *in vitro* and *in vivo*, while the so-called osteotropic cytokines IL-1, IL-6 and TNFα (see [42, 43] for review) all prevent osteoclast apoptosis *in vitro* [44]. The *in vitro* effect of oestrogen appears to be mediated by TGFβ [40], but the cellular source of this cytokine is not clear because the assay that we developed to study osteoclast apoptosis in these studies is a mixed bone marrow culture based on the Takahashi system [45] for generating osteoclasts from murine bone marrow precursors. Both oestrogen and tamoxifen increase TGFβ production by osteoblasts [46]; thus the effect of oestrogen to induce osteoclast apoptosis may well be indirect – a mechanism that would be supported by the finding of Arnett *et al.* [39] that oestrogen did not induce apoptosis of isolated neonatal osteoclasts.

On the basis of our findings, we [40] have proposed that oestrogen both before the menopause and when given after the menopause as hormone replacement therapy limits osteoclast life span on bone surfaces and prevents them from resorbing through the full thickness of trabeculae. In contrast, after the menopause, in the absence of oestrogen, increased cytokine production could not only increase generation of osteoclasts but also prolong their life spans, allowing osteoclasts to remain longer on bone surfaces and to erode through trabeculae with the resultant loss of connectivity characteristic of post-menopausal osteoporosis.

5.3.2 Cells in the osteoblast lineage

(a) Osteoblasts

Progress in the study of osteoblast apoptosis has been much slower than that of cell death in osteoclasts, and to date there have been few reports on the subject. There are several possible reasons for this and most are likely to be related to the technical difficulties associated with accurate *in vivo* detection of osteoblasts as they die. It has been recognized for many years that more osteoblasts appear to be recruited to the surface of bone remodelling units than the numbers that persist in the matrix as osteocytes or on the bone surface as lining cells [47]. The fate of the osteoblasts that disappear is unknown, but it may well be that they succumb to apoptosis. Variation in the number of osteoblasts recruited to each remodelling unit could be a determinant of the mean wall thickness of formed units, and recruitment of fewer osteoblasts in osteoporotic than in normal subjects could result in thinner trabeculae [48]. It is also possible that increased apoptosis resulting in decreased osteoblast life span could cause reduced wall thickness in osteoporotic patients, but as yet there are no data to support this hypothesis.

We have examined remodelling units in sections of human and murine bones for osteoblast apoptosis in a variety of circumstances, using both H&E and TUNEL staining (section 5.4.4), but have seen little or no evidence of it. However, apoptotic osteoblasts have been observed in an *in vitro* model of bone formation [49] and in cultured MC3T3-E1 osteoblastic cells following treatment with TNFα [50], an effect that appears to involve ceramide and nuclear translocation of NF-κB [51]. If apoptosis is the fate of some osteoblasts during bone remodelling and the percentage of osteoblasts undergoing apoptosis in bone remodelling units is similar to that reported in hepatocytes [52] or enterocytes [53], i.e. between one and five cells per thousand, the chances of detecting these few cells in bone sections are small, and thus our failure to detect them in tissue sections does not preclude this possibility.

(b) Osteocytes

In contrast to the experience with osteoblasts *in vivo*, there have been several reports of apoptosis in osteocytes in a variety of normal and pathological settings. During bone resorption, the matrix surrounding osteocytes is removed by osteoclasts and the fate of these osteocytes has been the subject of speculation for many years [54–56]. It now appears that some, if not all, of these cells undergo apoptosis and are then phagocytosed by osteoclasts in resorption lacunae [40, 57, 58]. These appear as mononuclear cells within vacuoles inside osteoclasts (Fig. 5.4). They have nuclear condensation and/or fragmentation typical of apoptosis, positive staining with TUNEL, but no cytoplasmic TRAP staining and, in our opinion, are not single osteoclast nuclei undergoing apoptosis individually.

Dunstan and colleagues [59] reported osteocyte death in trabecular bone in femoral heads from old subjects. They found significantly more empty osteocyte lacunae in bone from these individuals compared with young subjects, and although they did not comment on the mechanism of disappearance of the osteocytes, it is likely that they died by apoptosis. More recently, Noble *et al.* [60] described death of osteocytes in normal and patholgical bone in humans by a process which they considered to be similar, if not identical, to apoptosis. Death of osteocytes within bone could cause impairment of fatigue-related damage repair and further weakening of effete bone, particularly if, as proposed, viable osteocytes act as mechano- or trauma-receptors directing osteoclasts to foci of damaged bone for removal and subsequent replacement with new bone [47].

5.3.3 Chondrocytes

During endochondral bone formation, proliferating chondroblasts differentiate into hypertrophic chondrocytes in the growth plates of long bones. Cartilage matrix around these latter cells calcifies before being removed by chondroclasts and is replaced by new bone laid down by osteoblasts at the primary spongiosa. The hypertrophic chondrocytes are phagocytosed by chondroclasts along with the cartilage matrix [61], but prior to their removal they undergo apoptosis – a fate that has been confirmed by morphology and TUNEL staining [61, 62] and thus is similar to that of osteocytes.

Although little is known about the regulation of endochondral ossification, recent studies have indicated that the signalling protein, Indian hedgehog, and PTHrP negatively control the differentation of proliferating chondrocytes into hypertrophic chondrocytes [63, 64] and thus prevent accelerated chondrocyte differentiation and premature shortening of bones. Targeted over-expression of PTHrP to chondrocytes appears to increase the expression of the cell survival protein, Bcl-2, and to correspondingly decrease the expression of the cell death protein, Bax, in pre-hypertrophic chondrocytes both *in vitro* and *in vivo* [65]. This results in decreased apoptosis and accumulation of these cells and subsequent delay in endochondral bone formation.

5.4 METHODS FOR DETECTION OF APOPTOSIS

5.4.1 Gel electrophoresis

The fragmentation of a chromosomal DNA by endonucleases at nucleosomes results first in large 50–300 kilobase fragments that degrade subsequently into multiple oligonucleosomal pieces of DNA of varying sizes [66]. Because the number of nucleotides between nucleosomes is fairly constant (around 200), many of the oligonucleosomal DNA fragments are of similar length. The size of these fragments varies from 200 to several hundred nucleotides, depending upon the site and timing of endonuclease activity. Consequently, their electrophoretic mobility also varies, and when preparations of the DNA from apoptotic cells are run on a conventional agarose gel, a characteristic 'ladder' pattern is seen, with the smallest and lightest fragments travelling farthest to the bottom of the gel while the largest and heaviest fragments remain nearer the top. This pattern contrasts not only with that seen with viable cells, whose intact DNA remains at the top of the gel because of its low electrophoretic mobility, but also with that seen when viable cells are treated with DNase. This enzyme breaks up DNA into fragments with considerably more variation in size than that of apoptotic cells, resulting in a diffuse continuous band on electrophoresis.

Gel electrophoresis has been used to detect apoptotic cells for over 20 years and is still used extensively. It is a straightforward technique requiring equipment that is standard in most laboratories studying the biochemical or molecular regulation of cellular activities. The major strength of this technique is its fairly high specificity for distinguishing between apoptotic and necrotic cell death and between dead and viable cells. This is particularly useful in cultures of single cell types in which a proportion of the cells is undergoing apoptosis. Furthermore, it does not require the subjective judgment and training that are needed for recognition and accurate identification of apoptotic cells using morphological methods, especially in mixed cell populations.

Gel electrophoresis does have its limitations. For example, in mixed cell cultures it does not permit determination of which one cell type is undergoing apoptosis, nor does it allow quantitation of the percentage of a cell population that is apoptotic versus viable. Thus, it is of very limited use in most cultures of osteoclasts and stromal cells set up to determine whether particular compounds cause osteoclast apoptosis without affecting stromal cell viability and whether the effect has a dose–response relationship.

Attempts have been made recently to examine electrophoretic mobility of DNA extracted from individual cells undergoing apoptosis – the so-called comet assay [67], which was developed from an assay first described by Ostling and Johanson [68]. A comet-shaped electrophoretic pattern is observed when apoptotic cells embedded in agarose gel are lysed and the DNA is subjected to electrophoresis. To date there has been one report of this technique being used to examine apoptotic bone cells [36]. Because this technique requires individual cells to be removed from cultures and suspended in agarose gel, it is likely that the large size of osteoclasts may make them ideal cells for such study.

5.4.2 Electron microscopy

Ultrastructural analysis of cells is a very powerful technique for determination of the effects of agents on cellular organelles and was used to define in detail the sequential changes of apoptosis from the initial perinuclear membrane condensation of chromatin to the condensation and fragmentation of chromatin and cytoplasmic organelles to form the characteristic apoptotic bodies of disintegrated cells. It is, however, an expensive and relatively time-consuming technique with significant limitations for the study of apoptosis. For example, specimen size for transmission microscopy is limited to about 1 mm^3 and thus the sample volume per specimen is much smaller than that which can be submitted for light microscopy. Consequently, although each cell can be examined in great detail at very high magnification, the number of cells that can be examined per specimen is small.

Specimen size may not be a problem in tissues, such as some malignant tumours, that have a constitutively high rate of apoptosis or that have been subjected to chemotherapy and as a result have many cells undergoing apoptosis. However, it is a significant limitation in tissues with normal cell turnover, such as liver or intestine, in which the percentage of cells undergoing apoptosis is around 0.01–0.05%. In these instances, it would be necessary to examine many blocks of tissue to find the one apoptotic cell in 200 or 1000 viable cells, and in these circumstances the success rate is likely to be low. The problem of sample size and number of blocks required for detection of apoptotic cells is even greater in tissues, such as bone from normal or osteoporotic subjects, in which turnover not only occurs at a low rate but takes place at discrete remodelling sites which may occupy only 5–20% of the bone surface. Even within remodelling units, it is likely that apoptosis will be restricted to a small number of cells and occur at discreet times and places. Thus, electron microscopy would have little or no role in determining the frequency of apoptosis in osteoblasts or osteoclasts during normal bone remodelling.

Electron microscopy may have significant advantages over other techniques for detection and accurate identification of apoptotic cells in tissues, such as growth plate hypertrophic cartilage, in which cell death is common. Recent studies have indicated that chondrocytes in the hypertrophic zone of growth plates undergo apoptosis prior to resorption of the calcified cartilage around them by chondroclasts [61, 62]. The nuclei of these chondrocytes typically appear shrivelled and pyknotic in decalcified paraffin-embedded sections and thus it can be difficult to distinguish viable from apoptotic cells with certainty under light microscopy. In light of recent reports that some of these chondrocytes appear to become osteoblasts upon release from their lacunae by osteoclasts in chickens [62, 69], it is likely, given the limitations of light microscopy and of methods such as TUNEL (section 5.4.4), that documentation of the fate of these cells will be done best by using electron microscopy.

Ultrastructural analysis is also indicated for confirmation of apoptosis in cultured cells. For example, when non-adherent cells, such as thymocytes, are treated with glucocorticoids, some of them undergo apoptosis in the culture medium among their viable neighbours. These can be detected after all of the cells have been spun down to form a cell pellet which can be fixed, embedded and sectioned for ultrastructural analysis. Similarly, with adherent cells in culture, some of those undergoing apoptosis lose adherence to the culture plate and can be detected in the supernatant before they fragment into apoptotic bodies, while others can remain attached during the early stages of apoptosis.

Detailed ultrastructural analysis is particularly appropriate for the examination of cultured cells treated with agents which may cause cell death by apoptosis or by non-specific dose-dependent toxic mechanisms that result in cell death more closely resembling necrosis. The difference between toxicity and apoptosis may at first sight appear to be trivial in terms of the mechanism of action of a drug. However, the difference may be crucial to the drug's successful development. Apoptosis is a normal cellular process which can be harnessed specifically to cause the death of unwanted cells. Toxicity, in contrast, is a non-specific process which invokes a potentially harmful inflammatory reaction and is unlikely to have the regulatory controls that accompany apoptosis. Toxic cell injury may result in DNA fragmentation, possibly through partial activation of apoptotic mechanisms, and thus nick-end labelling techniques are likely to give a positive signal that will not discriminate between a toxic and apoptotic cell death.

5.4.3 Light microscopy

Light microscopy with standard staining techniques is the simplest and most accurate method for identifying apoptosis of individual cells, including bone cells, and can be used in a variety of ways with both tissue sections and cultured cells. However, recognition of apoptotic cells using morphology is not straightforward for those who are not experienced in this field and requires a significant amount of training. Furthermore, a very high technical standard is required in the preparation of tissue sections for the study of apoptosis. Poor fixation, processing, section cutting or staining, or suboptimal microscope optics can all compromise the histological detection of apoptosis. The principal advantage of light microscopy over other methods is that it permits direct examination of large numbers of cells using widely available inexpensive equipment and does not require the use of antibodies, labelled nucleotides or expensive reagents for visualization of apoptotic cells. The standard histological staining reagents are haematoxylin and eosin (H&E), which stain nuclear chromatin dark blue and cytoplasm pink. Cells undergoing apoptosis in tissue sections stained with H&E can be recognized readily by trained personnel, because their nuclear chromatin and cytoplasm condense with cell shrinkage and consequently stain more intensely than surrounding viable cells. In addition, apoptotic cells lose attachment to their neighbours and typically appear to have a clear space around them, which further highlights their presence.

Despite these characteristic appearances, it is difficult, if not impossible, to recognize all apoptotic cells in a single histological section 3–5 μm thick under light microscopy. The average mononuclear cell measures 10–20 μm on cross-section and most nuclei are between 5 and 10 μm in diameter. Consequently, in a typical section 3 μm thick most (but not all) nuclei of viable cells can be seen. As cell shrinkage and nuclear condensation and fragmentation proceed during apoptosis, the average size of nuclear and cytoplasmic profiles of dying cells in sections will decrease, and the chances of sectioning through occasional dying cells in normal tissues will inevitably be less than those of sectioning through both the nucleus and cytoplasm of viable cells. Furthermore, during the later stages of apoptosis, the cell breaks up into apoptotic bodies which can be less than 2–3 μm in diameter. These may be present between viable cells or inside them following phagocytosis and because of their small size are likely to go unrecognized in these circumstances, particularly if they do not contain condensed nuclear fragments. Even if they do contain a small nuclear fragment, these tiny bodies can be seen only after detailed scrutiny of sections. If they consist only of condensed cytoplasm, it is likely that most will go undetected. Thus, although in our opinion light microscopy is the most sensitive method for detection of apoptotic cells, it is still likely to underestimate significantly the frequency of apoptosis in both normal and disease states and, as with the nick-end labelling techniques described below, requires significant training to ensure that acquired data are accurate.

5.4.4 Nick-end labelling of apoptotic cell DNA

Several nick-end labelling methods have been developed to facilitate identification of apoptotic cells using methods developed many years ago to label chromosomes in metaphase. The first was described by Gavrieli *et al.* [70] and was called TUNEL (TdT-mediated dUTP-biotin nick end-labelling). Others, including ISNT (*in situ* nick translation) [71] and ISEL (*in situ* nick-end labelling) [72], followed quickly thereafter. They all work on the same principle, which is that single or double strand breaks at the ends of DNA fragments in apoptotic cells can have labelled nucleotides attached to them. Gavrieli *et al.* [70] used the enzyme terminal deoxynucleotidyl transferase to attach biotin-labelled nucleotides to the DNA and detected the biotin using a standard streptavidin and immunoperoxidase detection system and diaminobenzidine (DAB) as the chromagen. This system gives a positive brown signal in the nuclei of apoptotic cells (Fig. 5.5) whose cytoplasm can be counterstained with eosin. In theory, there should be no signal seen in the nuclei of viable cells and these can be counterstained lightly by haematoxylin or methyl green. Fluoroscein can also be attached to the nucleotides and visualized under fluorescence microscopy.

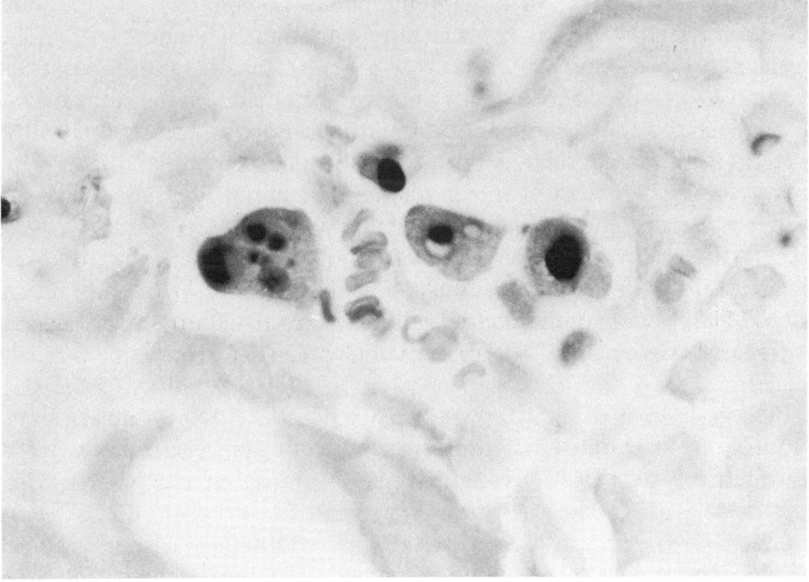

Figure 5.5 TUNEL staining of apoptotic osteoclasts in a TRAP-tag transgenic mouse. A strong dark signal is seen in some of the condensed nuclear fragments in four apoptotic osteoclasts that have retracted from the bone surface. Condensed nuclear chromatin of some apoptotic nuclei has remained unstained and these unstained nuclei are not distinguishable from the cytoplasm of the cells, because of the effect of the filter used to take this black-and-white picture. (Immunoperoxidase and methyl green staining, magnification ×845.)

Since their introduction, nick-end labelling methods have been used extensively, not only in tissue sections but also in cells cultured *in vitro*. Cells such as osteoblasts and osteoclasts, which adhere to culture plates, can be fixed and exposed to the nick-end labelling reagents, and cells with a positive signal can be identified and counted *in situ*. Alternatively, they can be removed by treatment with trypsin, fixed, exposed to the labelling reagents and then passed through a flow cytometer which can count total cell numbers as they pass through the detector and also identify and count those with a positive nuclear signal. The percentage of apoptotic cells can then be calculated automatically.

We have used both TUNEL and ISEL to identify apoptotic cells, predominantly in tissue sections but also in a modification of the murine bone marrow culture system that we developed as an *in vitro* assay to study osteoclast apoptosis. Using TUNEL and ISEL we have detected a strong positive signal in the nuclear fragments of osteoclasts with the classical morphological features of apoptotic cells, as illustrated in Fig. 5.5. However, we were disappointed to find that not all of the apoptotic nuclei of osteoclasts had a positive nuclear signal (Fig. 5.5) and that a significant number of apoptotic nuclei had no staining at all. When we counted osteoclasts with a positive TUNEL signal in tissue sections and expressed them as a percentage of all of the osteoclasts whose nuclei had the classical morphological features of apoptosis with a haematoxylin counterstain, we found that less than 50% of the apoptotic osteoclasts were detected using this method [73]. We found similarly low percentages of apoptotic hepatocytes staining positively with TUNEL in sections of liver from transgenic mice in which expression of large T antigen driven by the TRAP promoter is associated with hepatocyte neoplasia, dysplasia and increased apoptosis (unpublished observations).

In these TUNEL studies, we used commercially available TdT, biotin-labelled nucleotides and standard reagents for immunostaining. Nick-end labelling kits have been developed by several companies for the detection of apoptotic cells. These kits have the advantage that there is minimal reagent preparation required and the reaction product and times should be standardized for each experiment. It is more expensive per slide to use these kits than it is to purchase the reagents separately and they have the further disadvantage that the concentration of the reagents cannot be varied to examine the specificity and sensitivity of the reactions. Although the companies selling these apoptosis kits claim they have high specificity and sensitivity, their claims typically are based upon reproducibility of positive results with specific cells in specific circumstances, rather than on a comparison between nick-end labelling and standard morphological stains and criteria for detection of apoptosis.

We have also found a positive signal with TUNEL in the nuclei of cells that appear morphologically normal. For example, in bone sections from transgenic mice expressing large T antigen [22] in which less than 50% of morphologically apoptotic osteoclasts had a positive signal with TUNEL, we also detected a positive TUNEL signal in 20–30% of nuclei of cells in tendinous insertions. The nuclear chromatin in these cells was finely dispersed, the nuclei were not shrunk or contracted and the cytoplasm appeared normal. It seems most unlikely that, at the time of sacrifice of an otherwise healthy mouse, these cells could all have been in the early stages of apoptosis prior to nuclear condensation and fragmentation without many cells also being in the later stages of apoptosis. Furthermore, if these nuclear signals did indeed indicate a high rate of apoptosis, one would imagine that there would also be an equally high mitotic rate to maintain cell numbers. However, we did not observe any mitoses in cells lacking a positive TUNEL signal and, as far as we are aware, cells in these sites are long-lived and do not turn over. We do not know why so many of these cells have a positive TUNEL reaction, but speculate that we may be detecting cells with DNA damage which is repairable without the cell proceeding to apoptosis.

Detection of cells with DNA damage, rather than apoptosis, could also account for some of the recent reports of large numbers of TUNEL-positive osteocytes in sections of bone in which osteocytes appear otherwise normal. If large numbers of osteocytes in osteophytes or in bone from osteoporotic patients were undergoing apoptosis at the time of harvesting the specimens, it is likely that many of the other osteocytes would have undergone apoptosis previously, becoming shrunken and fragmented and disappearing from their lacunae, but this has not been observed in these recent reports. It could be argued that when osteocytes undergo apoptosis in their lacunae, they would remain there as detectable apoptotic bodies because these would be inaccessible to surrounding cells for phagocytosis – the typical fate of apoptotic bodies. However, we think it is more likely that they do eventually become undetectable by light microscopy due to progressive disintegration of the bodies into fragments that are too small to be detected under a light microscope. Empty osteocyte lacunae have been reported previously by Dunstan and colleagues [59] in samples of femoral heads from elderly subjects. The cause of the disappearance of the osteocytes from the lacunae in these elderly subjects was not examined by Dunstan and colleagues, but it is reasonable to speculate that they died by apoptosis.

A major advantage of purchasing the reagents used for nick-end labelling separately, rather than using a commercial kit, is that the incubation conditions can be controlled and altered, if necessary, should there be a suspicion of over- or underdetection of viable or apoptotic cells, respectively. The intensity of the staining reaction and the percentage of viable cells having a positive signal can vary depending upon a variety of factors, including concentrations of the proteinase K and TdT and the duration of the TdT incubation period. Furthermore, a positive signal can be detected in the nuclei of cells undergoing necrosis and these have to be distinguished from apoptotic cells morphologically. The typical recommended positive control for nick-end labelling reactions is the use of DNase which should give a positive signal in all cell nuclei because this enzyme splits all genomic DNA into oligonucleosomal fragments. We believe it is more important to use another positive control with each experiment, i.e. a section of tissue in which there are morphologically apoptotic cells and which has been used previously to establish optimal reagent concentrations and incubation times. This should help to reduce false positive and false negative reactions to a minimum and to avoid misinterpretation of staining reactions. This also illustrates the principle that, at present, standard light microscopic morphology should be used to validate techniques such as TUNEL, rather than vice versa.

We have no experience of the use of nick-end labelling techniques with flow cytometry for quantitation of apoptotic cells. However, from our reading of the literature and following discussions with investigators using this technique in disciplines other than bone, we believe that the problems of false positive and false negative staining are likely to be just as common with flow cytometry and TUNEL as they are in tissue sections. Flow cytometry has become a commonly used method for quantifying apoptotic cells, particularly in experiments designed to test the effects of drugs or compounds on cell survival *in vitro*. In addition to inducing apoptosis in some cells, many of these agents are likely also to cause reversible DNA damage to others and this damage may be detectable by nick-end labelling techniques. Investigators who use flow cytometry to detect cell death but who do not also examine their cells microscopically for the morphological features of apoptosis are likely to overestimate its occurrence.

5.4.5 Fluorescence techniques

Some fluorescent dyes can be used to detect apoptotic cells morphologically using fluorescence microscopy. There are several compounds that are used regularly (for example, Hoechst [74], acridine orange [17] and propidium iodide) and they all work by a similar mechanism. They bind to DNA and the intensity of the signal seen under fluorescence microscopy correlates with the density of the nuclear chromatin. Normal cells have finely dispersed, loosely packed chromatin which results in a light, granular fluorescent signal from the nuclei (Fig. 5.6). The signal detected in apoptotic cells differs from this and varies according to the stage each cell has reached during the apoptotic process. Thus, in cells in the early stages of apoptosis, bright fluorescence can be detected focally around the nuclear membranes corresponding to the sites of peripheral aggregation of chromatin. Later, intense fluorescence of condensed nuclear chromatin can be seen prior to nuclear fragmentation and disintegration of the cell (Fig. 5.6). The appearances are similar to those seen in the nuclei of cells stained with H&E and viewed under light microscopy.

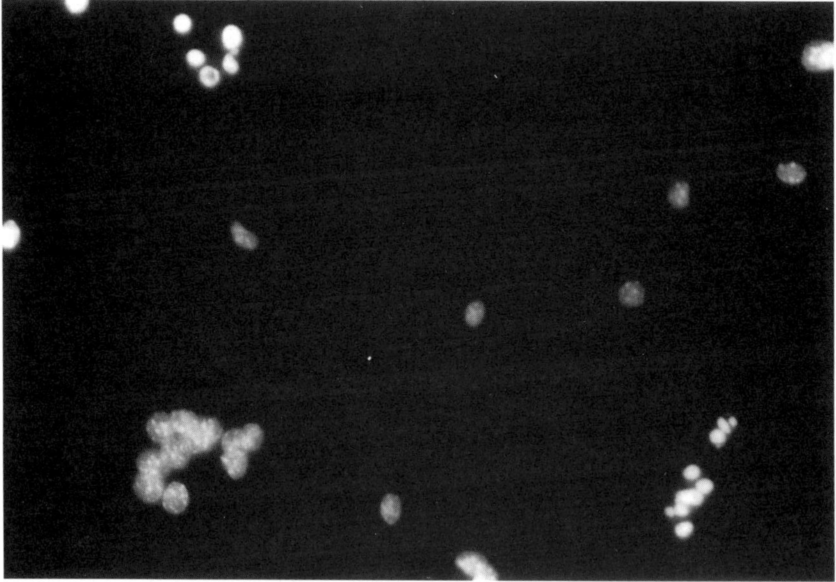

Figure 5.6 Hoechst staining of viable and apoptotic osteoclasts in culture. The characteristic dot-like pattern of fluorescence of viable nuclei is seen in the cluster of nuclei of a viable osteoclast in the bottom left-hand corner of this micrograph and in the nuclei of viable mononuclear cells elsewhere. In contrast, the uniformly bright fluorescence typical of apoptosis can be seen in the nuclei of apoptotic osteoclasts at the top left and bottom right. (Fluorescence microscopy, magnification ×845.)

The main advantage of these fluorescent techniques is that they permit ready identification of apoptotic cells based on only two morphological features, i.e. the intensity of the fluorescent nuclear signal and its distribution and shape. In cell culture preparations or tissue sections, the nuclear signal is much brighter than that seen in the cytoplasm and thus the number of apoptotic cells can be counted quickly and with relative ease.

These methods also have significant limitations. In our experience, Hoechst and acridine orange bind to a relatively small percentage of osteoclasts in the later stages of apoptosis in which nuclear condensation and fragmentation are readily detectable with haematoxylin counterstaining. In this respect, they behave similarly to the nick-end labelling techniques described above, but we do not know why binding fails to occur. We assume that the dye is unable to bind to nuclear chromatin in these circumstances, perhaps because the latter is too tightly packed or because something else binds to chromatin prior to its disintegration and prevents interaction. Whatever the mechanism involved, this failure to detect cells that are clearly apoptotic when stained with haematoxylin and viewed with light microscopy will result in underestimation of apoptosis.

Because there is a fluorescent signal in all viable nuclei, these techniques are unlikely to give the false positive signal that can be detected in some viable cells with nick-end labelling techniques. There is variability in the intensity of the fluorescence seen in viable cells and it is possible that nuclei with reversible DNA damage fluoresce more intensely than normal nuclei. We are unaware of any studies examining such phenomena, and in any event, provided that individuals using these methods define apoptosis on the basis of nuclear condensation and/or fragmentation, such cells should not be classified as being apoptotic.

Fluorescent microscopes are generally more expensive than light microscopes, but fluorescence attachments can be purchased along with the appropriate excitation filters and added to all but the basic microscope models in most manufacturers' ranges. They can be used in open multipurpose laboratories without the need to dim or switch off lights, provided that fluorescent signals are strong. In circumstances where it is necessary to observe weak fluorescence, they are best used in a room dedicated to this use and in which the lights can be turned off.

5.4.6 Detection of phosphatidylserine

During the early stages of apoptosis, phosphatidylserine is translocated from the inner to the outer surface of the plasma membrane, where it appears to be involved in signalling its demise to neighbouring cells and to professional phagocytes, which subsequently engulf it [75, 76]. Annexin V, a member of the Ca^{2+}-dependent phospholipid-binding annexin family of proteins, binds to phosphatidylserine and this property has been exploited to produce a new assay for the detection of apoptotic cells [77] that is available as a commercial kit. The assay appears to have been used predominantly with flow cytometry for detection of apoptotic cells cultured *in vitro*, but it can also be used on cells fixed on culture plates and on tissue sections with FITC or biotin-bound annexin V. We have no experience with this assay and do not know if it is more sensitive than nick-end labelling techniques for detection of cells in the later stages of apoptosis, when the cell contracts and breaks into apoptotic bodies.

5.5 SUMMARY

The study of apoptosis in bone cells, like that in most cell types, will be a growth area of research in the next few years. Like proliferation, apoptosis is a major determinant of the function of most cell types, and as such is likely to influence bone mass through its effects on the amount of bone that osteoblasts lay down and that osteoclasts remove during bone remodelling. Study of cell death in bone cells *in vivo* may be more challenging than that of many other cell types because the tissue is calcified and because cell numbers in bone and in bone remodelling units are low. Furthermore, the unavailability of osteoclast cell lines or of entirely pure populations of osteoclasts complicates study of osteoclast apoptosis *in vitro*. Nevertheless, existing techniques can be used not only to define the circumstances in which apoptosis of bone cells occurs, but also to examine the signalling pathways involved. The development of new methods that are applicable to the study of apoptosis in bone cells should enhance these efforts and, together with them, lead to a better understanding of bone cell turnover, of the mechanisms of action of drugs currently used to treat patients with bone diseases and possibly to the production of new drugs designed to maintain or increase bone mass by either promoting or preventing apoptosis of bone cells.

ACKNOWLEDGEMENTS

This work was supported by grants (AR43510, AR35929, CA40035 and DK45229) from the US National Institutes of Health, from the Center for the Enhancement of the Biology/Biomaterials Interface (CEBBI) and by an Eli Lilly/Medical Research Council of Great Britain Travelling Fellowship to DEH.

REFERENCES

1. Kerr, J.F.R., Wyllie, A.H. and Currie, A.R. (1972) Apoptosis: A basic biological phenomenon with wide-ranging implications in tissue kinetics. *British Journal of Cancer* **26**, 239–257.
2. Wyllie, A.H., Kerr, J.F.R. and Currie, A.R. (1980) Cell death: the significance of apoptosis. *International Reviews of Cytology* **68**, 251–306.
3. Zou, H. and Niswander, L. (1996) Requirement for BMP signaling in interdigital apoptosis and scale formation. *Science* **272**, 738–741.
4. Mooney, E.E., Ruis Peris, J.M., O'Neill, A. and Sweeney, E.C. (1995) Apoptotic and mitotic indices in malignant melanoma and basal cell carcinoma. *Journal of Clinical Pathology* **48**, 242–244.
5. Lee, J.M. and Bernstein, A. (1993) p53 Mutations increase resistance to ionizing radiation. *Proceedings of the National Academy of Sciences USA* **90**, 5742–5746.
6. Lowe, S.W., Ruley, H.E., Jacks, T. and Housman, D.E. (1993) p53-Dependent apoptosis modulates the cytotoxicity of anticancer agents. *Cell* **74**, 957–967.
7. Reed, J.C. (1994) Bcl-2 and the regulation of programmed cell death. *Journal of Cell Biology* **124**, 1–6.
8. Levine, B., Huang, Q., Isaacs, J.T. *et al.* (1993) Conversion of lytic to persistent alphavirus infection by the bcl-2 cellular oncogene. *Nature* **361**, 739–742.
9. Boyd, J.M., Malstrom, S., Subramanian, T. *et al.* (1994) Adenovirus E1B 19 kDa and Bcl-2 proteins interact with a common set of cellular proteins. *Cell* **79**, 341–351.

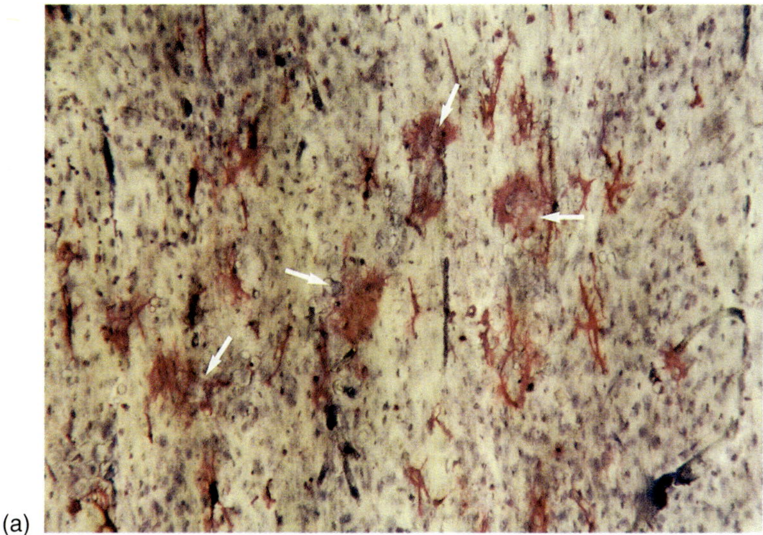

Plate 1 (a) A low-power photomicrograph, taken using transmitted light (×70), of bone marrow cells cultured on a bone slice in M-CSF. The strongly stained dendritic cells labelled immunohistochemically with monoclonal antibody 23c6 are large and partly cover osteoclastic excavations (arrows); the areas without osteoclastic excavations are devoid of 23c6-positive cells. (b) The corresponding bright-field reflected light photomicrograph emphasizes how closely the labelled cells are related spatially to the areas of bone resorption.

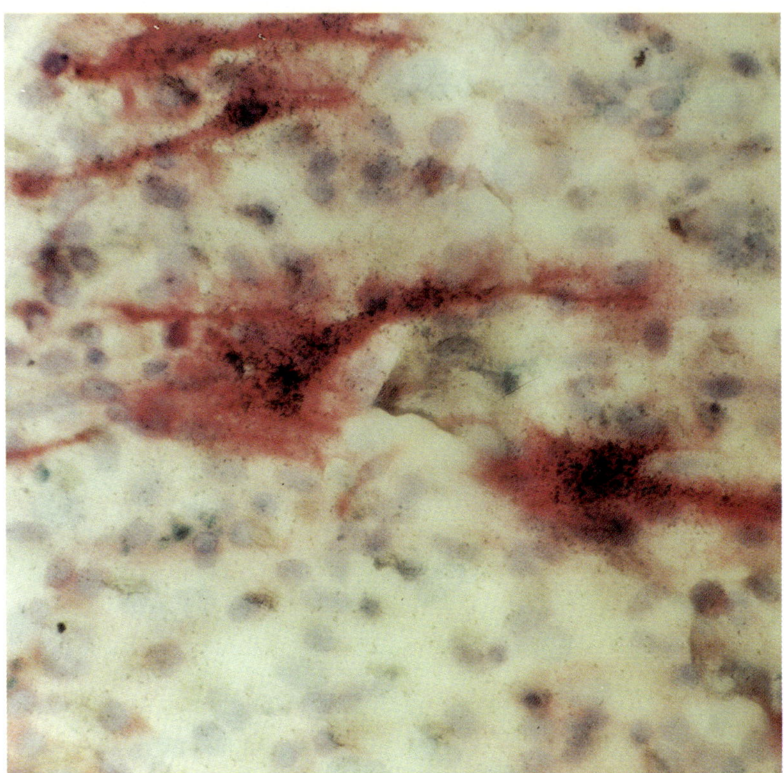

Plate 2 Autoradiograph of human bone marrow cells on a bone slice incubated with M-CSF (50 ng/ml) for 9 days before addition of ^{125}I-CT. The autoradiograph grains clearly label the strongly immunohistochemically labelled cells. Bone resorption is not evident because the bone slice was only weakly stained with toluidine blue, in order to emphasize the double-labelled cells.

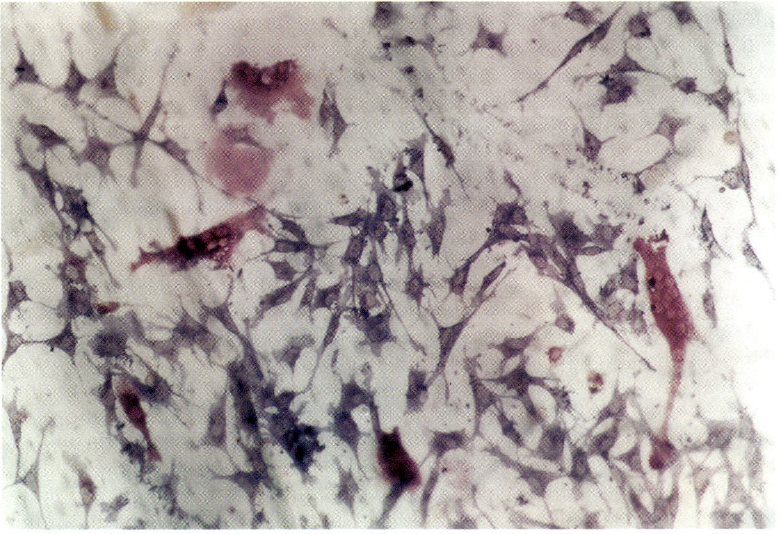

Plate 3 Osteoclasts (red) and osteoblasts (blue) as they appear on the bone slice at the end of the assay after double staining for TRAP and alkaline phosphatase. A pit is also visible, having stained weakly positive for TRAP, suggesting secretion of the enzyme onto the bone surface during resorption.

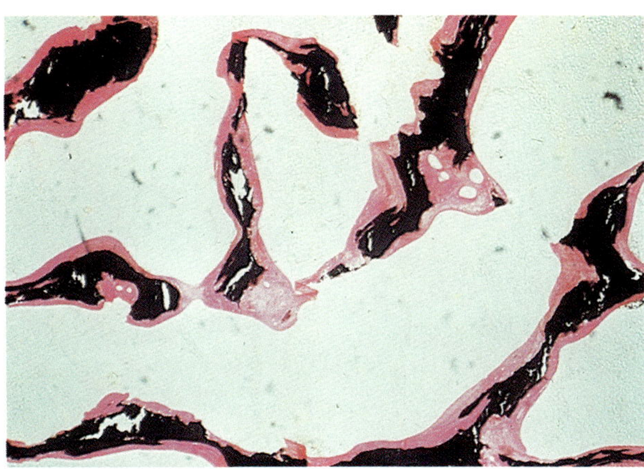

Plate 4 Section of iliac crest biopsy stained by the von Kossa technique to demonstrate mineralized bone (black) and osteoid (pink) in a patient with osteomalacia.

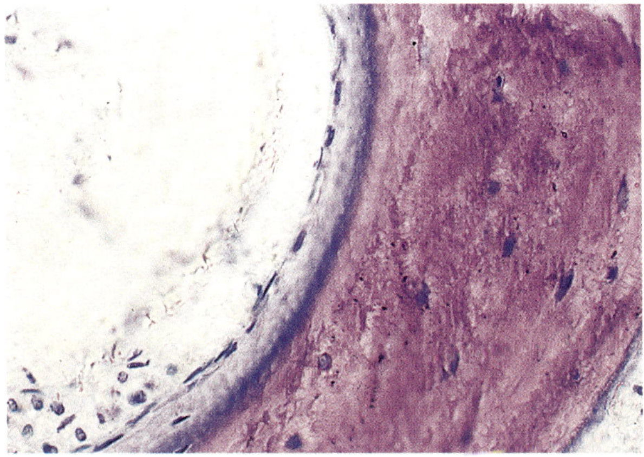

Plate 5 Toluidine blue-stained section of iliac crest cancellous bone showing mineralized bone (purple/blue) and osteoid (pale blue). The calcification front can be seen as a dark blue line at the interface of the osteoid and mineralized bone.

Plate 6 Image analysis system used for bone histomorphometry. The image of the biopsy section is captured by a CCD television camera onto a video monitor.

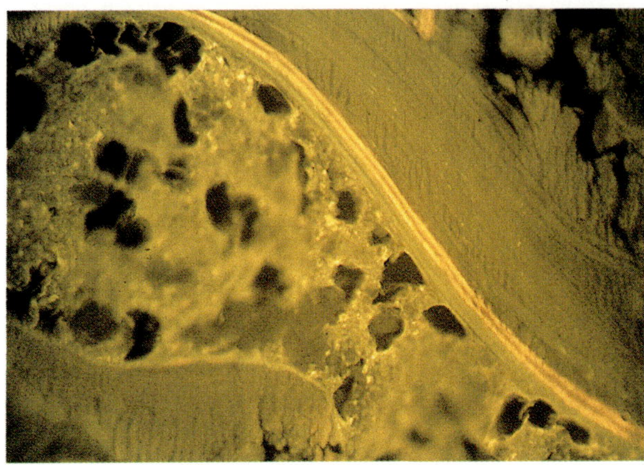

Plate 7 Unstained section of iliac crest cancellous bone viewed by fluorescence microscopy. The tetracycline labels are seen as double yellow fluorescent bands.

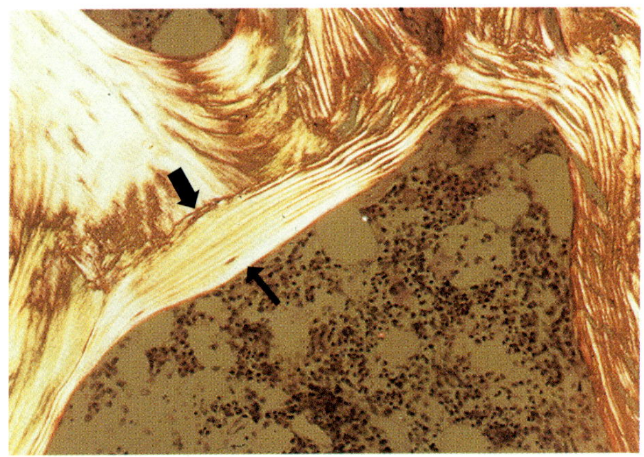

Plate 8 Section of iliac crest biopsy viewed under polarized light to show a completed bone structural unit bounded by the cement line (thick arrow) and mineralized bone surface (thin arrow).

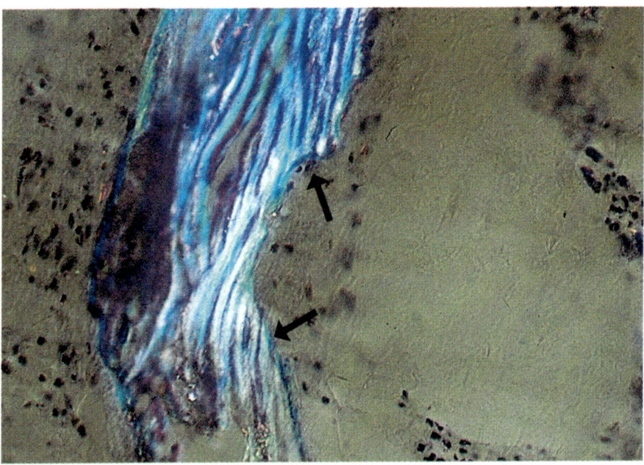

Plate 9 Resorption cavity viewed under polarized light. Note the cut-off collagen lamellae (arrowed) at the edges of the cavity.

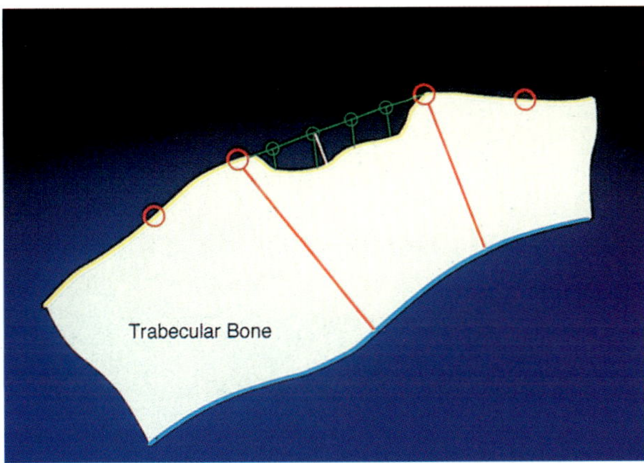

Plate 10 Measurement of erosion depth using a cubic spline to reconstruct the mineralized bone surface. Interactive graphics are superimposed over the image using the screen cursor controlled from a digitizer tablet. The two end-points of the cavity are identified using the cursor and two circles are drawn on the screen, centred on each end-point. The positions of the intersections between the circles and the trabecular surface are also entered using the cursor. A smooth continuous curve (cubic spline) is generated to pass between the four defined points. Mean erosion depth is measured as the mean distance from four approximately equidistant points (green lines). The trabecular width on either side of the cavity is also measured (red lines).

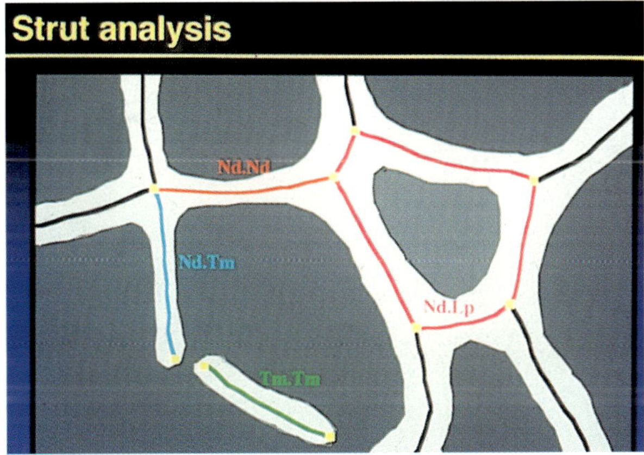

Plate 11 Diagrammatic representation of different strut types in strut analysis. Cancellous bone is shown in grey. Solid lines represent the skeletonized axis of the original bone profile. Solid squares represent nodes (Nd) and termini (Tm). Terminus-to-terminus (Tm.Tm), node-to-loop (Nd.Lp) and node-to-terminus (Nd.Tm) strut types are illustrated. (Reprinted with permission from *Journal of Bone and Mineral Research*, 1996 [54].)

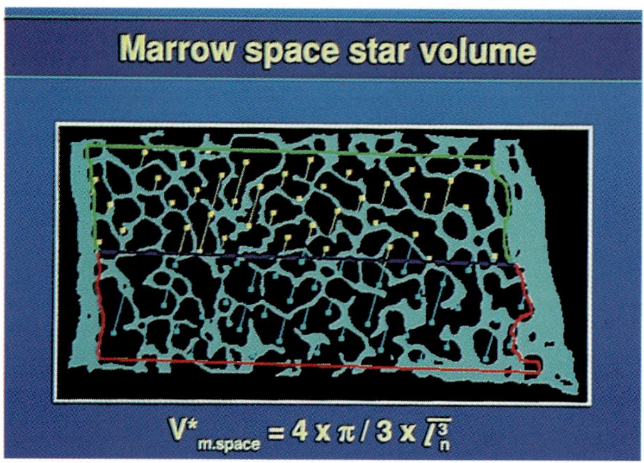

Plate 12 Binary image used in the assessment of marrow space star volume. The upper active region is defined by the green and blue lines and the lower region by red and blue lines. Blue and yellow solid squares represent grid points hitting the marrow space. Blue and yellow solid lines represresent grid lines intercepting trabeculae and the edge of the active region. The white arrow shows the direction of the vertical axis. C, cortex; CM, corticomedullary delineation. (Reprinted with permission from *Journal of Bone and Mineral Research*, 1996 [54].)

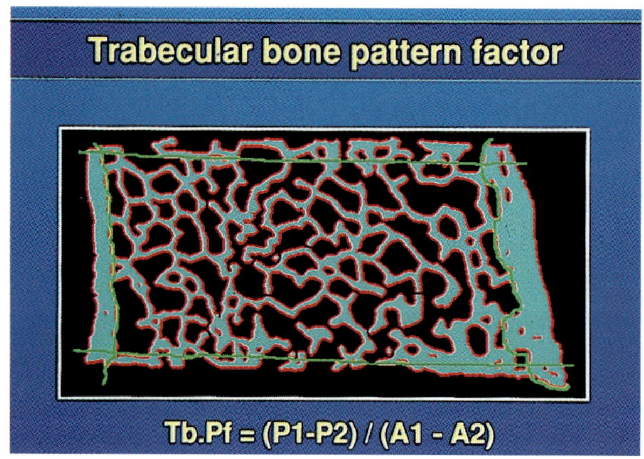

Plate 13 Binary image of cancellous bone used in the assessment of trabecular bone pattern factor. The active region is shown by the green line. The binary image is in grey and the dilated area is shown in red. (Reprinted with permission from *Journal of Bone and Mineral Research*, 1996 [54].)

Plate 14 Immunocytochemistry. Osteoclasts in cryostat sections (8 μm) of osteoclastoma (positive control tissue) and undecalcified human bone detected by (a)–(f) the alkaline phosphatase immunoenzymatic technique (sections are counterstained with Mayer's hematoxylin; (a)–(d), Fast Red TR post-coupling method; (e),(f), simultaneous coupling method) or (g) indirect immunofluorescence (counterstained with ethidium bromide). B, bone. The monoclonal antibodies C21, C22 and C35 are described in James *et al.* (1990) *Journal of Histochemistry and Cytochemistry* **39**, 905–914. (a) Reactivity with an osteoclast-selective Mab (C35).

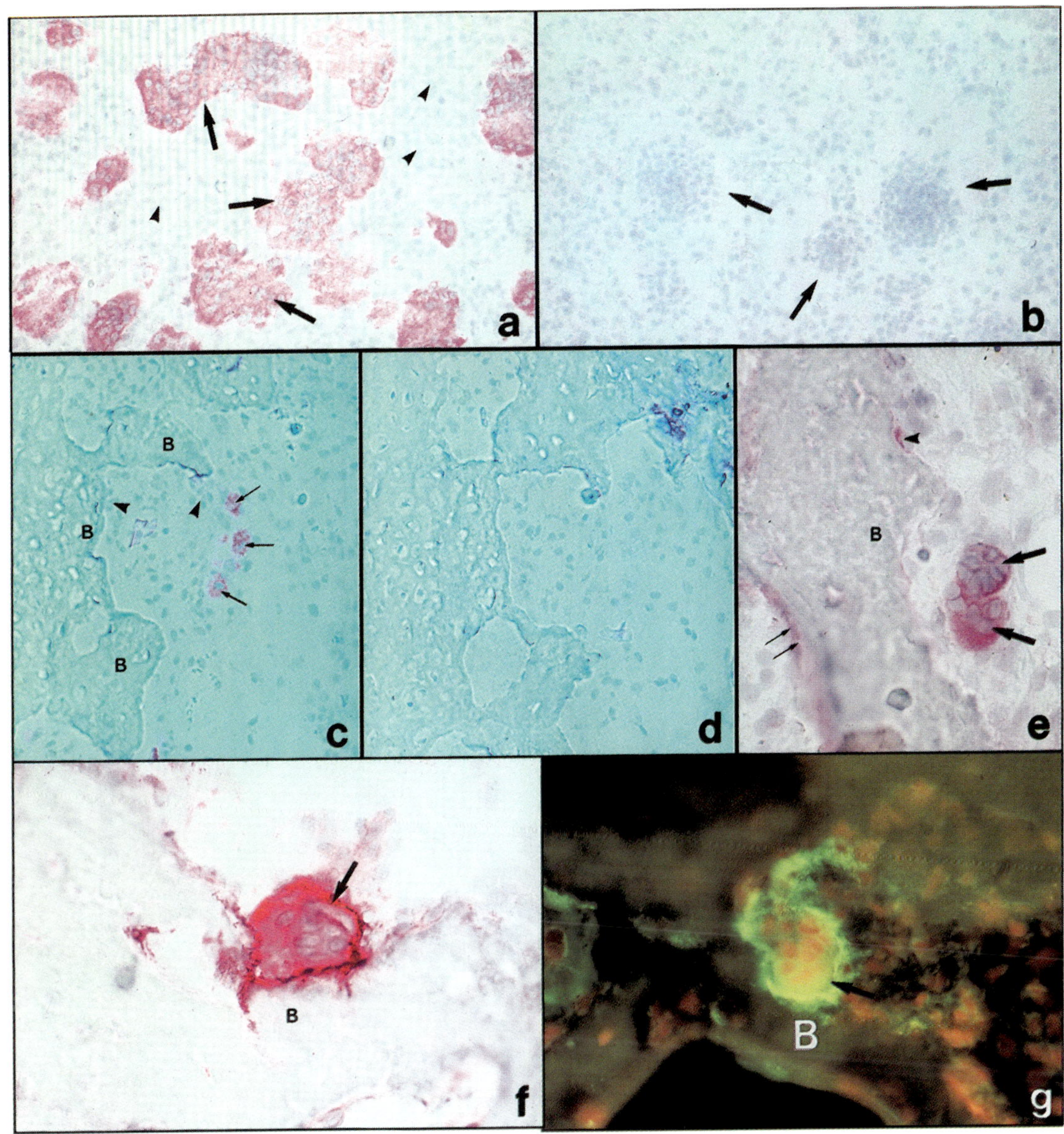

Plate 14 (cont) The cytoplasm of osteoclasts (arrows) and occasional mononuclear cell within human osteoclastoma tissue demonstrate positive red staining; stromal cells (arrowheads) are negative. (b) The serial section reacted with an isotype negative control shows no reactivity against osteoclasts (arrows). (c) A section of human osteophyte reacted with the Mab C35 demonstrates specific cytoplasmic reactivity against a cluster of osteoclasts (arrows); osteoblasts (arrowheads), osteocytes and marrow cells demonstrate no reactivity. Significantly, there is no apparent endogenous osteoblastic alkaline phosphatase activity. (d) The serial section reacted with an isotype negative control shows no reactivity against osteoclasts (arrows). (e) Osteoclasts (large arrows) within a resorption lacuna in a section of human osteophyte demonstrate cytoplasmic reactivity with C21, an osteoclast-selective Mab. Note the positive mononuclear osteoclast (arrowhead) and residual endogenous alkaline phosphatase activity within osteoblasts (small arrows) using the simultaneous method. (f) An osteoclast demonstrates a predominantly plasma-membrane reactivity with C22, a Mab reactive with the ß$_3$ chain of the vitronectin receptor. (g) Indirect immunofluorescence. An osteoclast (bright green fluorescencence, arrow) apposed to bone (dull green fluorescence, B) in a section of osteophyte demonstrates plasma membrane reactivity with Mab C22. Original magnifications: (a)–(d) ×50; (e)–(g) ×100.

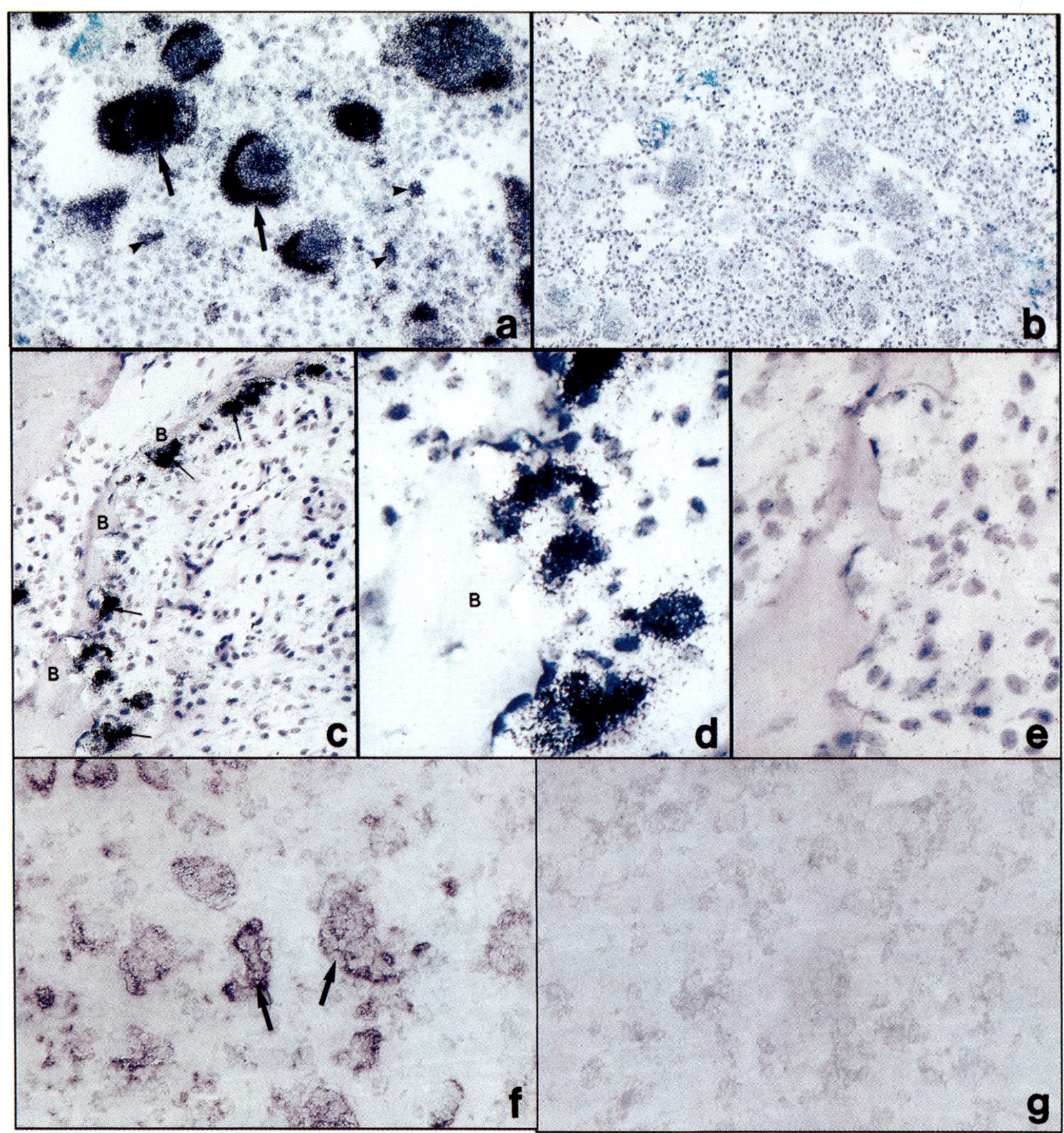

Plate 15 *In situ* hybridization. (a)–(g) ^{35}S-labelled and (h),(i) fluorescein-labelled riboprobes hybridized to sections of human osteoclastoma and undecalcified bone. Photomicrographs (a)–(e) are counterstained with methylene blue; (f) and (g) are counterstained with Fast Green. B, bone; arrows, osteoclasts. (a) Cathepsin K antisense probe on a section of human osteoclastoma. Osteoclasts and discrete populations of mononuclear cells (arrowheads) demonstrate strong cathepsin K mRNA expression. (b) Serial section probed for cathepsin B mRNA demonstrates negligible hybridization. (c) Cathepsin K mRNA expression in a section of human osteophyte. Osteoclasts resorbing or adjacent to bone demonstrate strong and selective cathepsin K mRNA expression. (d) High-power view of (c). (e) Serial section probed with the cathepsin K sense strand demonstrates no hybridization. (f) Osteoclast cathepsin K mRNA expression in a section of osteoclastoma hybridized with a fluorescein-labelled riboprobe. Compare the weak staining and poor tissue morphology with the excellent staining and morphology in (a). (g) The serial section hybridized with sense control. Original magnifications: (a),(c),(g) ×50; (d) ×125.

10. Henderson, S., Huen, D., Rowe, M. *et al.* (1993) Epstein–Barr virus-coded BHRF1 protein, a viral homologue of Bcl-2, protects human B cells from programmed cell death. *Proceedings of the National Academy of Science USA* **90**, 8479–8483.

11. Komiyama, T., Ray, C.A., Pickup, D.J. *et al.* (1994) Inhibition of interleukin-1β converting enzyme by the cowpox virus serpin CrmA. An example of cross-class inhibition. *Journal of Biological Chemistry* **269**, 19331–19337.

12. Alnemri E.S., Livingston, D.J., Nicholson, D.W. *et al.* (1996) Human ICE/CED-3 protease nomenclature. *Cell* **87**, 171.

13. Nagata, S. and Golstein, P. (1995) The Fas death factor. *Science* **267**, 1449–1456.

14. Steller, H. (1995) Mechanisms and genes of cellular suicide. *Science* **267**, 1445–1449.

15. Thompson, C.B. (1995) Apoptosis in the pathogenesis and treatment of disease. *Science* **267**, 1456–1462.

16. Fesus, L., Thomazy, V., Autuori, F. *et al.* (1989) Apoptotic hepatocytes become insoluble in detergents and chaotropic agents as a result of transglutaminase action. *Federation of Experimental Biology and Science* **245**, 150–154.

17. Arends, M.J., Morris, R.J. and Wyllie, A.H. (1990) Apoptosis: the role of the endonuclease. *American Journal of Pathology* **136**, 593–608.

18. Boyce, B.F., Windle, J.J., Reddy, S.V. *et al.* (1993) Targeting SV-40 T antigen to the osteoclast in transgenic mice causes osteopetrosis, transformation and apoptosis of osteoclasts. *Journal of Bone and Mineral Research* **8**, S118.

19. Fuller, K., Owens, J.M., Jagger, C.J. and Chambers, T.J. (1993) M-CSF suppresses osteoclastic apoptosis and switches function from bone resorption to migration/chemotaxis. *Journal of Bone and Mineral Research* **8**, S384.

20. Fuller, K., Owens, J.M., Jagger, C.J. *et al.* (1993) Macrophage colony-stimulating factor stimulates survival and chemotactic behavior in isolated osteoclasts. *Journal of Experimental Medicine* **178**, 1733–1744.

21. Williams, G.T., Smith, C.A., Spooncer, E. *et al.* (1990) Haemopoietic colony stimulating factors promote cell survival by suppressing apoptosis. *Nature* **343**, 76–79.

22. Boyce, B.F., Wright, K., Reddy, S.V. *et al.* (1995) Targeting simian virus 40 T antigen to the osteoclast in transgenic mice causes osteoclast tumors and transformation and apoptosis of osteoclasts. *Endocrinology* **136**, 5751–5759.

23. Lane, D.P. and Crawford, L.V. (1979) T antigen is bound to a host protein in SV40-transformed cells. *Nature* **278**, 261–263.

24. Lowe, S.W., Schmitt, E.M., Smith, S.W. *et al.* (1993) p53 is required for radiation-induced apoptosis in mouse thymocytes. *Nature* **362**, 847–849.

25. Liu, C.-C., Rader, J.I., Gruber, H. and Baylink, D.J. (1982) Acute reduction in osteoclast number during bone repletion. *Metabolic Bone Disease and Related Research* **4**, 201–209.

26. Liu, C.-C. and Howard, G.A. (1991) Bone-cell changes in estrogen-induced bone-mass increase in mice: dissociation of osteoclasts from bone surfaces. *Anatomical Record* **229**, 240–250.

27. Rowe, D.J. and Hausmann, R. (1976) The alteration in osteoclast morphology by diphosphonates in bone organ culture. *Calcified Tissue Research* **20**, 53–60.

28. Flanagan, A.M. and Chambers, T.J. (1989) Dichloromethylene bisphosphonate (C1$_2$MBP) inhibits bone resorption through injury to osteoclasts that resorb C1$_2$MBP-coated bone. *Bone and Mineral* **6**, 33–43.

29. Sato, M. and Grasser, W. (1990) Effects of bisphosphonates on isolated rat osteoclasts as examined by reflected light microscopy. *Journal of Bone and Mineral Research* **5**, 31–40.

30. Boyce, B.F., Aufdemorte, T.B., Garrett, I.R. *et al.* (1989) Effects of interleukin-1 on bone turnover in normal mice. *Endocrinology* **125**, 1142–1150.

31. Hughes, D.E., Wright, K.R., Mundy, G.R. and Boyce, B.F. (1995) Association of osteoclast apoptosis with loss of adhesion. *Journal of Bone and Mineral Research* **10**, S326.

32. Selander, K.S., Monkkonen, J., Karhukorpi, E-K. *et al.* (1996) Characteristics of clodronate-induced apoptosis in osteoclasts and macrophages. *Molecular Pharmacology* **50**, 1127–1138.

33. Wright, K.R., Hughes, D.E., Guise, T.A. *et al.* (1994) Osteoclasts undergo apoptosis at the interface between resorption and formation in bone remodelling units. *Journal of Bone and Mineral Research* **9**, S174.

34. Parfitt, A.M., Mundy, G.R., Roodman, G.D. *et al.* (1996) A new model for the regulation of bone resorption, with particular reference to the effects of bisphosphonates. *Journal of Bone and Mineral Research* **11**, 150–159.

35. Hughes, D.E., Wright, K.R., Uy, H.L. *et al.* (1995) Bisphosphonates promote apoptosis in murine osteoclasts in vitro and in vivo. *Journal of Bone and Mineral Research* **10**, 1478–1487.

36. Kameda, T., Ishikawa, H. and Tsutsui, T. (1995) Detection and characterization of apoptosis in osteoclasts in vitro. *Biochemical and Biophysical Research Communications* **207**, 753–760.

37. Kameda, T., Miyazawa, K., Yoshihisa, M. *et al.* (1996) Vitamin K_2 inhibits osteoclastic bone resorption by inducing osteoclast apoptosis. *Biochemical and Biophysical Research Communications* **220**, 515–519.

38. Lutton, J.D., Moonga, B.S. and Dempster, D.W. (1996) Osteoclast demise in the rat: physiological *versus* degenerative cell death. *Experimental Physiology* **81**, 251–260.

39. Arnett, T.R., Lindsay, R., Kilb, J.M. *et al.* (1996) Selective toxic effects of tomoxifen on osteoclasts: comparison with the effects of estrogen. *Journal of Endocrinology* **149**, 503–508.

40. Hughes, D.E., Dai, A., Tiffee, J.C. *et al.* (1996) Estrogen promotes apoptosis of murine osteoclasts mediated by TGF-β. *Nature Medicine* **2**, 1132–1136.

41. Hughes, D.E., Jilka, R., Manolagas, S. *et al* (1995) Sex steroids promote osteoclast apoptosis in vitro and in vivo. *Journal of Bone and Mineral Research* **10**, S150.

42. Horowitz, M.C. (1993) Cytokines and estrogen in bone: anti-osteoporotic effects. *Science* **260**, 626–627.

43. Pacifici, R. (1996) Estrogen, cytokines, and pathogenesis of postmenopausal ostteoporosis. *Journal of Bone and Mineral Research* **11**, 1043–1051.

44. Hughes, D.E., Wright, K.R., Mundy, G.R. and Boyce, B.F. (1994) TGFβ₁ induces osteoclast apoptosis *in vitro*. *Journal of Bone and Mineral Research* **9**, S138.

45. Takahashi, N., Yamana, H., Yoshiki, S. *et al.* (1988) Osteoclast-like cell formation and its regulation by osteotropic hormones in mouse bone marrow cultures. *Endocrinology* **122**, 1373–1382.

46. Oursler, M.J., Pederson, L., Fitzpatrick, L. and Spelsberg, T. (1996) Human giant cell tumors of bone (osteoclastomas) are estrogen target cells. *Proceedings of the National Academy of Science USA* **91**, 5227–5231.

47. Parfitt, A.M. (1994) Osteonal and hemi-osteonal remodeling: the spatial and temporal framework for signal traffic in adult human bone. *Journal of Cell Biochemistry* **55**, 273–286.

48. Parfitt, A.M., Villanueva, A.R., Foldes, J. and Rai, D.S. (1995) Relations between histologic indices of bone formation: implications for the pathogenesis of spinal osteoporosis. *Journal of Bone and Mineral Research* **10**, 466–473.

49. McCabe, L.R., Kockx, M., Lian, J. *et al.* (1995) Selective expression of fos- and jun-related genes during osteoblast proliferation and differentiation. *Experimental Cell Research* **218**, 255–262.

50. Kitajima, I., Nakajima, T., Imamura, T. *et al.* (1996) Induction of apoptosis in murine clonal osteoblasts expressed by human T-cell leukemia virus type I *tax* by NF-κB and TNF-α. *Journal of Bone and Mineral Research* **11**, 200–210.

51. Kitajima, I., Soejima, Y., Takasaki, I. *et al.* (1996) Ceramide-induced nuclear translocation of NF-κB is a potential mediator of the apoptotic response to TNF-α in murine clonal osteoblasts. *Bone* **19,** 263–269.

52. Bursch, W., Lauer, B., Timmermann-Trosiener, I. *et al.* (1984) Controlled death (apoptosis) of normal and putative preneoplastic cells in rat liver following withdrawal of tumor promoters. *Carcinogenesis* **5**, 453–458.

53. Lee, F.D. (1993) Importance of apoptosis in the histopathology of drug related lesions in the large intestine. *Journal of Clinical Pathology* **46**, 118–122.

54. Jande, S.S. and Bélanger, L.F. (1971) Electron microscopy of osteocytes and the pericellular matrix in rat trabecular bone. *Calcified Tissue Research* **6**, 280–289.

55. Tonna, E.A. (1972) An electron microscopical study of osteocyte release during osteoclasis in mice at different ages. *Clinical Orthopedics and Related Research* **87**, 311–317.

56. Aarden, E.M., Burger, E.H. and Nijweide, P.J. (1994) Functions of osteocytes in bone. *Journal of Cellular Biochemistry* **55**, 287–299.

57. Elmardi, A.S., Katchburian, M.V. and Katchburian, E. (1990) Electron microscopy of developing calvaria reveals images that suggest that osteoclasts engulf and destroy osteocytes during bone resorption. *Calcified Tissue International* **46**, 239–245.

58. Bronckers, A.L.J.J., Goei, W., Luo, G. *et al.* (1996) DNA fragmentation during bone formation in neonatal rodents assessed by transferase-mediated end labeling. *Journal of Bone and Mineral Research* **11**, 1281–1291.

59. Dunstan, C.R., Evans, R.A., Hills, E. *et al.* (1990) Bone death in hip fracture in the elderly. *Calcified Tissue International* **47**, 270–275.

60. Noble, B.S., Stevens, H., Loveridge, N. and Reeve, J. (1997) Identification of apoptotic changes in osteocytes in normal and pathological human bone. *Bone 20*, 273–282.

61. Lewinson, D. and Silbermann, M. (1992) Chondroclasts and endothelial cells collaborate in the process of cartilage resorption. *Anatomical Record* **233**, 504–514.

62. Roach, H.I., Erenpreisa, J. and Aigner, T. (1995) Osteogenic differentiation of hypertrophic chondrocytes involves cell divisions and apoptosis. *Journal of Cell Biology* **131**, 483–494.

63. Vortkamp A., Lee, K., Lanske, B. *et al.* (1996) Regulation of rate of cartilage differentiation by Indian hedgehog and PTH-related protein. *Science* **273**, 613–621.

64. Lanske, B., Karaplis, A.C., Lee, K. *et al.* (1996) PTH/PTHrP receptor in early development and Indian hedgehog-regulated bone growth. *Science* **273**, 663–666.

65. Amling, M., Neff, L., Tanaka, S. *et al.* (1997) Bcl-2 lies downstream of parathyroid hormone-related peptide in a signalling pathway that regulates chondrocyte maturation during skeletal development. *Journal of Cell Biology* **136**, 205–213.

66. Wyllie, A.H. (1994) Death from inside out: an overview. *Philosophical Transactions of the Royal Society of London* **345**, 237–241.

67. Olive, P.L, Wlodek, D. and Banath, J.P. (1991) DNA double-strand breaks measured in individual cells subjected to gel electrophoresis. *Cancer Research* **51**, 4671–4676.

68. Ostling, O. and Johanson, J.P. (1984) Microelectrophoretic study of radiation-induced DNA damages in individual mammalian cells. *Biochemical and Biophysical Research Communications* **123**, 291–298.

69. Galotto, M., Campanile, G., Robino, G. *et al.* (1994) Hypertrophic chondrocytes undergo further differentation to osteoblast-like cells and participate in the initial bone formation in developing chick embryo. *Journal of Bone and Mineral Research* **9**, 1239–1248.

70. Gavrieli, Y., Sherman, Y. and Ben-Sasson, S.A. (1992) Identification of programmed cell death in situ via specific labeling of nuclear DNA fragmentation. *Journal of Cell Biology* **119**, 493–501.

71. Gold, R., Schmied, M., Rothe, G. *et al.* (1993) Detection of DNA fragmentation in apoptosis: application of in situ nick translation to cell culture systems and tissue sections. *Journal of Histochemistry and Cytochemistry* **41**, 1023–1030.

72. Ansari, B., Coates, P.J., Greenstein, B.D. and Hall P.A. (1993) *In situ* end-labelling detects DNA strand breaks in apoptosis and other physiological and pathological states. *Journal of Pathology* **170**, 1–8.

73.   Wright, K.R., Story, B., Hughes, D.E. *et al.* (1995a) Standard morphology is more sensitive than TUNEL for identification of apoptosis in osteoclasts. *Journal of Bone and Mineral Research* **10**, S324.

74.   Arndt-Jovin, D.J. and Jovin, T.M. (1977) Analysis and sorting of living cells according to deoxyribonucleic acid content. *Journal of Histochemistry and Cytochemistry* **25**, 585–589.

75.   Fadok, V.A., Voelker, D.R., Campbell, P.A.S. *et al.* (1992) Exposure of phosphatidylserine on the surface of apoptotic lymphocytes triggers specific recognition and removal by macrophages. *Journal of Immunology* **148**, 2207–2216.

76.   Martin, S.J., Reutelingsperger, C.P.M., McGahon, A.J. *et al.* (1995) Early redistribution of plasma membrane phosphatidylserine is a general feature of apoptosis regardless of the initiating stimulus. Inhibition of overexpression of Bcl-2 and Abl. *Journal of Experimental Medicine* **182**, 1545–1557.

77.   Vermes, I., Haanen, C. and Reutelingsperger, C.P.M. (1995) A novel assay for apoptosis: Flow cytometric detection of phosphatidylserine expression on early apoptotic cells using fluorescein labelled Annexin V. *Journal of Immunologic Methods* **184**, 39–51.

CHAPTER SIX

Models for mechanical loading of bone and bone cells in vivo and in vitro

Timothy M. Skerry

6.1 INTRODUCTION

While the skeleton has many functions, the one that governs the shape and mass of each bone is the requirement to resist the effects of mechanical forces. When this requirement is not met, bones break. Since the disability associated with a broken bone affects the survival of an individual and therefore the species, there has been an evolutionary drive to minimize fracture incidence. Against this are the needs of efficiency, since an overengineered skeleton is excessively heavy to grow, maintain and use. The end result of these opposing requirements is that in each individual the mass of bone is adapted to habitual activity so that most normal and even abnormal loading incidents will not cause fracture.

Implicit in this ability to adapt to function is a response of bone to loading. This takes the form of an interaction of the effects of mechanical loading with other osteotropic stimuli, so that activity which in isolation might stimulate increased bone mass may, when combined with a drive towards resorption, only limit or moderate bone loss.

Because of the pervasive and potent influence of mechanical loading on bone mass, study of the effect of loading on the skeleton has been researched for many years. However, it is only recently that experiments have begun to shed light on the mechanisms behind the transduction mechanism of bone loading and its early intra- and intercellular consequences. This is the result of two separate factors. Firstly, long experience with different model systems has slowly allowed the refinement of experiments so that studies *in vivo* and *in vitro* can be accomplished without many of the problems that limited the scope of earlier work. Secondly, the application of sophisticated cellular and molecular biology techniques has led to the development of novel methods of investigation of mechanical responses.

The purpose of this chapter is to review models and methods of investigation of bone's loading responses. In order to do that, it is necessary (and interesting) to know a little of the early work in the field in order to put current developments into context. The importance of 'calibration' studies is emphasized in order to

Methods in Bone Biology. Edited by Timothy R. Arnett and Brian Henderson.
Published in 1997 by Chapman & Hall, London. ISBN 0 412 75770 2.

compare parameters between model systems. Considerations of space prevent the provision of fully detailed methods to replicate all the models described, but obvious pitfalls of different methods and models will be described, so that they can be avoided.

Because of the potent ability of mechanical stimulation to inhibit bone resorption and to stimulate formation, there is a largely untapped potential that research into the mechanisms involved may uncover novel targets for manipulation of bone mass. Such experiments could have profound clinical implications for treatment of the many diseases characterized by pathological changes in bone mass.

6.2 EARLY OBSERVATIONS OF FUNCTIONAL ADAPTATION

While there are allusions to the effects of different life styles on skeletal mass in Greek and Italian writing many hundreds of years ago, the beginnings of our understanding of the effects of loading on bone began in the late 19th century (reviewed by Roesler [1]). When the engineer Culmann saw drawings of trabecular arcades in the distal femur made by von Meyer, a German anatomist, he suggested that these were similar to the lines of principal stress in a curved model. Discussions between von Meyer, Culmann and Julius Wolff led to Wolff's Law, relating form to function, and the development of the idea of functional adaptation by Roux.

For nearly 100 years after that, there was little real advance over the concepts put forward by those workers. A large amount of phenomenological work catalogued differences in bone mass in individuals with differing activities, but advances in understanding of the mechanisms of the effect were limited by technical considerations.

6.3 LOADING AND STRAIN

One of the fundamental requirements of any loading study is that it should be clear what sort of stimulus is being applied to the cells. At the simplest level, it should be clear whether the applied forces cause physiological or pathological changes in the tissue. That is not to say that pathologically high levels of loading are invalid in studying bone formation, but that they almost certainly involve different processes from loading that has some close physiological counterpart. In order to determine these important fundamentals, it is necessary to understand the effect of loading of bone in a mechanical sense.

6.3.1 Bone strain

The way in which the effect of loading is measured is to describe the strain in the tissue. Strain is a simple ratio of the length before and after a load is applied – the deformation induced by a load (Fig. 6.1) Because strain is a ratio, it has no units. Compressive strains are usually given a negative sign, while tensile strains are positive.

Although preliminary investigations of strain in bone were performed in the 1940s [2, 3], the application of a strain gauge to the tibia of a live dog by F. Gaynor Evans in 1953 is regarded as the beginning of significant work on normal bone

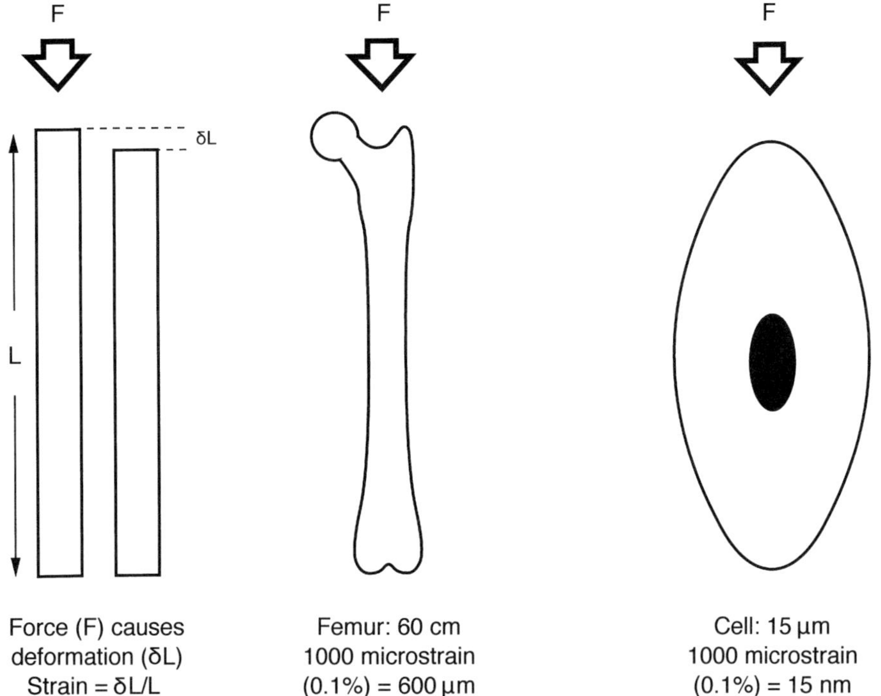

Figure 6.1   Strain is a ratio of the amount of deformation to the original length of a structure, and so it has no unit. Equal strains can be experienced by a bone and a cell, though the amount of deformation of one is greater than the other.

strains [4]. However, it was not until the availability of the cyanoacrylate super-glues over 10 years later that accurate bone strain measurements became possible. At that time in the late 1960s, Lanyon devised methods that made measurements of bone strain accurate, repeatable and routine [5–8]. These experiments were fundamental to current knowledge, as until then strains in bone had been calculated or measured *in vitro*.

Strains are measured by fixing strain gauges to the surface of a bone *in vivo* or *in vitro*. *In vivo* strain measurements have been made in a wide range of species, including humans, and the results are remarkably consistent. Most of the bones traditionally considered to be 'load-bearing' are deformed maximally by physiological activity by approximately 0.3% [9]. Strain in bone is more usually expressed in microstrain (strain $\times 10^{-6}$) so that physiological strains are up to ~3000 microstrain.

6.3.2 Effect of different attributes of strain

While the magnitude of strain applied to a bone is important (high strains are generally more osteogenic than low ones [10]), there are other important parameters of a loading regimen. *In vivo*, strain rates vary widely during different activities. Physiological activities generate strain rates in the human tibia between 10 000 (for example, during slow walking) and 120 000 microstrain/second when landing from a jump of over 1 metre [11]. Experimental data suggest that high strain rates (e.g. > 35 000 microstrain/second) are significantly more effective in stimulating new bone formation than low rates [12, 13].

Strain distribution or orientation is likely to be important in determining the effectiveness of a loading regimen *in vivo*. Naturally, the effects of non-habitual exercises which generate unusual strain distributions are more osteogenic than the same magnitudes of habitual strain distributions. There is also evidence that bone is more sensitive to high frequencies of strain than low ones, but the exact nature of this effect is not clear [14].

It is clear that the duration of loading has little effect once a threshold has been passed. In experimental studies, as few as 36 cycles of a highly unusual dynamic loading regimen have been shown to be effective in stimulating maximal new bone formation, but less unusual distributions of strain probably require somewhat longer stimuli [15].

Although most bone strains may reach 3000 microstrain as a result of physiological activity, there are some notable exceptions to this. The most important single site where strains are significantly lower is the calvariae of mammals, birds and humans, though strains in the tails of laboratory animals are similarly low. In those sites, strain magnitudes rarely exceed 50 microstrain with a rate of < 25 000 microstrain/second, unless there is a direct traumatic episode. The normal strains within a tissue should be considered when deciding on a loading regimen to mimic the effects of physiological loading, whether it is *in vivo* or in an explant or in cells.

6.4 EFFECT OF LOADING ON BONE

6.4.1 Cells affected by loading

Before describing the specific effects of loading on bone and the many model systems available, it is helpful to provide some knowledge of the sequences of events that follow or accompany loading of bone. This requires consideration of which cells may be affected either directly or indirectly.

It is probable that almost all bone cells stylized in Fig. 6.2 are affected indirectly by mechanical loading. If formation is to occur in response to loading, osteoblast precursors in the marrow differentiate and move to the bone surface, where, in addition, existing osteoblasts and lining cells will become activated to produce and mineralize matrix. In disuse, osteoclast precursor production in the marrow is stimulated and the cells migrate to the relevant site, fuse into active multinucleated osteoclasts and resorb bone. While either formation or resorption is proceeding, changes in bone loading have the ability to enhance, moderate or inhibit those processes as appropriate.

When the direct effects of loading are concerned, it is likely that the cells affected are the osteocytes and osteoblasts/lining cells [16]. This is because the ability to sense the effects of load in bone almost certainly requires close attachment to the bone. Logically, it is not unreasonable that osteocytes should be involved in the response of bone to loading. The cells are distributed throughout the matrix and communicate with each other [17, 18], and so they form an ideally located population of strain sensors, capable of providing information to the bone surface on the mechanical environment of a large region of bone. A significant body of circumstantial evidence supports this hypothesis. Osteocytes have a number of

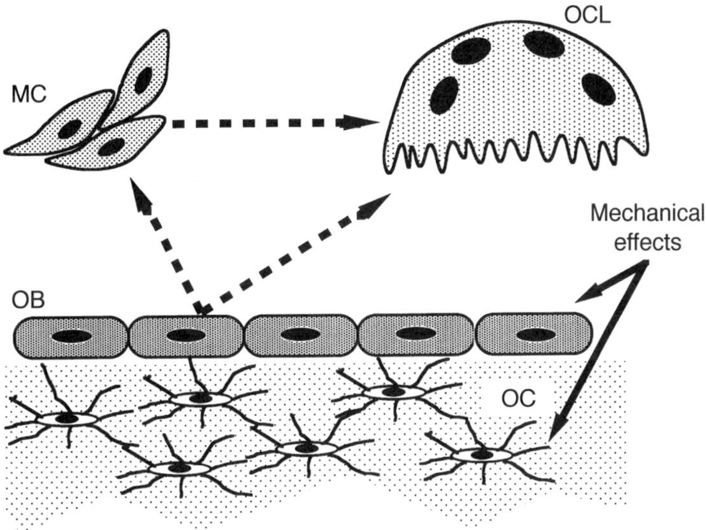

Figure 6.2 Bone cells affected by loading. Osteocytes (OC) and osteoblasts (OB) are affected directly by deformation of the bone matrix. Mechanical forces also influence recruitment and activity of osteoblast or osteoclast precursors in the marrow cells (MC), and can influence osteoclasts (OCL) directly.

responses to loading that are consistent with such a role [16, 19–22]. In addition, the orchestrated influence of loading on periosteal, endosteal and intracortical remodelling makes unlikely the possibility that it is controlled simply by bone surface cells.

While it is clear that loading or disuse can also alter marrow cell activity, both *in vivo* and *in vitro* [23, 24], it is not certain that this is a direct effect. Pressure changes within the medullary cavity during loading could affect marrow cells directly, but it is unlikely that these are high [25]. What seems more likely is that changes at the bone surface are involved in the formation, recruitment or attraction of marrow stromal cells to the sites where their action is needed.

6.4.2 Specific effects of loading on bone cells

When a bone is loaded cyclically, a sequence of changes occurs which can culminate in formation of new bone on the surface. The earliest change so far recorded *in vitro* in osteoblast-like cells 100 msec after loading is a rise in intracellular calcium in response to a single stretch cycle [26]. Within a few minutes, cascades of second messenger pathways, early response gene activation and enzyme activities have been observed [27, 28]. In addition, there is increased expression of nitric oxide and different prostaglandins [29]. *In vivo*, the earliest recorded change is a rapid increase in activity of the enzyme glucose 6-phosphate dehydrogenase (G6PD) in osteocytes, within 6 minutes of the start of loading [16]. Subsequent changes in mRNA expression in osteocytes are followed by changes on the bone surface which culminate in either inhibition of markers of bone resorption, or stimulation of formation, in association with cellular proliferation and early markers of matrix expression.

Since it is impossible to assess the effects of loading on bone strength except in long-term studies *in vivo*, it is necessary to calibrate *in vitro* and short-term *in vivo* studies so that their results can be viewed in a physiological context. Table 6.1 therefore details the sequences of changes seen after loading of bones, explants and cells, to highlight common responses.

6.5 MECHANISMS OF LOADING RESPONSES

While the fact of responses to load in bone is beyond question, the mechanism by which it occurs is not clear. Since numerous different models exist to mimic effects of loading by different means, it is helpful to consider briefly the numerous possible transduction mechanisms.

6.5.1 Deformation of cells

One reason that direct deformation of cells by loading has been held to be the stimulus for functional adaptation is that strain has been demonstrated to correlate well with the results of loading, whether in terms of area of matrix formed *in vivo* [30], or (for example) collagen mRNA expression *in vitro* [31]. It is not clear how cellular deformation stimulates bone cells, but numerous mechanisms involved in other early responses of cells to stimuli have been suggested [32, 33]. These include stretch-activated membrane ion channels [34], G protein-dependent pathways [32], and linkage between the cytoskeleton and phospholipase C or protein kinase C pathways [26]. It is entirely likely that loads which deform cells may also cause changes in fluid flow or electrical potential so that many of the proposed transduction mechanisms may be linked.

Table 6.1 Calibration parameters: sequential changes in bone cells and matrix after loading

Time after loading	Effect	System	Reference
100 /msec	↑ i/c calcium	c	[26]
5 seconds to 15 minutes	↑ phospholipase A_2 activation	c	[26, 27]
< 5 seconds	↑ protein kinase C activation	c	[26]
< 6 minutes	Matrix proteoglycan orientation changes	a,b	[78, 121]
6–20 minutes	↑ enzyme activity (G6PD, alkaline phosphatase)	a,b,c	[16, 20, 22]
5–15 minutes	↑ PG expression	b,c	[122, 123]
< 15 minutes	↑ NO production	b,c	[29]
1–18 hours	↑ gene expression (c-fos, TGFβ, IGF-1, type I collagen)	a,b,c	[28, 31, 82]
8–24 hours	↑ $_3$H-uridine incorporation	a,b	[20, 124]
48 hours	↓ TRAP activity, ↑ BrdU incorporation	a	[86] and Turner (personal communication)
48 hours	Osteoblast proliferation and matrix synthesis	a,b	[77, 90]
72 hours	↑ OC and type I collagen mRNA expression	a	[21]
> 96 hours	Mineralized bone formation	a	[30, 77, 81, 86]

a, *in vivo*; b, explant; c, cell culture.

6.5.2 Fluid flow and streaming potentials

It has been suggested that deformation of bone affects cells by means of fluid flow through the interstices of the cortex. This may be due to direct deformation of the cells by fluid movement as is thought to occur in endothelial cells [35], but it is harder to conceive of this happening in osteocytes, which are confined in three dimensions, than in endothelial cells anchored to a basement membrane and more free to deform. Surface osteoblasts may be more able to deform in response to fluid flow, but there does not appear to be any information on fluid flow in the periosteum. There is no doubt that intracortical fluid flow occurs in response to load [36], and that fluid flow models cause changes in bone cell activity [37], but the mechanism of fluid flow effects within bone is not clear.

Since the interstitial fluid in bone contains ions, this fluid flow has also been linked to streaming potentials – movement of charged particles past cells, inducing electrical effects [36]. For this reason, there has been considerable investigation of the effects of electrical stimuli of different kinds on bone remodelling and repair, and these experiments are often stated to be linked with mechanical effects, because deformation of bone induces electrical potentials. The study of electrical phenomena in bone is outside of the scope of this chapter but can be found elsewhere [38].

6.5.3 Hydrostatic pressure

Hydrostatic forces are those applied via a surrounding medium, such as air or liquid, and they act equally in all directions. Because of this, and because the cytoplasm of a cell is almost incompressible, hydrostatic forces experienced physiologically in bone do not induce deformation of cells which can alter remodelling activity if applied directly. This is quite different from other more compressible tissues, such as cartilage, which do experience significant hydrostatic pressures *in vivo* [39]. However, that is not to say that hydrostatic pressures are not capable of influencing bone cells.

Following studies to investigate the effect of hydrostatic compression on chondrocytes [40], a similar model was devised to apply compressive force to bone explants *in vitro*, as a result of gas pressure above the culture medium. Although it has been shown that shear strains are induced in the cartilage of the growth plate [41], the nature of the effect may be different from bone adaptation. It is clear that the proposed direct effects of compression on osteoblasts and osteoclasts [42] are unlikely to have physiological relevance. This is because the effects seen in culture are initiated by intermittent compression of 130 mbar at 0.3 Hz. For example, swimmers' bone would experience greater intermittent compressive forces than this during most strokes, yet swimmers have bone mass which is no greater than non-exercising control subjects [43, 44].

6.5.4 Effects of loads mediated through other tissues

Some model systems have been described for the study of responses of bone to loading, where loads are applied indirectly to the tissues. Examples of these techniques are the use of orthodontic devices to move teeth through alveolar bone *in vivo* [45] and the application of forces to calvarial sutures *in vitro* [46]. Tooth movement mediated by changes in periodontal ligament cells and bone resorption induced by stretching of sutures are valid subjects in the study of processes associated with formation or resorption, but the model systems have more in common with the response of bone to direct periosteal pressure than an adaptive change.

6.6 ALTERED LOADING MODELS *IN VIVO*

Methods of alteration of loading of bones *in vivo* fall into a number of groups, relating to the nature of the change in loading – whether it involves reduced or increased loading, the whole body or just one region, and whether the loading method involves surgical implantation of devices to allow loads to be applied or is a non-invasive method. Some models fit into more than one category, so that surgical preparation may lead to bone loss due to disuse, while loading of the functionally isolated segment will allow investigation of positive adaptive changes.

6.6.1 Osteotomy models

Local disuse can be induced by isolating a section of a bone from its extremities, so that loads are not passed through it. This procedure works effectively where pairs of bones are present together, so that the surgery on one does not completely prevent the use of the limb in question.

In early studies, Lanyon used osteotomy models to stimulate bone formation due to overload. In these experiments, the mid section of the ulna was excised. Since the radius and ulna share the loads on the forearm, this procedure increased loads on the remaining bone and stimulated a hypertrophic response. In both young pigs and adult sheep [47, 48], the removal of the ulnar shaft increased the strains on the radius during walking by over 20%. One year later, the radial cross-section had increased by approximately the area of the section of ulna removed, and the walking strains had fallen to less than those before surgery (Fig. 6.3). Similar studies in dogs later replicated these findings [49].

These studies were comprehensive. The strains were measured on the bones before and after the surgery and also after the bone formation had occurred, when they had returned to near normal levels. Naturally, the trauma of surgery complicates this type of methodology, and it is limited to quadrupeds with approximately equal sized bones in the forearm. Where there is a great disparity in the strengths of two bones in a limb, such as with the tibia and fibula, removal of the weaker has little effect; removal of the stronger prevents normal weightbearing for a prolonged time and therefore affects the experiment.

Osteotomy methods were also developed and used in the most widely practised of these models. One that has contributed significantly to the understanding of bone responses to load has been the isolated avian ulna model [50]. In this model, the mid section of the ulna of a rooster or turkey is isolated from the load path by cutting through the bone at the ends. Since the nutrient artery enters the ulna in this mid section, the procedure does not devitalize the bone. Pins are inserted into the ends of the free-floating segment (Fig. 6.4). These allow either protection of the segment from loads by external clamps, or the application of loads through the pins. Using this model, Rubin and Lanyon showed that over 6 weeks profound disuse bone loss was caused, by both endosteal resorption and incompletely infilled secondary remodelling. Static loads were shown to be ineffective in preventing that

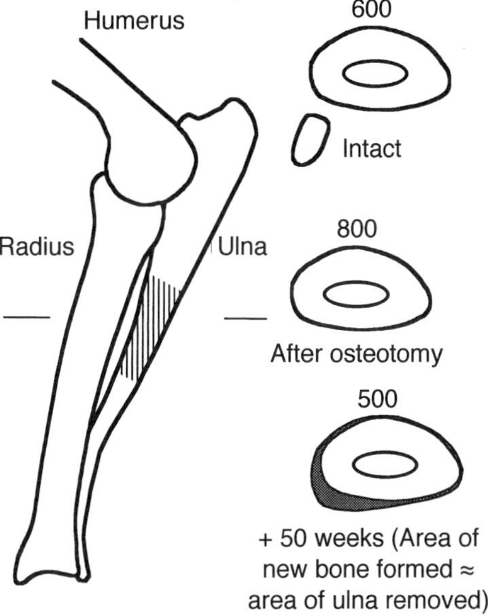

Figure 6.3 Osteotomy model to induce overstrains. Removal of the central section of the ulna (hatched area) increases strains during walking from 600 to 800 microstrain. One year later, new bone formed on the radius (shaded area) strengthens the bone, and strains during walking are 500 microstrain.

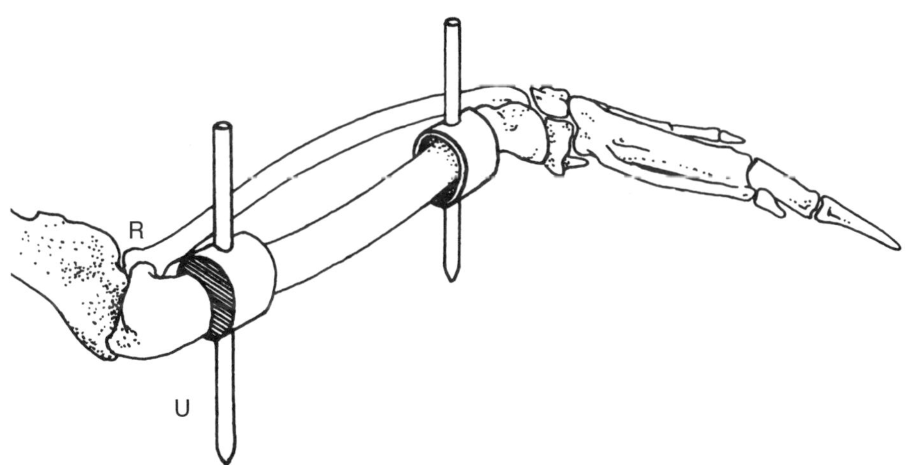

Figure 6.4 Isolated avian ulna model, allowing a large section of the ulna (U) to be isolated from loading, while the radius (R) continues to support the limb normally. If the pins emerging from the wing are fixed with external clamps, bone is lost due to disuse, while alternatively it can be loaded to stimulate formation.

loss, but small numbers of cyclical loads, if repeated daily, produced a strain-dependent increase in bone mass [10].

Osteotomy models have been informative in the past, but are technically demanding and use species that are not very convenient for use of standard antibody and molecular reagents. These models generally allow long-term effects of mechanical perturbations to be studied, but are relatively cumbersome for studies of more acute effects. For these reasons there is little use of these models now.

6.6.2 Casting/bandaging

One of the simplest disuse models involves casting or bandaging a limb to prevent normal loading. When rigid support is applied to a limb, there is loss of bone mass over the course of a number of weeks [51, 52]. Studies using these methods in dogs [51, 53, 54] have shown that in young animals, prolonged limb support causes reduction in growth and cortical thickness. In adults, bone loss occurs under the same circumstances, showing that both modelling and remodelling activity shift to a negative balance in disuse. Remobilization of the limbs after a year of disuse showed that net bone formation was induced by the return of loading, but after 6 months there was incomplete restoration of bone mass.

In rats, studies in which one hindlimb was bandaged to the body to prevent weightbearing [55] show similar progressive bone loss. It has been suggested that the non-bandaged leg is overloaded [56] and that the bone formation which results can be used as a model for adaptation to increased loading. That suggestion is reasonable but suffers from the same flaws as the disuse aspect of the model – that the difference in strain during the procedure is unquantified and rather variable in different individuals. In addition, there may be a brief period of increased bone mass during the initial stages of the immobilization, while the animal acclimatizes to the restrictions of the support. In response to bandaging (or sciatic neurectomy) of the rat hindlimb, measured strain magnitudes during the first day after the procedure drop by less than 10% of those in normal animals, although the distribution of strain is altered markedly (Skerry, unpublished).

However, the method is simple and cheap, though it requires close supervision. Some animals tolerate bandaging well, while others resent it greatly and remove casts and dressings with ease. Animals must be observed twice daily, and re-bandaged or re-cast as necessary.

6.6.3 Neurectomy

Other methods that cause local bone loss due to disuse include sectioning of a regional nerve trunk, or one or more tendons crossing a joint. The most common neurectomy model is to section the sciatic nerve in the hindlimb of the rat as it passes round the greater trochanter of the femur, a process that stimulates a rapid progressive loss of bone from the tibial metaphysis [57, 58]. Neurectomy induces rapid bone loss, which is detectable within 30 hours of surgery and highly significant within 72 hours [59]. In very young rats that are not weaned, the procedure is effective [60] and provides a good model for reduced bone formation and increased bone resorption.

Neurectomy is simple to perform and is well tolerated. The nerve is easily identified. After surgery the gait is altered, as the extensor muscles of the hock joint are inactive and rats walk with the foot flat and twisted out. There is also a lack of sensation on the outer side of the foot. Occasionally, phantom sensations in the foot can cause self-mutilation, but this is relatively infrequent. Sectioning of nerves to the forelimbs is much less effective in inducing bone loss (A.G. Torrance, personal communication), despite similarly effective alterations in gait.

Tenotomy of the straight patellar tendon/ligament has similar effects [58, 61], both in the time scale and magnitude of bone loss, but does not appear to offer significant advantages.

6.6.4 Generalized bone loss

A more radical method of reducing loading of the entire skeleton is to fly animals in space [62]. During orbit, free fall induces weightlessness, and the bones of both experimental animals and human space pilots have shown the dramatic consequences of this generalized disuse. Very few researchers have access to material from these experiments. The forces induced at landing provide an unquantifiable but opposite stimulus to the disuse induced by the weightlessness.

In order to mimic the effects of weightlessness, rats can be suspended by their tails, so that the hindlegs are lifted clear of the ground and therefore relieved of normal loading [63]. This model has been used extensively in the United States, but it can be hard to obtain Home Office licence for its use in the UK. The system is expensive, as it requires individual accommodation for each animal due to the height of the mast from which the animal's tail is suspended.

6.6.5 Increased loads – whole body

It is possible to increase loading of individual bones or the entire skeleton by exercising the subjects. Studies of this sort include methods where human subjects [64–67] or experimental animals [68, 69] are exercised on tracks, on treadmills, in water, or by specific sport programmes. In general these studies have shown that exercise programmes have a beneficial effect on the skeleton, either by stimulating increased bone mass or by inhibition of bone loss. Perhaps the more important result of these studies is not the information they reveal about the mechanisms of loading responses (which is, by the nature of the studies, fairly limited) but the nature of exercises that are most effective in elevating bone mass or reducing loss in humans.

These studies are usually limited to long-term investigations, as the early effects of the exercise regimen are usually variable between individuals to such an extent that rapid changes are not discernible above individual variation. In the case of human studies, results are limited to parameters that can be measured by non-invasive means, or by methods that are tolerated both by the subjects and by the ethical approval body concerned. This varies considerably in different countries.

In human studies, compliance can be a problem. Animal studies are often very time consuming, as the subjects have to be trained to perform the activities. In many cases, groups of animals self-select into those that will and those that will not learn the activity successfully. It is not safe to assume that those groups are equivalent, and so in order to perform a meaningful study the subjects for both control and exercised groups should be taken from an exercise-compliant subset of the original population.

6.6.6 Regional alterations in physiological activity

Where some form of asymmetrical exercise is performed, loads may be increased on one bone in the skeleton. This has the advantage that the opposite bone can be used as a control, and increases the power of the methodology considerably.

The simplest of these studies have been human studies where the bone mass of tennis players has been investigated [70]. Radiographic measurements revealed that the cortical thickness of the humerus of the dominant arm was 30% thicker than that of the opposite one. More recently, it has been shown that these differences are achieved if vigorous loading exercises begin before skeletal maturity. Where loading starts after most of the skeleton has formed, much less difference is seen [71]. Longitudinal studies of human responses are possible, but are often protracted.

While most human studies are of long duration and are non-invasive, there is one exception. In a somewhat extreme experiment performed on humans, the tibial cortices of six volunteers were microdialysed via implanted catheters [72]. In response to a 5-minute period of heel lifts with hard impacts on a solid floor, there was a 2.5 to 3-fold increase in expression of PGE_2 for over 8 hours after the end of the exercises. While this result is not surprising in the light of numerous previous animal studies, it serves as an illustration of a calibration between models and perhaps more importantly between experimental model systems and human physiology.

6.6.7 Applied mechanical loading via implants

Modern studies on the effects of alterations of measured applied loading on bone mass began in the 1960s with work by Hert, who devised a method for application of mechanical loads to the tibiae of rabbits *in vivo* [73]. Pins were implanted across the cortices of the bone, protruding from the skin, so that static and dynamic forces could be applied to them and therefore the bone on a daily basis. These studies provided the basis for most of our current understanding of the effects of loading on bone, as they preceded even the studies by Lanyon (section 6.6.1). In a series of studies, Hert and his co-workers showed that dynamic but not static loads were effective in altering bone mass [74], that functional responses are independent of peripheral innervation [73] and that tension and compression are equally effective in influencing bone mass [75]. While these studies were flawed because bone strains were calculated, not measured, they form the starting point to other methods of applied loading.

Since the studies by Hert, several model systems have used similar methods to apply loads to bones. In studies in sheep [76] and in birds [12], loads applied via pins in otherwise intact bones were shown to induce bone formation which was dependent on the magnitude and rate of the loading. In an attempt to remove the uncontrolled loads applied by the use of the bone by the animal, Lanyon's group developed the functionally isolated avian ulna preparation (section 6.6.1), which has since been used for more acute experiments showing the early cellular responses to loading [77], and the matrix and cellular changes that are induced early in the response to loading [16, 78] (section 6.4.2). Since then, use of those models has diminished considerably as more convenient alternatives have been developed.

One of the difficulties with *in vivo* models in sheep and birds is that the availability of reagents for those species is very limited in comparison with rodents. For these reasons, rat models for mechanical loading have been developed in several laboratories [79–81]. Of these, two are non-invasive, requiring no surgery, but the model developed by Chambers requires surgical placement of pins into two vertebrae in the tail of a rat. Loads applied to those pins are used to compress the vertebra between those containing the pins, and to induce bone formation accordingly. While inflammatory changes may colour the results of these studies, the model has shown similar responses to loading in other earlier experiments [77, 82], inducing bone formation and IGF-1 mRNA expression after a single loading period [21]. Despite some drawbacks with this model, it does offer the possibility of studying cancellous bone responses to loading, which is not easy with other rat models.

With all surgical methods there is significant induction of surgical trauma, which is likely to cause the regional acceleratory phenomenon (RAP) leading to stimulation of local bone formation as a result of trauma or inflammation. The effects of this are likely to be more significant in short-term studies than long ones, where the tissues sampled are close to the points of application of force, and in the rat, in which the effect is very pronounced. Attempts to minimize these problems are made by fixing the pins at the extremities of the long bones, or in adjacent vertebrae to the one studied in the case of the rat model. While these precautions minimize the effects of trauma or vascular disturbance, it is unlikely that they overcome them completely, and the possibility that some inflammatory stimuli are involved in responses must be considered.

6.6.8 Non-invasive applied loading

In order to overcome the problems of trauma associated with surgical methods of preparation of bones for loading *in vivo*, two groups devised methods to load rat bones non-invasively *in vivo*. One of these involved application of four point-bending forces to the tibiae of rats, by cushioned pads pressed against the leg of an anaesthetized animal [79]. Because the leg rests on two pads spaced 11 mm apart, and loads are applied via pads spaced 8 mm apart, the tibia is bent, and an area between the inner pads is not directly affected by periosteal pressure. This method is in fairly widespread use, and has been shown to confirm many responses shown in other systems [83–85]. In addition, the model has been used to study changes in periosteal gene expression after brief periods of loading [82], showing rapid increase or induction of c-fos, TGFβ and IGF-1, with reductions in osteopontin, osteocalcin and alkaline phosphatase mRNA expression. Some questions have been raised regarding the sites of application of pressure and the possible induction of inflammatory responses, and it has been suggested that the endosteal responses of the model represent the best controlled functionally adaptive response.

The second non-invasive method of loading rat bone *in vivo* was devised by Torrance *et al.* [81] (Fig. 6.5). This involved the application of compressive forces between the carpus and elbow of the rat, which caused the already curved bone to bend more. In a series of studies, this model has been used to characterize responses to different magnitudes and rates of loading, and to study the reduced effectiveness of loading in ovariectomized animals compared with intact females [13]. Because the sites of application of load are well away from the sample region, there appears to be insignificant complication with inflammatory responses. In old animals loading is capable of inducing bone formation on a quiescent surface [81], but in younger rats, where one periosteal surface undergoes resorption as part of a modelling drift, the loading induces a sequence of changes, beginning with inhibition of formation and culminating with formation, mimicking the events of the reversal phase of remodelling [86]. Because of the very specific zones of response

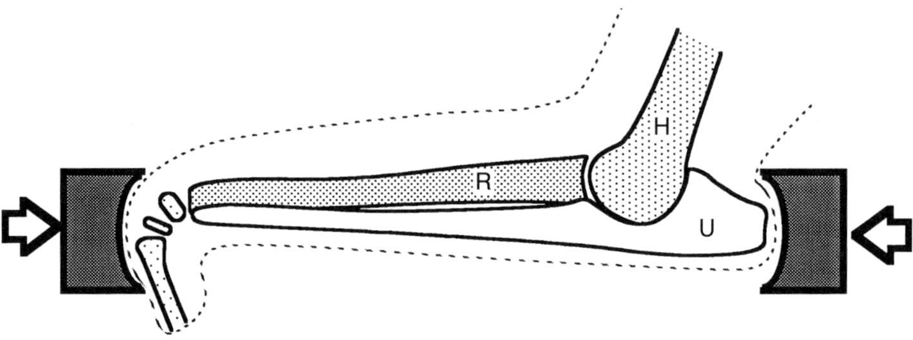

Figure 6.5 Rat ulna loading model. Loads are applied to the flexed carpus and elbow, causing the curved ulna (U) to bend more. The sites of load are distant from the midpoint of the bone, where responses are measured, and the loads necessary to cause mild overstrains are low, and do not cause detectable inflammatory effects at the sites of the compression. R, radius; H, humerus.

to loading in this model, it has been possible to use it to determine the effects of loading on osteocyte gene expression and has led to novel discoveries of regulated excitatory amino acid transport by bone cells in response to loading [87, 88]. Subsequent studies have also shown that if the bone is fatigue loaded on a single occasion, instead of loading within the physiological range, it is possible to induce intracortical Haversian remodelling in the ulnar cortex, providing a suitable rodent model for that process [89].

6.6.9 Specific problems with *in vivo* models

It is important to differentiate between the effects of physiological loads and those that would not be experienced normally. This requires that some measurement is made of the strains experienced by a bone *in vivo*. Although in ideal circumstances it is preferable to make measurements of strains induced in the particular region of the specific bone to be used, that would influence any remodelling responses and so strain measurements should be made in bones of equivalent dimensions to those that will be loaded. *In vivo* strain measurement is beyond the scope of this chapter; it is relatively specialized and is best performed in collaboration with one of the groups experienced in these techniques.

The limb bones of most species (including humans, many other mammals and birds) experience approximately 1000 microstrain during moderate activity, and extremes of physiological activity may induce 3000 microstrain. If a model is to be used which involves sampling for somewhere other than the shaft of a long bone, such assumptions are very unsafe. For example, in the skull and tail vertebrae of the rat, normal strains are less than 50 microstrain [11], so that loads inducing many hundreds of microstrain could not be considered physiological. In such cases, *in vivo* strain measurements in a number of animals are a prerequisite for meaningful experiments.

In models where surgery is performed, trauma can be associated with new bone formation which has no connection with the adaptive response. Even non-invasive models can suffer from this problem to a lesser extent. To control for this, sham surgery should be performed to demonstrate the effect of the preparation method without the loading, or with loading which does not induce dynamic strain change. A useful control would be to apply a static compressive load for the time that dynamic loading is applied in the experimental groups.

Where models do not deprive subjects of the ability to use the bone in question normally, outside of an experimental period of loading, then the effects of the subject's own use must be controlled. This is important in humans and animals where locomotion is unrestricted for most of the time. If there are significant differences in the non-experimental use of the bone between groups, the results will be affected. This point raises another associated issue. Where experimental groups, particularly in animal studies, are acquired and housed for a study, their previous exercise history must be considered. If a group of fit, active, non-cage-housed animals are brought into a laboratory, then they may undergo some degree of disuse-related bone loss as a result of this change in activity. To overcome this, it is necessary to acclimatize animals to the experimental environment. The time this will take depends on the species, age, previous activity and current activity of the groups. In the case of large mammals, particularly older ones, this can be protracted, and preliminary studies may be necessary to establish the time taken for bone mass to reach a plateau, if a new model is being established.

The influence of other osteotropic influences on exercise/load-induced changes in bone must also be considered. Of these, hormonal status, nutrition, day length, diet and concurrent illness are major points.

While it is undesirable to alter hormonal status in many cases, ovariectomy may reduce the spread of results where cycling in females is profoundly different from the target species. An example of this is in the sheep, where prolonged anoestrous is associated with low circulating oestrogen for many months. If experiments are to be performed on non-ovariectomized animals, then it is important that different groups start and end at the same time of the year, unless complete environmental control of day length and temperature is available.

Even mild concurrent illness has the ability to influence bone mass, as a result of the systemic acute phase response to local inflammation. Changes in circulating acute phase proteins and cytokines have known effects on bone metabolism, underlining the necessity for regular observation and good standards of animal care. Where infection or inflammation occur at or near the experimental site, results should be treated with great caution.

6.7 *IN VITRO* MODELS

Because of the numerous difficulties associated with animal models of bone loading, and because there are many questions which it is either difficult or impossible to answer with those methods, there has been a profusion of methods devised to mimic those effects *in vitro*. These fall into two categories: explant methods and cell culture methods. While these methods offer many clear advantages over *in vivo* studies, they have problems of their own, both in methodology and interpretation of data. The primary need for any *in vitro* study is that it should be calibrated in relation to other systems, so that the results can be linked realistically to the processes which, *in vivo*, lead to increased bone strength. This is important because nearly all cells have responses to mechanical stimulation of some sort, and it is not safe to assume that all consequences of deforming or loading of bone cells are associated with the functionally adaptive response *in vivo*. Unless it can be demonstrated that the regimen used to stimulate the cells induces some changes common with other systems, particularly those *in vivo*, interpretation of data is very hard.

6.7.1 Loaded explants

Experiments to determine the effects of loading on bone explants require rigorous characterization in order to determine that the culture conditions maintain the cells in the tissue in a viable and therefore responsive state. It is usual that following removal from the animal, tissues are washed and then placed in culture, either at a medium/air interface or immersed in medium. The responses of the cells must be studied at regular time points after explanting in order to determine the length of time for which the system remains viable. This is necessary so that experiments can be conducted in the window available after the cells have equilibrated to their new conditions, and before they begin to die or respond abnormally to stimuli. Suitable assessment criteria might include morphology of cells assessed by electron microscope, and various markers of function such as lactate dehydrogenase activity and PTH-stimulated cAMP expression. Morphological signs that the tissue is of poor viability include pyknotic nuclei and empty osteocyte lacunae.

Several models have been developed and used over recent years. One relatively easy criterion to determine whether the system has significant difficulties is to examine the publications relating to that model. Where the model is used for studies in a single publication, and not again, or where use is confined to a single laboratory, usually the one where it was devised, there may be easier systems to use.

If a model system is capable of maintaining viable cells in a section of bone, then it is necessary to apply the same rigorous criteria to the modes of loading of the bone as were discussed in the context of *in vivo* studies. Strains applied to the tissues should mimic those experienced *in vivo*, and ideally the loads applied should be related to measured strains in equivalent explants. The advent of miniature strain gauges allows measurement of strain in very small bones, and such data have considerably more credibility than calculated tissue strains.

Some of the problems of *in vitro* explant models for loading are illustrated by the studies of Lozupone *et al.* [90, 91]. In this model, the effects of loads calculated to be double those experienced *in vivo* were studied on 18-day-old rat metatarsals. Unfortunately, in the non-loaded control bones 65% of the osteocytes were non-viable, suggesting that the culture method was not suitable, even though loading increased the numbers of viable cells.

Better viability was seen in a model in which 17-day-old chick tibiotarsal bones were cultured and loaded [20, 31]. In response to defined strains in the bone, increases were detected in enzyme activity and in type I collagen mRNA synthesis, in common with other studies *in vivo*.

To study responses of cancellous bone to loading *in vitro*, samples were taken from the distal femora of adult dogs [92, 93]. After washing and equilibration, they were loaded in a system in which medium was perfused through the tissue. Some of the same changes that had been demonstrated previously *in vivo* were confirmed in response to calculated bulk strains, and it was also shown that cAMP and prostaglandin E2 were released into the medium as a result of loading. In a later study, samples of media from those studies were analysed and shown to contain high levels of nitrite, consistent with nitric oxide synthase (NOS) induction by loading [29]. Difficulties with this model are connected with the samples of bone used. In the studies published, impounded stray dogs of differing age, sex and history were the source, and the experiments would have been prohibitively expensive if samples were removed from laboratory bred animals.

In another explant model, the ulnae were cultured from young rats (<100 g body weight) and loaded *in vitro* [94–96]. Because the same bones can be loaded *in vivo* [86], the model has the advantage that direct comparisons can be made between the two systems. In this model, loading has several similarities with other studies: namely, increased G6PD activation, alkaline phosphatase and NOS activity and prostaglandin expression [29, 96].

6.7.2 Hydrostatically compressed explants

Numerous experiments have used intermittent hydrostatic compressive forces to study their effects on bone cells and rudiments [42, 97–101]. These studies show a range of effects ranging from enhancement of enzyme activity to alterations of calcium incorporation and release by the explants used. As described in section 6.5.3, there are unanswered questions regarding the nature of the hydrostatic stimulus. The low pressures and rates are more than exceeded by many physiological activities with no ability to affect bone mass. For this reason, it is important to ensure that in these model systems considerable care is taken to control for gas partial pressure changes which could affect media in compressed cultures. Although the mechanism of these effects is not fully understood, the devices used for these studies are simple and inexpensive and the effect is clear and reproducible.

6.7.3 Loaded cell cultures

Cell culture methods have some significant advantages over explant studies, as the viability of the cells is rarely in doubt. However, since the cells have been removed from their normal organization within a three-dimensional matrix, there are questions regarding the extrapolation of effects of load on monolayer cultures to changes in bone mass. Since numerous workers have studied effects of deformation on bone cells by these methods, over a long period, the answers that can be obtained from these models do provide significant insights into bone metabolism.

Several issues must be considered in these studies. The cells used, the method of applying the mechanical stimulus and the amount of strain will all affect results. In general, all the parameters that affect cell culture studies of other sorts will be important in loading experiments. Passage number, origin, effects of transformation if a clonal cell line, culture substrate and medium used will all alter cell behaviour.

Numerous studies have investigated the response of bone cells to stretching after culture on substrates ranging from collagen ribbons and plastic films of various sorts to glass [26, 102].

One finding which is apparent is that the method of application of load is important. Where cells are cultured on circular substrates deformed by air pressure changes under the base of the dish, the strains within the chamber vary widely (Fig. 6.6a). If low strains are applied to culture plates with a relatively inflexible base [24, 103], then this variation in strain across the plate may be acceptable for some studies. For investigation of mechanical effects, it is important to use systems that are appropriate to bone cells. Systems designed for use with cells experiencing much higher strains are not ideal for bone cell experiments. Where very deformable culture dish bases are used, and very high strains (~24 000 microstrain) are applied, results may not relate to effects in bone [104].

One advance on those systems has been the development of a device that presses a ring-shaped indenter against the undersurface of the dish (Fig. 6.6b) [105]. This system induces more uniform biaxial strains over the surface of the dish than the previous models.

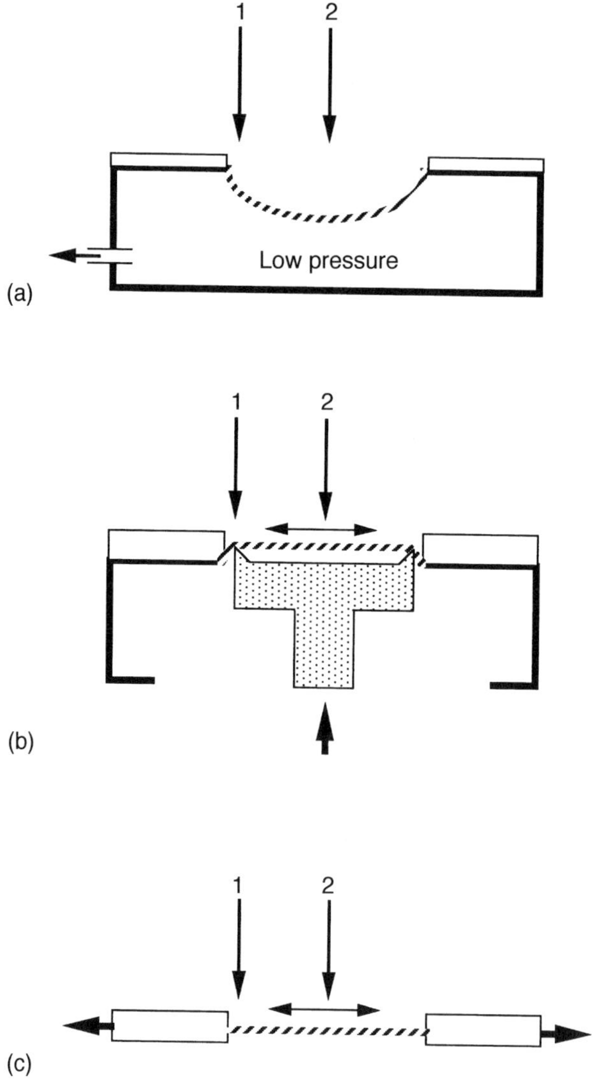

Figure 6.6 Loading of cultured cells. (a) Where cells are cultured on dishes with flexible bases, air pressure changes under the dish deform the base and stretch the cells. Strains are very different across the whole plate, but especially near the edge (1) and in the centre (2). (b) A better system uses similar flexible bases, but stretches them with an annular indenter. Although the strains are very high at a small zone near the edge (1), they are relatively uniform throughout the rest of the plate (2). (c) Direct stretching (or bending) of a substrate gives approximately uniform strains at zones 1 and 2.

The most controlled way to apply strain to cultured cells is to deform the substrate directly (Fig. 6.6c). Models for bending glass and plastic plates have been developed and used by the groups of Jones [106], Lanyon [107] and others [108, 109] and have shown a range of early effects (section 6.4.2) [26, 27]. These methods of bending or stretching a flat substrate have the advantage that they induce strains in the material which can be shown to be relatively uniform over the area used for culture [29, 106–109]. Such studies have been among those which have given rise to the earliest consequences of loading [29], and currently represent the most well characterized systems appropriate for use in studies of bone cell responses to direct mechanical strain. As with any system, there are problems. Bending of the substrate causes fluid flow, inducing shear on the cells, and this may complicate the results. In all models of this sort, there are technical difficulties associated with maintaining a seal between the substrate and the wall which retains the culture medium, but these problems are soluble [29, 106–108].

6.7.4 Fluid flow culture models

Fluid flow models were developed originally for investigation of endothelial cell responses to blood flow [89]. Load-induced flow of fluid through cortical bone was suggested to be a transduction mechanism for loading, and so studies using similar models have been used in several laboratories. Early studies showed that fluid flow was able to stimulate profound increases in prostaglandin-dependent cAMP expression, which was shear stress dependent, but not related to streaming potentials. Later studies [37] compared the responses of cells to pulsatile fluid flow and hydrostatic compression, and showed similar responses in prostaglandin production in both systems. Others have developed similar systems [110, 111], and these methods are generally simpler to set up than models in which controlled stretching is applied to the substrate of the cells.

6.7.5 Direct application of force to single cells

Where single cells can be visualized microscopically, direct pressure can be applied to the cells by means of a micropipette. Poking the cells in this way has been shown to stimulate rapid rises in intracellular calcium (H.J. Donahue, unpublished) which are consistent with the changes seen after stretching cells [26]. Such methods are very appropriate for optical studies, or those where electrophysiological methods are to be used to study responses of cells, as gross movement of the sort needed to stretch the substrate or even to flow fluid past the cells would break the electrode seal. In this context, it is possible to stimulate cells mechanically by electrophysiological methods [112]. If a patch clamp electrode is applied to the surface of the cell, suction can activate stretch-activated channels, either in cell-attached or isolated patches or membrane.

A modification of the method of direct mechanical stimulation of single cells is to coat a pipette tip or ferromagnetic microbead with a specific cell adhesion molecule, and use this to investigate the roles of specific ligands and receptors in cell matrix interactions and mechanosensing [35]. These methods allow the most specific dissection of responses of cells to mechanical stimulation. While the range of techniques that can be applied to single cells has been limited, single-cell PCR is possible and therefore these methods are likely to increase in importance in the future.

6.7.6 Other models

Because monolayer cultures do not allow the complex interactions of bone cells *in vivo*, several workers have developed methods for growing bone cells in three-dimensional culture in agarose, collagen and other gels [113–115]. There are no publications relating to the use of these models to study effects of loading on bone cells, but such studies would be appropriate and offer advantages over monolayer systems. One study has been performed in which cells were cultured on porous beads before loading, where calculated strains were associated with increased RNA synthesis [116].

In plants, models have been developed to rotate germinating seeds in order that they should not experience the force of gravity in a single direction as usual. This device, the clinostat, has been modified to allow cultures of cells to be rotated in order to mimic the lack of gravitational pull in a single direction experienced in weightless conditions. While the rotation alters cell growth rates and alkaline phosphatase activity [117], the mechanism is not clear. Cells are not exposed to fluid flow, as the wells are completely full, but they are not weightless as they would be in orbit. Further work is in progress with this model.

6.7.7 Problems with *in vitro* models

Both explant and cell culture models offer considerable advantages over *in vitro* models of loading in the effects that can be discriminated. However, there are numerous factors to consider in designing such studies.

Explant models may be complicated because the tissues may experience some disuse reaction. In models where tissues are removed from animals capable of movement, this may exert a significant effect on the response of the tissue to loading. Where embryonic tissues are used, there is less likelihood that the bones have been subjected to significant loading before explanting, but that assumption is relatively untested.

The range of cell types used for culture studies is very large, and detailed description is beyond the scope of this chapter, but it is important to consider the same principles which relate to all cell culture experiments. Where clonal cell lines are used, the transformation process may affect responsiveness to stimuli and, given the poor understanding of mechanotransduction in bone cells, it is possible that cell lines may be significantly more or less sensitive to deformation than untransformed cells. When primary cells are used, it is vital to consider their origin. Numerous studies have used cells derived from calvariae. However, it is hard to interpret such results in the light of the extremely low functional strains recorded in calvariae compared with long bones [11] and the recent observation of differences in strain responses of cultured calvarial and long bone derived cells [107]. Culture of either long bone derived cells or minimally transformed clonal cells on a deformable substrate that is stretched uniformly probably represent the best methods for investigation of mechanical effects on groups of cells in monolayer culture.

Wherever possible, the strains induced by the loading should be measured. The application of strain gauges to a culture container of the type to be used will allow estimates of strain to be made, though it is important to realize that fixing a strain gauge to a flexible surface may reinforce it and reduce the measured strains for a given load in comparison with those induced when no gauge is present.

6.8 METHODS OF APPLICATION OF LOADS

Where loads are to be applied to bones, explants or cells, the method of application of force affects the response to loading [118]. The frequency, rate and magnitude of loading will affect the response, while the duration is probably less important once a threshold has been reached. This is considered to be of the order of 50 cycles, though many workers exceed that number by an order of magnitude to be certain of inducing a response. The method of application of load is therefore important in controlling those variables that do affect the response.

6.8.1 Manual loading

In some studies *in vitro* and *in vivo*, forces applied to the bones were controlled by hand [80,119]. This method is simple and cheap, and if the force of a weight is lifted to remove load from the system, rather than to apply the force, the magnitude of each load cycle will be equal. However, the frequency and, most important, the rate will vary greatly between cycles and experiments, and these methods should be avoided.

6.8.2 Mechanical devices

A substantial improvement of manual application of force is to use a weight or spring to apply loading, which is removed by the action of a motorized system. Cams driven by motors are used widely [95, 120] and this method has been used in the later experiments on loading of rat tail vertebrae. Pneumatic cylinders have also been used [12, 50], because they are able to apply higher rates of strain and are relatively inexpensive.

6.8.3 Electromagnetic devices

Electromagnetic actuators can provide rapid and controllable loading, and were used in Hert's original studies. However, there may be effects of the electromagnetic fields generated by these machines, and it may be difficult to shield those effects.

6.8.4 Computer-controlled servohydraulics

Without doubt, the most controllable way to load whole bones, explants and cell cultures is to use a computer-controlled servohydraulic testing machine. These machines were used originally for *in vivo* loading, where the forces required were relatively high, but they are now capable of very precise loading, beyond the ability of pneumatic or mechanical devices. They are now used both for *in vivo* loading and more recently for loading of cultured cells [108, 109]. Waveforms can be specified precisely, so that accurate studies on the effects of strain rate (for example) can be performed without varying the frequency of loading, which is necessary with mechanical devices.

These machines are expensive, but they are significantly more controlled than other methods, and the reduction in spread of data which they permit can justify their use if serious commitment to mechanical loading studies is envisaged.

6.9 CHOICE OF MODEL SYSTEM

In a functionally adaptive response to loading *in vivo*, bone will model or remodel in order to resist applied loads optimally. This means that the structure being loaded will adjust its mass and architecture to become stronger than it was before loading began. Altered bone strength is the logical and indisputable goal of loading responses but it will almost certainly be impossible to obtain information on bone strength with most models designed to answer specific questions on the short-term effects of load. This is not a barrier to the use of models where bone strength is not altered. Strength changes will inevitably be preceded by alterations of cellular activity, and with a knowledge of these changes in models where the consequences of loading alterations on strength are also known, it is possible to provide a series of 'calibration' parameters that allow valid interpretations of earlier changes to be made (Table 6.2).

For example, if a model is chosen for its ability to allow discrimination of transcriptional regulation by mechanical loading of cultured cells, it is possible to validate the loading method by use of parameters that are shared with other systems. In this case, the use of early markers of loading-related responses such as cellular proliferation, increased enzyme activity and a response to relevant strains of the culture substrate provides the necessary links with a functionally adaptive response in whole bone.

6.10 CONCLUSIONS

Mechanical loading has a potent ability to increase bone mass, so the study of its mechanisms is a large and growing field. No experimental method of loading of bone cells *in vivo* or *in vitro* is free from flaws, but methods that attempt to mimic physiology, that are well characterized and that show similar responses to other systems are to be favoured. Dissection of the mechanosensing mechanism of bone cells and the early downstream consequences of loading will generate new information and targets to allow manipulation of bone mass by activation of mechanically related signalling pathways without undergoing exercise, as well as providing information on the types of exercise which benefit bone strength during life.

Table 6.2 Changes that can be investigated by *in vivo*, explant and cell culture models

Change	In vivo	Explant	Cell culture
Bone strength changes	+	−	−
Bone formation	+	+	?
Bone resorption	+	+	+
Cell/matrix interactions	+	+	−
Cell proliferation	+	+	+
Enzyme activity	+	+	+
Second messengers	−	?	+
Protein expression	+	+	+
Gene expression	+	+	+
Real time imaging	−	−	+

+, Possible/relevant; −, not possible; ?, difficult/questionable relevance.

REFERENCES

1. Roesler, H. (1981) Some historical remarks on the theory of cancellous bone structure (Wolffs law), in *Mechanical Properties of Bone* (ed. S.C. Cowin), ASME, New York, pp. 27–42.
2. Gurdjian, E.S. and Lissner, H.R. (1944) Mechanism of head injury as studied by the cathode ray oscilloscope. *Journal of Neurosurgery* **1**, 393–399.
3. Gurdjian, E.S. and Lissner, H.R. (1945) Stresscoat methods in studies on skull fractures. *Surg, Gynec, Obstet* **81**, 679–687.
4. Evans, F.G. (1953) Method of studying biomechanical significance of bone form. *American Journal of Physical Anthropology* **11**, 355–360.
5. Lanyon, L.E. and Smith, R.N. (1970) Bone strain in the tibia during normal quadrupedal locomotion. *Acta Orthopaedica Scandinavica* **42**, 238–248.
6. Lanyon, L.E. (1971) Strain in sheep lumbar vertebrae recorded during life. *Acta Orthopaedica Scandinavica* **42**, 102–112.
7. Lanyon, L.E. (1972) In vivo bone strain recorded from the thoracic vertebrae of sheep. *Journal of Biomechanics* **5**, 277–281.
8. Lanyon, L.E. (1973) Analysis of surface bone strain in the calcaneus of the sheep during normal locomotion. *Journal of Biomechanics* **6**, 41–49.
9. Rubin, C.T. and Lanyon, L.E. (1984) Dynamic strain similarity in vertebrates; an alternative to allometric limb bone scaling. *Journal of Theoretical Biology* **107**, 321–327.
10. Rubin, C.T. and Lanyon, L.E. (1985) Regulation of bone mass by mechanical strain magnitude. *Calcified Tissue International* **37**, 411–417.
11. Hillam, R.A., Mosley, J.M. and Skerry, T.M. (1994) Regional differences in bone strain. *Bone and Mineral* **25(S1)**, 32 (abstr.).
12. O'Connor, J.A., Lanyon, L.E. and McFie, H.F. (1982) The influence of strain rate on adaptive bone remodelling. *Journal of Biomechanics* **15**, 767–781.
13. Mosley, J.M. (1996) The influence of mechanical load and oestrogen on the development of long bone architecture. PhD Thesis, University of London.
14. Rubin, C.T. and McLeod, K.J. (1994) Promotion of bony ingrowth by frequency-specific, low-amplitude mechanical strain. *Clinical Orthopaedics and Related Research* **298**, 165–174.
15. Rubin, C.T. and Lanyon, L.E. (1987) Osteoregulatory nature of mechanical stimuli: function as a determinant for adaptive remodeling in bone. *Journal of Orthopaedic Research* **5**, 300–310.
16. Skerry, T.M., Bitensky, L., Chayen, J. *et al.* (1989) Early strain-related changes in enzyme activity in osteocytes following bone loading in vivo. *Journal of Bone and Mineral Research* **4**, 783–788.
17. Doty, S.B. (1981) Morphological evidence of gap junctions between bone cells. *Calcified Tissue International* **33**, 509–512.
18. Jones, S.J., Gray, C., Sakamaki, H. *et al.* (1993) The incidence and size of gap-junctions between the bone-cells in rat calvaria. *Anatomy and Embryology* **187**, 343–352.
19. Pead, M.J., Suswillo, R.F.L., Skerry, T.M. *et al.* (1988) Increased 3H-uridine levels in osteocytes following a single short period of dynamic bone loading in vivo. *Calcified Tissue International* **43**, 92–96.
20. Dallas, S.L., Zaman, G., Pead, M.J. *et al.* (1993) Early strain-related changes in cultured embryonic chick tibiotarsi parallel those associated with adaptive modeling in vivo. *Journal of Bone and Mineral Research* **8**, 251–259.
21. Lean, J.M., Jagger, C.J., Chambers, T.J. *et al.* (1995) Increased insulin-like growth factor I mRNA expression in rat osteocytes in response to mechanical stimulation. *American Journal of Physiology – Endocrinology and Metabolism* **268**, 318–327.
22. Dodds, R.A., Ali, N.N., Pead, M.J. *et al.* (1993) Early loading-related changes in the activity of glucose 6-phosphate dehydrogenase and alkaline phosphatase in osteocytes and periosteal osteoblasts in rat fibulae *in vivo*. *Journal of Bone and Mineral Research* **8**, 261–267.

23. Keila, S., Pitaru, S., Grosskopf, A. *et al.* (1994) Bone-marrow from mechanically unloaded rat bones expresses reduced osteogenic capacity in-vitro. *Journal of Bone and Mineral Research* **9**, 321–327.

24. Thomas, G.P. and ElHaj, A.J. (1996) Bone-marrow stromal cells are load responsive in-vitro. *Calcified Tissue International* **58**, 101–108.

25. Wozasek, G.E., Simon, P., Redl, H. *et al.* (1994) Intramedullary pressure changes and fat intravasation during intramedullary nailing – an experimental study in sheep. *Journal of Trauma* **36**, 202–207.

26. Jones, D.B. and Bingmann, D. (1991) How do osteoblasts respond to mechanical stimulation. *Cells And Materials* **1**, 329–340.

27. Binderman, I., Zor, U., Kaye, A.M. *et al.* (1988) The transduction of mechanical force into biochemical events in bone cells may involve activation of Phospholipase A2. *Calcified Tissue International* **42**, 261–266.

28. Inaoka, T., Lean, J.M., Bessho, T. *et al.* (1995) Sequential analysis of gene expression after an osteogenic stimulus – c-fos expression is induced in osteocytes. *Biochemical and Biophysical Research Communications* **217**, 264–270.

29. Pitsillides, A.A., Rawlinson, S.C.F., Suswillo, R.F.L. *et al.* (1995) Mechanical strain-induced no production by bone-cells – a possible role in adaptive bone (re)modeling. *FASEB Journal* **9**, 1614–1622.

30. Lanyon, L.E. (1992) Control of bone architecture by functional load bearing. *Journal of Bone and Mineral Research* **7**, S369–S375.

31. Zaman, G., Dallas, S.L. and Lanyon, L.E. (1992) Cultured embryonic bone shafts show osteogenic responses to mechanical loading. *Calcified Tissue International* **51**, 132–136.

32. Banes, A.J., Tsuzaki, M., Yamamoto, J. *et al.* (1995) Mechanoreception at the cellular-level – the detection, interpretation, and diversity of responses to mechanical signals. *Biochemistry and Cell Biology* **73**, 349–365.

33. Stamenovic, D., Fredberg, J.J., Wang, N. *et al.* (1996) A microstructural approach to cytoskeletal mechanics based on tensegrity. *Journal of Theoretical Biology* **181**, 125–136.

34. Duncan, R.L. and Turner, C.H. (1995) Mechanotransduction and the functional response of bone to mechanical strain. *Calcified Tissue International* **57**, 344–358.

35. Wang, N. and Ingber, D.E. (1995) Probing transmembrane mechanical coupling and cytomechanics using magnetic twisting cytometry. *Biochemistry and Cell Biology* **73(7/8)**, 327–335.

36. Turner, C.H., Forwood, M.R. and Otter, M.W. (1994) Mechanotransduction in bone – do bone-cells act as sensors of fluid- flow. *FASEB Journal* **8**, 875–878.

37. KleinNulend, J., van der Plas, A., Semeins, C.M. *et al.* (1995) Sensitivity of osteocytes to biomechanical stress in vivo. *FASEB Journal* **9**, 441–445.

38. McLeod, K.J., Rubin, C.T. and Donahue, H.J. (1995) Electromagnetic-fields in bone repair and adaptation. *Radio Science* **30**, 233–244.

39. Oloyede, A. and Broom, N.D. (1994) Complex nature of stress inside loaded articular cartilage. *Clinical Biomechanics* **9**, 149–156.

40. van Kampe, G.P.J., Veldhuijzen, J.P., Kuijer, R. *et al.* (1985) Cartilage response to mechanical force in high density chondrocyte cultures. *Arthritis and Rheumatism* **28**, 419–424.

41. Wong, M. and Carter, D.R. (1990) Theoretical stress analysis of organ culture osteogenesis. *Bone* **11**, 127–131.

42. KleinNulend, J., Veldhuijzen, J.P., de Jong, M. *et al.* (1987) Increased bone formation and decreased bone resorption in fetal mouse calvaria as a result of intermittent compressive force in vitro. *Bone and Mineral* **2**, 441–448.

43. Grimston, S.K., Willows, N.D. and Hanley, D.A. (1993) Mechanical loading regime and its relationship to bone-mineral density in children. *Medicine and Science in Sports and Exercise* **25**, 1203–1210.

44. Taaffe, D.R., Snow Harter, C., Connolly, D.A. *et al.* (1995) Differential effects of swimming versus weight-bearing activity on bone-mineral status of eumenorrheic athletes. *Journal of Bone and Mineral Research* **10**, 586–593.

45. Michaeli, Y., Shamir, D., Weinreb, M. *et al.* (1994) Effect of loading on the migration of periodontal fibroblasts in the rat incisor. *Journal of Periodontal Research* **29**, 25–34.

46. Miekle, M.C., Sellers, A. and Reynolds, J.J. (1980) Effect of tensile mechanical stress on the synthesis of metalloproteinases by rabbit coronal sutures in vitro. *Calcified Tissue International* **30**, 77–82.

47. Lanyon, L.E., Magee, P.T. and Baggott, D.G. (1979) The relationship of functional stress and strain to the process of bone remodelling: an experimental study in the sheep radius. *Journal of Biomechanics* **12**, 615–622.

48. Goodship, A.E., Lanyon, L.E. and McFie, H. (1979) Functional adaptation of bone to increased stress. *Journal of Bone and Joint Surgery* **61A**, 53.

49. Burr, D.B., Schaffler, M.B., Yang, K.H. *et al.* (1989) Skeletal change in response to altered strain environments: is woven bone a response to elevated strain? *Bone* **10**, 223–233.

50. Rubin, C.T. and Lanyon, L.E. (1984) Regulation of bone formation by applied dynamic loads. *Journal of Bone and Joint Surgery* **66A**, 397–402.

51. Jaworski, J.F.G., Liskova-Kiar, M. and Uhthoff, H.K. (1980) Effect of long term immobilisation on the pattern of bone loss in older dogs. *Journal of Bone and Joint Surgery* **62B**, 104–110.

52. Li, X.J., Jee, W.S.S., Chow, S.Y. *et al.* (1990) Adaptation of cancellous bone to aging and immobilization in the rat a single photon absorptiometry and histomorphometry study. *Anatomical Record* **227**, 12–24.

53. Uhthoff, H.K. and Jaworski, Z.F.G. (1978) Bone loss in response to long term immobilisation. *Journal of Bone and Joint Surgery* **60B**, 420–429.

54. Jaworski, Z.F. and Uhthoff, H.K. (1986) Reversibility of nontraumatic disuse osteoporosis during its active phase. *Bone* **7**, 431–439.

55. Li, X.J. and Jee, W.S.S. (1991) Adaptation of diaphyseal structure to aging and decreased mechanical loading in the adult rat – a densitometric and histomorphometric study. *Anatomical Record* **229**, 291–297.

56. Jee, W.S.S. and Li, X.J. (1990) Adaptation of cancellous bone to overloading in the adult rat: a single photon absorptiometry and histomorphometry study. *Anatomical Record* **227**, 418–426.

57. Goshi, N., Ogihara, A. and Ohno, H. (1989) Histological alterations in rat tibia caused by the sciatic neurectomy. *Journal of Electron Microscopy* **38**, 219.

58. Weinreb, M., Rodan, G.A. and Thompson, D.D. (1989) Osteopenia in the immobilized rat hind limb is associated with increased bone resorption and decreased bone formation. *Bone* **10**, 187–194.

59. Weinreb, M., Rodan, G.A. and Thompson, D.D. (1991) Immobilization-related bone loss in the rat is increased by calcium deficiency. *Calcified Tissue International* **48**, 93–100.

60. Weinreb, M., Rodan, G.A. and Thompson, D.D. (1991) Depression of osteoblastic activity in immobilized limbs of suckling rats. *Journal of Bone and Mineral Research* **6**, 725–731.

61. Thompson, D.D. and Rodan, G.A. (1988) Indomethacin inhibition of tenotomy-induced bone resorption in rats. *Journal of Bone and Mineral Research* **3**, 409–414.

62. Spengler, D.M., Morey Holton, E.R., Carter, D.R. *et al.* (1983) Effects of spaceflight on structural and material strength of growing bone. *Proceedings of the Society for Experimental Biology and Medicine* **174**, 224–228.

63. Wronski, T.J. and Morey Holton, E.R. (1982) Skeletal abnormalities in rats induced by simulated weightlessness. *Metabolic Bone Disease and Related Research* **4**, 69–75.

64. Smith, E.J., Smith, P.E., Ensign, C.J. *et al.* (1984) Bone involution decrease in exercising middle-aged women. *Calcified Tissue International* **36(S1)**, 129–138.

65. Barlet, J.P., Coxam, V. and Davicco, M.J. (1995) Physical exercise and the skeleton. *Archives of Physiology and Biochemistry* **103**, 681–698.

66. Thorsen, K., Kristoffersson, A. and Lorentzon, R. (1996) The effects of brisk walking on markers of bone and calcium- metabolism in postmenopausal women. *Calcified Tissue International* **58**, 221–225.

67. Ayalon, J., Simkin, A., Leichter, I. *et al.* (1987) Dynamic bone loading exercises for postmenopausal women – effect on the density of the distal radius. *Archives of Physical Medicine and Rehabilitation* **68**, 280–283.

68. Woo, S., Kuei, S.C., Amiel, D.,*et al.* (1991) The effect of prolonged physical training on the properties of long bone: a study of Wolff's law. *Journal of Bone and Joint Surgery* **63A**, 780–787.

69. Raab, D.M., Crenshaw, T.D., Kimmel, D.B. *et al.* (1991) A histomorphometric study of cortical bone activity during increased weight-bearing exercise. *Journal of Bone and Mineral Research* **6**, 741–749.

70. Jones, H.H., Priest, J.D., Hayes, W.C. *et al.* (1977) Humeral hypertrophy in response to exercise. *Journal of Bone and Joint Surgery* **59A**, 204–208.

71. Kannus, P., Haapasalo, H., Sankelo, M. *et al.* (1995) Effect of starting age of physical-activity on bone mass in the dominant arm of tennis and squash players. *Annals of Internal Medicine* **123**, 27–31.

72. Thorsen, K., Kristoffersson, A. and Lorentzon, R. (1995) Differences in effect of moderate endurance exercise on calcium and bone metabolism in young women early and late postmenopausal women. *Journal of Bone and Mineral Research* **10**, S454.

73. Hert, J., Skelenska, A. and Liskova, M. (1971) Continuous and intermittent loading of the tibia in rabbit. *Folia Morphologica* **19**, 378–387.

74. Hert, J., Liskova, M. and Landgrot, B. (1969) Influence of the longterm continuous bending on bone: an experimental study on the tibia of the rabbit. *Folia Morphologica* **17**, 389–399.

75. Liskova, M. and Hert, J. (1971) Periosteal and endosteal reaction of tibial diaphysis in rabbit to intermittent loading. *Folia Morphologica* **19**, 301–317.

76. Churches, A.E. and Howlett, C.R. (1982) Functional adaptation of bone in response to sinusoidally varying controlled compressive loading of the ovine metacarpus. *Clinical Orthopaedics and Related Research* **168**, 265–280.

77. Pead, M.J., Skerry, T.M. and Lanyon, L.E. (1988) Direct transformation from quiescence to bone formation in the adult periosteum following a single brief period of bone loading. *Journal of Bone and Mineral Research* **3**, 647–656.

78. Skerry, T.M., Bitensky, L., Chayen, J. *et al.* (1988) Loading-related reorientation of bone proteoglycan in vivo: strain memory in bone tissue? *Journal of Orthopaedic Research* **6**, 547–551.

79. Turner, C.H., Akhter, M.P., Raab, D.M. *et al.* (1991) A noninvasive, in vivo model for studying strain adaptive bone modeling. *Bone* **12**, 73–79.

80. Chambers, T.J., Evans, M., Gardner, T.N. *et al.* (1993) Induction of bone formation in rat tail vertebrae by mechanical loading. *Bone and Mineral* **20**, 167–178.

81. Torrance, A.G., Mosley, J.M., Suswillo, R.F.L. *et al.* (1994) Noninvasive loading of the rat ulna in vivo induces a strain related modeling response uncomplicated by trauma of periosteal pressure. *Calcified Tissue International* **54(3)**, 241–247.

82. RaabCullen, D.M., Thiede, M.A., Petersen, D.N. *et al.* (1994) Mechanical loading stimulates rapid changes in periosteal gene expression. *Calcified Tissue International* **55**, 473–478.

83. Turner, C.H., Woltman, T.A. and Belongia, D.A. (1992) Structural changes in rat bone subjected to long-term, in vivo mechanical loading. *Bone* **13**, 417–422.

84. RaabCullen, D.M., Akhter, M.P., Kimmel, D.B. *et al.* (1994) Periosteal bone formation stimulated by externally induced bending strains. *Journal of Bone and Mineral Research* **9**, 1143–1152.

85. Turner, C.H., Forwood, M.R., Rho, J.Y. *et al.* (1994) Mechanical loading thresholds for lamellar and woven bone-formation. *Journal of Bone and Mineral Research* **9**, 87–97.

86. Hillam, R.A. and Skerry, T.M. (1995) Inhibition of bone resorption and stimulation of formation by mechanical loading of the modeling rat ulna in vivo. *Journal of Bone and Mineral Research* **10(5)**, 683–689.

87. Mason, D.J., Suva, L.J., Genever, P.G. *et al.* (1997) Mechanically regulated expression of a neural glutamate transporter in bone. A role for excitatory amino acids as osteotropic agents? *Bone* **20(3)**, 1–3.

88. Mason, D.J., Hillam, R.A. and Skerry, T.M. (1996) Constitutive in vivo mRNA expression by osteocytes of β-actin, osteocalcin, connexin-43, IGF I, c-fos and c-jun, but not TNFα or tartrate resistant acid phosphatase. *Journal of Bone and Mineral Research* **11(3)**, 350–357.

89. Hsieh, H.J., Li, N.Q. and Frangos, J.A. (1991) Shear-stress increases endothelial platelet-derived growth-factor messenger-rna levels. *American Journal of Physiology* **260**, H642–H646.

90. Lozupone, E., Favia, A. and Grimaldi, A. (1992) Effect of intermittent mechanical force on bone tissue in vitro: preliminary results. *Journal of Bone and Mineral Research* **7(S2)**, 407–409.

91. Lozupone, E., Palumbo, C., Favia, A. *et al.* (1996) Intermittent compressive load stimulates osteogenesis and improves osteocyte viability in bones cultured in-vitro. *Clinical Rheumatology* **15**, 563–572.

92. ElHaj, A.J., Minter, S.L., Rawlinson, S.C.F. *et al.* (1990) Cellular responses to mechanical loading in vitro. *Journal of Bone and Mineral Research* **5(9)**, 923–933.

93. Rawlinson, S.C.F., ElHaj, A.J., Minter, S.L. *et al.* (1991) Loading-related increases in prostaglandin production in cores of adult canine cancellous bone in vitro: a role for prostacyclin in adaptive bone remodeling? *Journal of Bone and Mineral Research* **6**, 1345–1351.

94. Ming, Z., Zaman, G., Rawlinson, S.C.F. *et al.* (1996) Mechanical loading and sex-hormone interactions in organ-cultures of rat ulna. *Journal of Bone and Mineral Research* **11**, 502–511.

95. Ming, Z., Zaman, G. and Lanyon, L.E. (1993) Estrogen enhances the osteogenic effects of mechanical loading and exogenous prostacyclin, but not prostaglandin-E2. *Journal of Bone and Mineral Research* **8(S1)**, 151.

96. Ming, Z., Zaman, G. and Lanyon, L.E. (1994) Estrogen enhances the stimulation of bone collagen synthesis by loading and exogenous prostacyclin, but not prostaglandin E2, in organ cultures of rat ulnae. *Journal of Bone and Mineral Research* **9**, 805–816.

97. Quinn, R.S. and Rodan, G.A. (1981) Enhancement of ornithine decarboxylase and Na,K ATPase in osteoblastoma cells by intermittent compression. *Biochemical and Biophysical Research Communications* **100**, 1696–1702.

98. KleinNulend, J., Veldhuijzen, J.P. and Burger, E.H. (1986) Increased calcification of growth plate cartilage as a result of compressive force in vitro. *Arthritis and Rheumatism* **29**, 1002–1009.

99. Bagi, C.M., Vinter, I., Filipovic, I. *et al.* (1990) Intermittent compressive force stimulates sulphate metabolism in fetal mouse metatarsal bone rudiments. *Periodicum Biologorum* **92**, 377–384.

100. KleinNulend, J., Veldhuijzen, J.P., van Strien, M. *et al.* (1990) Inhibition of osteoclastic bone resorption by mechanical stimulation in vitro. *Arthritis and Rheumatism* **33**, 66–72.

101. KleinNulend, J., Semeins, C.M., Veldhuijzen, J.P. *et al.* (1993) Effect of mechanical stimulation on the production of soluble bone factors in cultured fetal mouse calvariae. *Cell and Tissue Research* **271**, 513–517.

102. Yeh, C.K. and Rodan, G.A. (1984) Tensile forces enhance prostaglandin E synthesis in osteoblastic cells grown on collagen ribbons. *Calcified Tissue International* **36**, S67–S71.

103. Sandy, J.R., Meghji, S., Scutt, A.M. *et al.* (1989) Murine osteoblasts release bone-resorbing factors of high and low molecular weights: stimulation by mechanical deformation. *Bone and Mineral* **5**, 155–168.

104. Buckley, M.J., Banes, A.J., Levin, L.G. *et al.* (1988) Osteoblasts increase their rate of division and align in response to cyclic, mechanical tension in vitro. *Bone and Mineral* **4**, 225–236.

105. Schaffer, J.L., Rizen, M., Litalien, G.J. *et al.* (1994) Device for the application of a dynamic biaxially uniform and isotropic strain to a flexible cell-culture membrane. *Journal of Orthopaedic Research* **12**, 709–719.

106. Jones, D.B., Leivseth, G. and Tenbosch, J. (1995) Mechano-reception in osteoblast-like cells. *Biochemistry and Cell Biology* **73**, 525–534.

107. Rawlinson, S.C.F., Mosley, J.R., Suswillo, R.F.L. *et al.* (1995) Calvarial and limb bone cells in organ and monolayer culture do not show the same early responses to dynamic mechanical strain. *Journal of Bone and Mineral Research* **10**, 1225–1232.

108. Murray, D.W. and Rushton, N. (1990) The effect of strain on bone cell prostaglandin E2 release: a new experimental method. *Calcified Tissue International* **47**, 35–39.

109. Fermor, B., Emerton, M., Urban, J.P.G. *et al.* (1996) Application of defined uniform dynamic strain to human bone cells in-vitro. *Journal of Bone and Mineral Research* **11**, 45.

110. Owan, I., Burr, D.B., Turner, C.H. *et al.* (1996) A system to study the anabolic effects of fluid-flow and mechanical stretch in osteoblasts. *Journal of Bone and Mineral Research* **11**, M334.

111. Smalt, R., Mitchell, F.T., Howard, R.L. *et al.* (1996) Mechanical strain versus wall shear-stress as the stimulus to bone-cells in mechanical loading. *Journal of Bone and Mineral Research* **11**, M336.

112. Duncan, R. and Misler, S. (1989) Voltage-activated and stretch-activated Ba^{2+} conducting channels in an osteoblast-like cell line (UMR 106). *FEBS Letters* **251**, 17–21.

113. Highfill, J.G., Todd, P. and Kompala, D.S. (1993) A 3-dimensional culture system for murine bone-marrow. *Abstracts of Papers of the American Chemical Society* **205**, 134-BIOT.

114. Benayahu, D., Kompier, R., Shamay, A. *et al.* (1994) Mineralization of marrow-stromal osteoblasts mba-15 on 3-dimensional carriers. *Calcified Tissue International* **55**, 120–127.

115. Gerstenfeld, L.C., Uporova, T., Schmidt, J. *et al.* (1996) Osteogenic potential of murine osteosarcoma cells – comparison of bone-specific gene-expression in in-vitro and in-vivo conditions. *Laboratory Investigation* **74**, 895–906.

116. Shelton, R.M. and ElHaj, A.J. (1992) A novel microcarrier bead model to investigate bone cell responses to mechanical compression in vitro. *Journal of Bone and Mineral Research* **7(S2)**, 403–405.

117. Alajmi, N., Braidman, I.P. and Moore, D. (1995) Effect of clinostat rotation on differentiation of embryonic bone in- vitro. *Advances In Space Research* **17**, 189–192.

118. Skerry, T.M. (1997) Mechanical loading and bone: what sort of exercise is beneficial to the skeleton? *Bone* **20(3)**, 1–3.

119. Jones, D.B., Nolte, H., Scholubbers, J.G. *et al.* (1991) Biochemical signal transduction of mechanical strain in osteoblast like cells. *Biomaterials* **12**, 101–110.

120. Turner, C.H., Forwood, M.R., RaabCullen, D.M. *et al.* (1994) On animal models for studying bone adaptation. *Calcified Tissue International* **55**, 316–318.

121. Skerry, T.M., Suswillo, R.F.L., ElHaj, A.J. *et al.* (1990) Load-induced proteoglycan orientation in bone tissue in vivo and in vitro. *Calcified Tissue International* **46**, 318–326.

122. Somjen, D., Binderman, I., Berger, E.H. *et al.* (1980) Bone remodelling induced by physical stress is prostaglandin E2 mediated. *Biochem et Biophys Acta* **627**, 91–100.

123. Rawlinson, S.C.F., ElHaj, A.J., Minter, S.L. *et al.* (1991) Loading-related increases in prostaglandin production in cores of adult canine cancellous bone in vitro: a role for prostacyclin in adaptive bone remodeling? *Journal of Bone and Mineral Research* **6**, 1345–1351.

124. Lanyon, L.E. (1987) Functional strain in bone tissue as an objective, and controlling stimulus for adaptive bone remodelling. *Journal of Biomechanics* **20**, 1083–1093.

CHAPTER SEVEN

Bone histomorphometry

Juliet Compston

7.1 INTRODUCTION

Bone histomorphometry is the means by which bone remodelling, modelling and structure may be quantitatively assessed. It provides information that is not currently available from other approaches, such as bone densitometry and biochemical markers of bone turnover, and enables a more precise characterization of disease states and their response to treatment than can be obtained from qualitative examination of bone histology. In recent years there have been significant advances in histomorphometric techniques, most notably the use of computerized rather than manual techniques and the development of sophisticated approaches to the assessment of bone microarchitecture. The application of these techniques has been particularly valuable in determining the cellular pathophysiology of different forms of osteoporosis and in defining the mechanisms by which drugs affect bone. This chapter reviews current methodology with particular reference to its application in humans; however, the techniques described are also widely used in animal research, especially in the context of preclinical testing of drugs for osteoporosis.

7.2 BONE BIOPSY

7.2.1 Procedure

The iliac crest is the usual site for bone biopsy in humans. Most investigators favour the transverse approach, in which a biopsy containing two cortices and intervening cancellous bone is obtained (Fig. 7.1) in contrast to vertical biopsies, which contain only one cortical plate. A number of specially designed trephines are commercially available; ideally, for bone histomorphometry, the internal diameter of the specimen should be at least 6 mm.

In most cases, the biopsy is performed as an out-patient procedure under mild sedation. For a trans-iliac biopsy, the patient lies in the supine position and the specimen is obtained approximately 2.5 cm below and behind the anterior superior iliac spine. The biopsy should be performed under sterile conditions.

The area around the anterior superior iliac spine is infiltrated with 1% lignocaine and the inner and outer periosteum are then anaesthetized. The author uses a small

Methods in Bone Biology. Edited by Timothy R. Arnett and Brian Henderson.
Published in 1997 by Chapman & Hall, London. ISBN 0 412 75770 2.

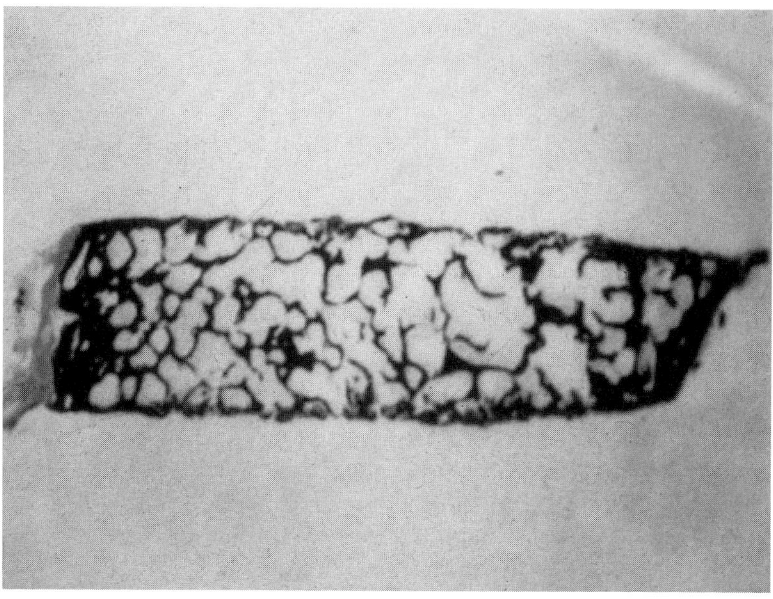

Figure 7.1   Section of trans-iliac biopsy obtained with an 8 mm internal diameter trephine. The biopsy contains inner and outer cortical plates and intervening cancellous bone.

trocar and stilette which is driven with a weighted instrument until it lies just in the outer cortex; lignocaine is infiltrated under the periosteum and the trocar and stilette are then advanced through the iliac crest to the inner cortex, where the procedure is repeated. Alternatively, the inner cortex may be anaesthetized by introducing a needle through the skin from opposite the biopsy site. A small skin incision is then made and a hollow cannula with a serrated edge is introduced and placed firmly on the outer periosteum. A smaller, hollow cannula is then inserted through the larger cannula and the biopsy is obtained by advancing the serrated edge of this cannula through the iliac crest until it has reached the outer surface of the inner cortex. The cannula is then withdrawn after rotation through 360° (to ensure that the core of bone has been freed from adjoining tissues) and the biopsy is removed from the cannula using a metal rod. The incision is then sutured and dressed and the patient is instructed to lie on the side of the biopsy to apply pressure to the site and reduce the risk of bruising. Ideally, this position should be maintained for 2 hours and thereafter the patient should be advised to rest for 24 hours.

7.2.2 Adverse effects

Bone biopsy is safe and generally well tolerated, though there is often some discomfort after the procedure – usually lasting between 24 and 48 hours. The morbidity is low Bond mainly due to haematoma, which may occasionally be extensive; this is most likely to occur in obese subjects or in patients with bleeding diatheses. Other reported complications include infection, transient femoral nerve palsy, avulsion of the superior ramus of the iliac crest, fracture of the iliac crest and osteomyelitis. It should be stressed that these are extremely rare and, overall, the incidence of all complications is less than 1% [1].

7.3 PREPARATION OF BONE SECTIONS FOR HISTOMORPHOMETRY

The preparation of high quality histological sections is an essential prerequisite for bone histomorphometry. Assessment of bone mineralization requires the preparation of undecalcified sections, in which the bone specimen is embedded in a hard resin prior to cutting. A number of different protocols have been described for the fixation and embedding of bone specimens; these depend, to some extent, on local circumstances and research interests. In particular, existing regimens are being modified to enable successful application of *in situ* hybridization and immunolocalization techniques to bone histological sections. These include low temperature embedding in glycolmethacrylate, the use of frozen sections and other modifications of conventional approaches.

Trans-iliac biopsies should ideally contain both inner and outer cortical plates and the intervening cancellous bone; regular sharpening of the trephine ensures that 'bone dust' along the borders of the biopsy specimen is kept to a minimum and that artefactual breakage of individual trabeculae is avoided. The use of sharp knives in the microtome is also of key importance in this respect. Unfortunately, particularly in subjects with osteoporosis, it is not always possible to obtain a complete biopsy and, where bone loss is severe, fragmentation of the specimen may occur. Even in such cases, however, it may be possible to make some assessment of bone mineralization and turnover, though most techniques for structural analysis require the presence of both cortices and relatively intact cancellous bone.

7.3.1 Embedding and fixation

For the preparation of undecalcified sections for histological examination, the author uses fixation in ethanol at increasing concentrations from 70 to 100% over 8 days, followed by embedding in LR White Resin, which is polymerized at a temperature of 60°C for 2–4 hours. Alternatively, methylmethacrylate may be used; however, this necessitates the use of a fume cupboard and requires a longer time for polymerization than does LR White Resin. Sections between 5 and 10 μm thick are then cut on a heavy duty microtome using hard steel or tungsten carbide knives.

7.3.2 Staining procedures

A number of different staining procedures may be used to demonstrate the histological features of bone. Differentiation of osteoid from mineralized bone is best achieved using the von Kossa stain with a van Geison counterstain (Plate 4), solochrome cyanin B, or Goldner's trichrome stain. For examination of bone cell morphology and polarized light microscopy, 1% toluidine blue (pH 4.2) is a useful stain and also enables visualization of the calcification front (Plate 5). Examination of tetracycline fluorescence is achieved using unstained sections viewed under blue light (365 nm).

7.4 HISTOMORPHOMETRY

7.4.1 Theoretical considerations

The conventional histological sections on which bone histomorphometry is generally performed are viewed as a two-dimensional image in which profiles of three-dimensional structures are seen. Extrapolation of two-dimensional data to three-dimensional quantities necessitates the application of stereological formulae, which are based on the assumptions that sampling is unbiased and random and, for most applications, that the structure is isotropic, i.e. evenly dispersed and randomly orientated in space [2]. None of these conditions is totally fulfilled in the case of bone and the expression of histomorphometric indices as three-dimensional quantities is thus subject to some error. In practice, some histomorphometrists prefer to express data as two-dimensional values, whereas others apply stereological formulae to convert these to three-dimensional quantities. The approach adopted will affect absolute values and, hence, accuracy but should not influence comparisons between patient groups or the diagnostic value of histomorphometry, provided that the approach adopted is consistent.

The use of vertical sections to obtain unbiased stereological estimates from histological sections was first described by Baddeley *et al.* [3] and subsequently applied to bone [4]. The cycloid test system employed has the same axis as the sections and the test lines are defined in relation to the axis (sine-weighted). The bone biopsy is randomly rotated around its long axis before embedding and all sections are cut parallel to this axis (the latter also being the case for conventionally prepared sections); the vertical axis of the sections must be kept parallel to the vertical axis of the test system. This procedure has clear theoretical advantages over the non-random sampling and biased measurement procedures used by the majority of histomorphometrists, though the practical significance of these has not been established. Because of the need for random rotation before embedding, vertical sections as described above cannot be prepared from archival material. However, since random rotation is effectively (if unintentionally) performed prior to the embedding procedure, the application of the system described above to archival specimens appears justifiable.

7.4.2 Grids

Grids are arrangements of lines and points that may be inserted into the eyepiece of the microscope or outlined on a flat surface onto which the microscope image can be projected. A number of different systems have been described. Some of these contain horizontal and/or vertical lines with points – for example, the Zeiss Integrationsplatte 1 and 11; random test line orientation, which is theoretically superior, can be achieved using alternating hemispherical lines or, for vertical sections, a cycloid test grid. Alternatively, random orientation of the test lines may be achieved by random rotation of the graticule between each field of measurement [5]. Areas are measured by counting grid points which fall on a profile, and perimeters by using intersection by test lines. Distances such as osteoid seam and wall width are measured using an eyepiece micrometer. In order to avoid systematic sampling errors in width measurements, Kragstrup *et al.* [6] have described a method in which a measuring grid is moved over the section in equidistant steps and orthogonal intercepts on the structures undergoing width measurements are sampled. This method provides a systematic sampling procedure which takes into account the surface extent of the structure.

The advent of interactive computerized systems for bone histomorphometry has superseded the use of grids and micrometers. These systems are much less labour intensive and tedious for the operator and can perform complex measurements such as cancellous bone structural analysis, which cannot easily be achieved using non-computerized techniques. Several purpose-built systems are now commercially available; alternatively, in-house systems can be designed using relatively simple hardware, provided that programming expertise is available.

An example of such a system is illustrated in Plate 6. The system is developed around a PC386 computer (Elonex UK Ltd). Direct measurements are made using a digitizing tablet and cursor with an LED point light source and a binocular trans-mitted light microscope with a drawing attachment. Sections are viewed directly through the microscope, the cursor light being superimposed on the section. The field of measurement is defined by a square etched on the eyepiece graticule; the square is mapped to correspond with an active area on the tablet and an active drawing area on a video monitor. The system is calibrated for each magnification used. Perimeter measurements are made by tracing with the cursor LED and distance measurements are made either by dotting with the cursor on either side of the structure at appropriately spaced points or by tracing the outline of the struc-ture (for example, a bone structural unit), distances between the outer and inner boundaries then being made automatically at four equidistant points. For cancel-lous bone structural analysis [7, 8] the system used incorporates a frame-store board to capture images from a CCD television camera on a video monitor.

7.4.3 Terminology

In 1987 the American Society of Bone and Mineral Research Histomorphometry Nomenclature Committee proposed a standardized system for the description of histomorphometric indices derived from two-dimensional histological sections [9] (Tables 7.1–7.3). The proposed nomenclature has become widely accepted and has provided a much needed clarification of the sometimes arbitrary and often unin-telligible formats used previously. All data are expressed in the format of source (the structure on which the measurement is made; for example, bone tissue or surface), the measurement (primary or derived) and the referent (area or perimeter in two-dimensional and volume or surface in three-dimensional terminology). The recommended order and format to be used (including punctuation) is: source – measurement/referent. In practice, because only one source is often used in any one study, it is unnecessary to specify this once it has been defined. The most commonly used source in bone histomorphometry is cancellous bone (Cn); other possible sources include cortical bone (Ct), endocortical surface (Ec), periosteal surface (Ps) and transitional zone (Tr.Z). It should be noted that although the term 'cancellous' is preferred to 'trabecular' when referring to the source, 'trabecular' rather than 'cancellous' is used in descriptions of structural indices. Abbreviations or symbols are constructed from the first letters in the order of the words of the descriptive term, a single capital letter generally being used for the more commonly used terms and an additional lower case letter being used for less frequently used terms. The use of a single lower case letter is stated to denote terms related to time; here the system is somewhat inconsistent, since words such as hit, single, double and active are included in this category.

The primary measurements are referred to as area, perimeter and width when expressed in two dimensions and volume, surface and thickness in three dimensions. The fourth type of primary measurement, number, cannot be extrapolated to a three-dimensional value, unless serial sections are examined. Absolute area and perimeter measured in two dimensions have no three-dimensional equivalent but if three-dimensional nomenclature is adopted these quantities are referred to as volume and surface, though the absolute values are identical. Values for width can be converted to thickness by dividing width by $4/\pi$ (1.273) for isotropic structures or 1.2 for human iliac cancellous bone [2, 10]; the one exception to this rule is cortical width and thickness, which are numerically equal. Measurements of width may be obtained directly, or indirectly by calculation from area and perimeter; the former method has the advantage of providing information on the frequency distribution of widths within a biopsy.

Table 7.1   Referents commonly used in bone histomorphometry (from [9])

Referent (3D/2D)	Abbreviation (3D/2D)
Bone surface/perimeter	BS/B.Pm
Bone volume/area	BV/B.Ar
Tissue volume/area	TV/T.Ar
Core volume/area	CV/C.Ar
Osteoid surface/perimeter	OS/O.Pm
Eroded surface/perimeter	ES/E.Pm
Mineralized surface/perimeter	Md.S/Md.Pm
Osteoblast surface/perimeter	Ob.S/Ob.Pm
Osteoclast surface/perimeter	Oc.S/Oc.Pm

Table 7.2   Primary histomorphometric indices of bone remodelling

Name	Abbreviation	Units
Bone area	B.Ar/T.Ar	%
Osteoid area	O.Ar/T.Ar or O.Ar/B.Ar	%
Osteoid perimeter	O.Pm/B.Pm	%
Osteoblast perimeter	Ob.Pm/B.Pm	%
Osteoid width	O.Wi	μm
Interstitial width	It.Wi	μm
Trabecular width	Tb.Wi	μm
Eroded perimeter	E.Pm/B.Pm	%
Osteoclast perimeter	Oc.Pm/B.Pm	%
Mineralizing surface	M.Pm/B.Pm	%
Mineral appositional rate	MAR	μm/d
Wall width	W.Wi	μm
Erosion depth	E.De	μm
Erosion length	E.Le	μm
Erosion area	E.Ar	μm^2
Cavity number	N.Cv./B.Pm or /TA	No./mm or /mm^2

Table 7.3 Derived histomorphometric indices of bone remodelling

Name	*Abbreviation*	*Units*[b]
Adjusted appositional rate	Aj.AR	μm/d
Bone formation rate	BFR/B.Pm	μm²/μm/d
	BFR/B.Ar	%/y
Erosion rate	ER	μm/d
Mineralization lag time	Mlt	d
Osteoid maturation period	Omt	d
Formation period	FP	d
Active formation period	FP(a+)	d
Resorption period	RP	d
Reversal period	Rv.P	d
Quiescent period	QP	d
Remodelling period	Rm.P	d
Total period	Tt.P	d
Activation frequency	Ac.f	/y
Trabecular separation[a]	Tb.Sp	μm or mm
Trabecular number²	Tb.N	/mm

[a] May also be measured directly.
[b] d, day; y, year.

Area, perimeter and number have meaning only if expressed in terms of a referent, whereas distance measurements do not require any referent. Areas and perimeters (two-dimensional) or volumes and surfaces (three-dimensional) are used as referents for most measurements. When measurements with a surface referent are expressed as a percentage, the referent is termed bone surface (or a subdivision of it) and is equivalent to two-dimensional perimeter values. However, in the new nomenclature, the term 'bone surface' is also used specifically to denote bone surface to tissue volume ratio or surface density, a three-dimensional quantity expressed in units of mm²/mm³. This and the other surface/volume and volume/volume ratios are derived from the corresponding two-dimensional perimeter/area ratios using a multiplication factor of $4/\pi$ or 1.2 (see above).

Bone histomorphometry is most commonly applied to cancellous bone, which has a considerably higher turnover than cortical bone and hence a greater capacity to exhibit alterations in bone remodelling. Cancellous bone may be subdivided into cortico-endosteal and pure mid-cancellous regions; changes at the cortico-endosteal region are particularly important as they may affect both cortical and trabecular bone mass and structure. Cortical bone is largely ignored by most histomorphometrists in spite of its predominance in the skeleton and its importance as a determinant of fracture risk. Frost [11] pioneered the application of histomorphometric techniques to cortical bone and more recently, detailed analysis of the bone remodelling cycle in iliac crest biopsies has been reported [12].

7.4.4 Limitations of bone histomorphometry

Bone histomorphometry has a number of limitations. Some of these are inherent in the restrictions imposed by a single biopsy site and disease heterogeneity whilst others reflect imperfections in measurement techniques and difficulties in identification of some of the key processes in remodelling.

The large measurement variance associated with bone histomorphometry has been demonstrated by many groups and arises from a number of sources, including intra- and inter-observer variation, sampling variation and methodological factors [13–15]. Intra- and inter-observer variation reflect the subjective approach to identification of many of the histological features assessed – for example, resorption cavities and osteoid seams. Methodological factors include the criteria used for cortico-medullary differentiation (which are often arbitrary), the staining method used and the magnification at which measurements are made. The technique used for quantitation may also affect the values obtained, as a result of different sampling procedures and variations in the number of sampling points utilized. Many of these sources of variance can be minimized by standardization of staining, cortico-medullary delineation and magnification and the employment of criteria for identification of osteoid seams, resorption cavities and newly formed bone structural units.

The important issue of how closely bone remodelling in the iliac crest resembles that at other skeletal sites has not been fully resolved. Some metabolic bone disorders – for example, osteomalacia and some forms of renal osteodystrophy – appear to affect the whole skeleton and in such cases an iliac crest biopsy is representative. In osteoporosis, however, there is clear evidence of disease heterogeneity and it is well documented that bone volume in iliac crest biopsies is a poor indicator of bone loss elsewhere in the skeleton [16]. There is also some evidence for variations in bone turnover at different skeletal sites [17] but the demonstration of clear abnormalities of bone remodelling in iliac crest bone obtained from patients with osteoporosis indicates that changes responsible for the disease process are reflected, at least to some extent, in bone from this site. Similarly, changes in bone turnover and remodelling balance have been shown in patients with treated osteoporosis and have generally been consistent with the observed changes in bone mass at sites such as the spine and hip.

Finally, current histomorphometric techniques are seriously limited by the lack of reliable markers for activation and resorption. Dynamic indices related to these processes are at present calculated from bone formation rates, based on the assumptions that bone resorption and formation are coupled temporally and spatially and that bone remodelling is in a steady state. These are unlikely to be tenable in the presence of untreated or treated osteoporosis [18].

7.5 ASSESSMENT OF MINERALIZATION

7.5.1 *In vivo* tetracycline labelling

The administration of two time-spaced doses of a tetracycline derivative before bone biopsy enables measurement and calculation of dynamic indices of bone formation and, by extrapolation, bone resorption [19]. Various regimes have been described; most involve a 10 day gap between the two doses, bone biopsy being performed 3–5 days after the last dose. The regime used by the author is as follows.

- Days 1 and 2: 150 mg demeclocycline twice daily
- Days 3–10: no demeclocycline
- Days 13 and 14: 150 mg demeclocycline twice daily.

The bone biopsy is performed 3–5 days after day 14.

Adverse effects of demeclocycline and related compounds are rare but include diarrhoea and other gastrointestinal symptoms. Occasionally, skin rashes occur; these are sometimes severe and may exhibit photosensitivity.

There is evidence that different tetracycline compounds differ with respect to their uptake by mineralizing bone. Parfitt *et al.* [20] reported that demeclocycline labelling resulted in a greater surface extent of fluorescence than oxytetracycline, significantly affecting values for dynamic indices of bone formation. These differences should be borne in mind when comparing data between centres and, in particular, when using control data obtained from other sources.

In animals, other fluorochromes such as alizarin red and calcein, which are too toxic for human use, may also be used as markers of active mineralization and enable multiple labelling regimes to be administered in order to measure short-term changes in bone formation rates.

7.5.2 Measurement of osteoid

Primary measurements of osteoid include area, perimeter and seam width; assessment of dynamic indices of mineralization – for example, mineral appositional rate, mineralization lag time and osteoid maturation rate – requires double tetracycline labelling prior to biopsy. Calcification fronts along osteoid seams can be demonstrated by staining with toluidine blue (Plate 5), Sudan black B or thionin but the surface extent assessed in this way does not always correlate well with the mineralizing perimeter as measured using tetracycline uptake [21].

Osteoid seam width may be measured directly or calculated from osteoid area and perimeter. However, when relatively small amounts of osteoid are present, the latter approach is inaccurate since the value for osteoid area, expressed as a percentage of bone area, is influenced not only by the seam width but also by the mineralized bone area [22]. Direct measurement of seam width can be made using an eyepiece micrometer and calculating the mean of several equidistant measurements of width along each seam. Alternatively, using interactive systems, the width may be calculated automatically at equidistant points along each seam.

Measurements of osteoid, in particular its perimeter, are strongly influenced by the magnification used. At high magnifications, it becomes difficult to distinguish osteoid seams from the thin endosteal membrane covering the quiescent bone surface and for this reason it is preferable to use defined criteria; for example, all seams less than one lamella (approximately 3 μm) are excluded. Osteoid measurements may also be affected by the staining procedure used, and poor differentiation of osteoid from mineralized bone in images used for semi-automatic techniques may reduce the accuracy of measurements; thus values obtained using image analysis are generally lower than those generated by manual measurements [14]. Finally, the delineation of the cortico-medullary junction may affect the values obtained for primary measurements of osteoid, since osteoid amount tends to be greater in the cortico-endosteal region than in pure cancellous bone.

7.5.3 Dynamic indices of mineralization

The administration of time-spaced tetracycline labels prior to biopsy enables calculation of dynamic indices of matrix formation and mineralization. Tetracycline binds to calcium and becomes permanently incorporated into the mineralization fronts at sites of active mineralization (Plate 7). In this respect, the timing of the biopsy in relation to the labelling regime is crucial, a period of 3–5 days after the last label enabling deposition of a sufficiently thick layer of new mineral to retain the tetracycline label [23].

Tables 7.2 and 7.3 give abbreviations and units for primary and derived histomorphometric indices used in the following sections.

(a) Mineral appositional rate

The mineral appositional rate (MAR) (Table 7.2) is calculated as the distance between two time-spaced tetracycline labels divided by the time between the administration of these labels. Measurements are made from the mid-point of each label at approximately equidistant points along the labelled surface and the inter-label period is calculated as the number of days between the mid-points of the two labelling periods [24]. MAR is used in the calculation of many derived indices of bone formation and accurate measurement is thus of key importance. In the absence of tetracycline uptake, mineral appositional rate and derived indices should be treated as missing data; in biopsies in which only single labels can be detected, the finite lower limit of 0.3 μm/day for mineral appositional rate should be used for the calculation of derived indices [25].

(b) Adjusted appositional rate

The adjusted appositional rate (Aj.AR) (Table 7.3) represents the mineral appositional rate of the bone formation rate averaged over the osteoid surface. In the absence of a mineralization defect the apposition of matrix and mineral, whilst not synchronous, can be assumed to occur at the same rate, and under such circumstances the adjusted appositional rate is equivalent to the osteoid or matrix appositional rate. It is calculated as follows:

Aj. AR = MAR × M.Pm/O.Pm

From the above formula it is clear that Aj.AR is usually less than MAR and cannot exceed it.

(c) Mineralization lag time and osteoid maturation time

The mineralization lag time (Mlt) (Table 7.3) is the interval between deposition and mineralization of a given amount of osteoid, averaged over the life span of the osteoid seam. It is calculated as follows:

Mlt = O.Wi/Aj.AR

The osteoid maturation time (Omt) (Table 7.3) is the period between the deposition and onset of mineralization of a given amount of osteoid and results from processes such as collagen cross-linking which are necessary before mineralization can proceed. In humans it is usually shorter than Mlt and never exceeds it. It is calculated as follows:

Omt = O.Wi/MAR

7.6 ASSESSMENT OF BONE TURNOVER

Bone turnover describes the tissue level of bone resorption and formation, a key determinant of which is activation frequency. The uptake of tetracycline derivatives at sites of actively forming bone enables the rate of mineralization of osteoid seams and the surface extent of bone formation to be assessed and a number of derived indices, including activation frequency, to be calculated (Table 7.3).

7.6.1 Mineralizing perimeter

The extent of bone perimeter or surface which exhibits tetracycline fluorescence is an important primary measurement from which many dynamic indices are derived. When a double tetracycline label has been administered, both double and single labels will be seen; this reflects the labelling escape error, caused by initiation of mineralization before the first label or its termination between administration of the two labels [26] and probably also the switch from an active to resting state in a minority of osteoid seams (the so-called 'on/off' phenomenon) [2]. Because of the former, the extent of double-labelled surface underestimates the actively mineralizing surface and the double plus half the single label is therefore used to estimate the mineralizing perimeter. In cases where only single labels can be detected, it has been suggested that the mineralizing perimeter should be expressed as half the single-labelled perimeter [25].

In the absence of tetracycline administration prior to biopsy, the osteoid perimeter may provide some indication of bone turnover, increased osteoid perimeter being characteristic of high turnover states. An increase in the extent of perimeter occupied by resorption cavities does not necessarily imply increased bone turnover, however, since these may not represent active resorption but rather reflect failure of formation to occur in previously resorbed cavities.

7.6.2 Bone formation rates

Bone formation rates are usually expressed in terms of the bone perimeter or area. In the former case, either the osteoid perimeter may be used as the referent (adjusted appositional rate, Aj.AR) or the total bone perimeter (tissue-based bone formation rate, BFR/BS). They are calculated as follows:

Aj.AR = MAR × M.Pm/O.Pm

BFR/BS = MAR × M.Pm/B.Pm

The bone formation rate/bone area (BFR/B.Ar) is calculated as MAR × M.Pm × (B.Pm/B.Ar) × 100.

7.6.3 Bone resorption rates

Because of the lack of markers of active resorption analagous to the use of tetracycline to identify actively forming bone, bone resorption rates can only be calculated indirectly, from bone formation rates, based on the assumptions discussed earlier. Erosion rate (ER) is expressed in μm/day and calculated as follows:

ER = E.De/EP

(Calculation of the erosion period, EP, is shown below.)

7.6.4 Remodelling periods

The average duration of a single remodelling cycle is described as the remodelling period, which can be further divided into quiescent, erosion, reversal and formation remodelling periods [2]. The formation period (FP) can be further divided into the active (FPa+) and inactive formation period (FPa–) [27]; the latter is a measure of the 'off-time', which accounts for the discrepancy between osteoid and mineralizing perimeter after correction for label escape [26]. Formation periods are calculated as follows:

FP = W.Wi/Aj.AR

FPa+ = W.Wi/MAR

FPa– = FP – FPa+

Quiescent (QP), erosion (EP) and reversal (Rv.P) periods are calculated as follows:

QP = Q.Pm/B.Pm × FP

EP = E.Pm/B.Pm × FP

Rv.P = Rv.Pm/B.Pm × FP

The mean time between initiation of two successive remodelling cycles at the same site is defined as the total period (Tt.P) and calculated as follows:

Tt.P = Rm.P + QP

7.6.5 Activation frequency

Activation frequency (Ac.F) is a key determinant of bone mass in the adult skeleton and increased activation frequency, resulting in high bone turnover, forms quantitatively the most important mechanism of bone loss in osteoporosis [28]. At present, there are no *in situ* markers of activation and hence direct assessment of activation frequency cannot be made. Rather, it is calculated as the frequency with which a given site on the bone surface undergoes new remodelling, as follows:

Ac.f = 1/Tt.P or (BFR/B.Pm)/W.Wi

These formulae define activation frequency as the reciprocal of the time taken from the initiation of one remodelling cycle to initiation of a new one at that site, thus implying its dependence on the life span of previously activated units. Since activation may occur at any site on the bone surface, however, there seems no *a priori* reason why this should be so, particularly in non-steady states [18]. In addition, the use of indices of bone formation to calculate activation frequency relies on assumptions about the coupling of bone resorption and formation which are unlikely to be tenable in many disease states.

7.7 ASSESSMENT OF REMODELLING BALANCE

7.7.1 Bone formation

Within the individual bone remodelling unit, the amount of bone formed is termed the wall width (W.Wi) [29] and is measured as the mean width of completed bone structural units, identified under polarized light (Plate 8) or by stains such as toluidine blue or thionin [30] which demonstrate the cement line. Completed bone structural units are identified by the absence of resorption lacunae or osteoid.

Investigation of the effect of a disease or its treatment on wall width (and calculated dynamic indices for which wall width is required) requires differentiation of units formed during the period of observation from those formed prior to this time. This can only be achieved by identification of uncompleted bone structural units, which have a covering of osteoid and hence can be presumed to represent current or recent remodelling activity. Reconstruction of these forming sites can then be performed [31]; however, the number of such units that can be identified in any one biopsy is usually very small.

7.7.2 Bone resorption

There are several problems associated with accurate assessment of the amount of bone resorbed during each remodelling cycle. The identification of resorption cavities is often difficult and to some extent subjective; in addition, it is difficult to identify those cavities in which resorption has been completed. The use of polarized light microscopy to demonstrate cut-off lamellae at the edges of the cavity assists recognition [32] (Plate 9), as does the presence of osteoclast-like cells within the cavity. More precise identification of osteoclasts can be achieved by histochemical techniques which demonstrate the presence of tartrate-resistant acid phosphatase (TRAP), though this is not specific to osteoclasts [33, 34]. For further discussion of osteoclast identification, see Chapters 1 and 2.

Direct measurement of erosion depth was first reported by Eriksen *et al.* [35], using a method in which the number of lamellae eroded beneath the bone surface is counted and cavities are characterized according to the presence of osteoclasts, mononuclear cells and preosteoblastic cells on the assumption that these cell types are specifically associated with increasing stages of completion of the resorptive phase. This approach depends critically on the accurate identification of the different cell types within resorption cavities; even in high quality histological sections this can be difficult and in the authors' hands 24% of resorption cavities were excluded from measurement because of failure to classify by cell type or inability to define and count the eroded lamellae. This method has not been widely adopted by other groups, though some have used a simplified approach in which the eroded lamellae are counted without subdivision of cavities by cell type [36]. The latter technique provides an estimate of the mean depth of cavities in all stages of completion and thus underestimates the final resorption depth.

The computerized method developed by Garrahan *et al.* [37] involves reconstruction of the eroded bone surface by a curve-fitting technique (cubic spline) and provides measurements of mean and maximum erosion depth together with the area, surface extent and number of cavities (Plate 10). Since all resorption cavities are included in the measurements, the mean values for mean and maximum depth and for area considerably underestimate the final resorption depth and area. This method may also be performed using interactive reconstruction of the eroded bone surface [38]. Another modification is to include for measurement only those cavities that contain a thin layer of osteoid, ensuring that the final resorption depth has been achieved [38]; however, the number of such cavities that can be identified in a single biopsy is usually extremely small. Interestingly, the values reported using this approach in normal human biopsies were approximately 17% lower than those obtained in the same biopsies by the technique of counting eroded lamellae and were more consistent with reported age-related changes in trabecular width [39, 40].

Further work is required to improve existing techniques for the measurement of erosion depth. The approaches currently used most widely assess the depth of all resorption cavities and therefore underestimate the completed erosion depth, whereas measurements made using Eriksen's method are likely to overestimate the true value [41]. Although these limitations preclude accurate assessment of remodelling balance at present, measurement of erosion depth has generated valuable information about the pathophysiology of bone loss in untreated and treated disease states.

7.8 ASSESSMENT OF BONE STRUCTURE

The importance of cortical and cancellous bone structure as a determinant of bone strength is well established and there has been increasing interest over recent years in the quantitative assessment of bone microarchitecture. Structural determinants of the mechanical strength of bone include cortical width and porosity and, in cancellous bone, trabecular size, shape, connectivity and anisotropy. Early approaches to the quantitative assessment of cancellous bone structure were based on direct or indirect measurements of trabecular width, separation and number [42–44]. In recent years more sophisticated methods have been developed which enable assessment of connectivity and, in some cases, anisotropy.

Since all of the structural determinants of cancellous bone strength are three-dimensional characteristics, their assessment on conventional histological sections provides only indirect information about these qualities and three-dimensional images are required for direct measurement of connectivity, anisotropy and trabecular size and shape [45]. Although further studies are required to examine the relationship between structural indices obtained using two- and three-dimensional approaches, there are several lines of evidence to support the contention that measurements from two-dimensional sections are representative of three-dimensional structure [46, 47].

7.8.1 Assessment of trabecular width, separation and number

Direct measurements of trabecular width can be made using an eyepiece graticule or grid but these are time-consuming, provide values for width only at relatively widespread points along the trabeculae and usually omit measurements at nodes [42–44]. More recently, computerized techniques have been described [48, 49]; these enable rapid measurements of width to be made at intervals of the operator's choice and include width measurements at the nodes. One such technique involves skeletonization of the trabecular bone and subsequent generation of circles around the symmetric axis which are simultaneously tangential to the two opposing boundaries. Measurements are performed automatically on stored, edited binary images [49].

Alternatively, trabecular width can be calculated from area and perimeter measurements [50]; the validity of this approach depends on the assumption that the width of the measured structures is small relative to length. This is generally, though not always, true of trabecular bone in the iliac crest and trabecular width calculated in this way shows a good correlation with directly measured width.

Trabecular separation may also be measured directly using an eyepiece graticule or grid [43, 44]; these measurements are made between the mid-points of the trabeculae and hence are unaffected by trabecular thinning. Trabecular spacing, also defined as the distance between mid-points, can be calculated as the reciprocal of trabecular number [50]. More commonly, trabecular separation is calculated as follows [50]:

$$\text{trabecular separation (μm)} = \text{trabecular width (μm)} \times \frac{100}{(\text{TV/BV} - 1)}$$

or

$$\text{trabecular separation (μm)} = (1000/\text{trabecular number}) - \text{trabecular width (μm)}$$

Trabecular number or density can be assessed directly by counting but is usually calculated according to the parallel plate model as follows [50]:

$$\text{Trabecular number} = (\text{BV/TV})/\text{trabecular width}$$

These calculations of trabecular separation and number are based on the assumption that cancellous bone consists of parallel trabecular plates and are therefore not strictly applicable to the iliac crest.

7.8.2 Strut analysis

In 1986 Garrahan *et al.* [7] described a new technique for the quantitative analysis of cancellous bone structure, based on the topological classification of trabeculae or struts and on the definition of free ends (termini) and nodes (junctions). Using a computerized image analysis system, the number of termini and nodes are counted automatically and a node/terminus ratio derived. The different strut types are classified as shown in Plate 11; node-to-loop and node-to-node strut types are positively related to connectivity, whereas struts with termini are inversely related. The total length of each individual strut type within a section is measured automatically and expressed as a percentage of the total strut length within the section or per mm^2 of section. Whilst termini may clearly be created artefactually by the process of sectioning, nodes, node-to-node and node-to-loop struts are true indices of connectivity although their number may be underestimated. Termini may also result from sectioning through a trabecular window in a connected structure.

This technique was originally based on expensive hardware (Ibas II image analyser) but has now been adapted for use on simpler and more widely used systems. Indices such as nodes, free ends and different strut types may also be counted manually [51], but this approach is relatively labour intensive.

7.8.3 Star volume

The star volume is defined as the mean volume of solid material or empty space that can be seen unobscured from a point of measurement chosen at random inside the material [52]. The application of this parameter to bone was described by Vesterby [53] using the vertical section technique [3] and a cycloid test system. The method provides an unbiased stereological estimate of the mean size of the trabeculae and of the marrow space and can be used both for measurements of trabecular width (trabecular star volume) and trabecular separation or connectivity (marrow space star volume); it was initially applied to assessment of vertebral bone structure but, with some modifications, can also be used in iliac crest bone. The method involves the generation of intercepts from random sampling points, the cubed length of intercepts being used in the calculation of star volume (Plate 12). Values generated by marrow space star volume measurements are significantly influenced by biopsy size, particularly in poorly interconnected cancellous bone, in which a large proportion of the measured intercepts may hit the boundary rather than bone, resulting in underestimation of star volume [54].

The requirement for randomly orientated 'vertical sections' for assessment of star volume is probably not absolute, since random rotation of the biopsy core is likely to occur during the normal embedding process. Hence the technique may be applicable to archival material, provided that the cycloid test system is used.

7.8.4 Trabecular bone pattern factor

This index is related to trabecular connectivity and is based on the concept that patterns or structures can be defined by the relationship between convex and concave surfaces [55], convexity indicating poor connectivity and concavity being associated with structural integrity. Convexity and concavity are assessed by the measurement of bone perimeter and area before and after dilatation of a structure; thickening of convex structures increases the perimeter whilst the reverse applies to concave structures. The change in bone perimeter after dilatation thus depends on the ratio of convex to concave structures. Computer-based dilatation is carried out on binary images (Plate 13) and the trabecular bone pattern factor (TBPf) is calculated as follows:

$$\text{TBPf} = (P1 - P2)/(A1 - A2)$$

where P1 = first perimeter measurement, P2 = second perimeter measurement, A1 = first area measurement and A2 = second area measurement.

Values obtained for trabecular bone pattern factor may be influenced by the computer-based smoothing technique used, the degree of dilatation employed and the magnification at which the measurements are performed.

7.8.5 Fractal analysis

Fractal objects are characterized by scale invariance or self-similarity over a wide range of magnifications so that any one piece of a fractal, if magnified sufficiently, resembles the intact object [56]. Fractal analysis has been described in a number of biological systems, including the bronchial tree and vascular networks; in bone, it has been applied to both radiological images and histological sections [57]. The value obtained for the fractal dimension depends critically on the magnification used for measurement and the relationship between fractal dimension and connectivity in cancellous bone has not been established. Multidirectional fractal analysis can be used to assess structural anisotropy [58].

7.8.6 Three-dimensional approaches

In recent years a number of techniques have been used to generate three-dimensional images of bone. These include reconstruction of serial sections, scanning and stereo microscopy, volumetric, high resolution and microcomputed tomography and magnetic resonance imaging [59, 60]. The potential for imaging techniques such as magnetic resonance imaging and computed tomography to provide information about bone structure *in vivo* is an important and active area of current research. At present, such application of these approaches is restricted by limited resolution, partial volume effects and noise; nevertheless, such approaches provide a means whereby direct assessment of connectivity, anisotropy and trabecular size and shape may be directly assessed.

(a) High resolution microcomputed tomography
This technique generates a three-dimensional reconstruction array from which bone area, bone perimeter, trabecular width, separation and density can be determined and structural anisotropy examined [46]. Connectivity is assessed using the Euler number, which is calculated from the number of nodes and the number of branches. Breaking a single connection increases the Euler number by 1, whilst the creation of a connection increases it by the same amount. An alternative approach to the production of three-dimensional images is reconstruction of serial sections, for which automated techniques have recently been described [47].

Connectivity is independent of trabecular shape, size or orientation and its ability to predict the mechanical strength of bone is thus limited; Goldstein *et al.* [61] reported that bone volume and anisotropy accounted for 90% of the variance in elastic modulus and ultimate strength in human metaphyseal bone. Correlations have been demonstrated between a number of two-dimensional structural indices and mechanical strength. Thus Dempster *et al.* [62] reported a significant relationship between trabecular thickness and ultimate compressive strength in the second lumbar vertebra and structural indices obtained by strut analysis have also been shown to correlate with mechanical strength at this site [51]. The marrow space star volume of vertebral bone is also correlated with compressive strength [63]; however, in all these studies, bone volume overshadowed structure as a predictor of strength.

(b) Euler number and the Conn-Euler
The Euler number is a topological property based on the number of holes and connected components in an object and provides a direct measure of connectivity. The Conn-Euler, described by Gundersen *et al.* [64], is a method in which the Euler number is calculated on projections through parallel thin section pairs, or disectors, spaced 10-40 μm apart. The Euler number is calculated from the numbers of islands, holes and bridges identified in the disectors.

7.9 FUTURE DEVELOPMENTS

Despite the current limitations of bone histomorphometry, its value in defining pathophysiological processes responsible for disease and its unique ability to demonstrate the mechanisms by which drugs affect bone are increasingly recognized. The rapid advances which have occurred in molecular biological techniques and in our knowledge of bone cell biology should soon enable better identification of the processes of activation and resorption *in situ*, leading to a better understanding of mechanisms of bone loss and bone gain. Finally, the rapid advances which are being made in the *in vivo* assessment of cancellous bone architecture will provide new insights into structural determinants of bone strength and the ability of anabolic skeletal agents to restore bone architecture in patients with advanced bone loss.

REFERENCES

1. Rao, D.S. (1983) Practical approach to bone biopsy, in *Bone Histomorphometry: Techniques and Interpretation*, (ed. R. Recker), CRC Press, Boca Raton, Florida, pp. 3–11.
2. Parfitt, A.M. (1983) The physiological and clinical significance of bone histomorphometric data, In *Bone Histomorphometry: Techniques and Interpretations*, (ed. R. Recker), CRC Press, Boca Raton, Florida, pp. 143–224.
3. Baddeley, A.J., Gundersen, H.J.G. and Cruz Orive, L.M. (1986) Estimation of surface area from vertical sections. *Journal of Microscopy* **142**, 259–276.
4. Vesterby, A., Kragstrup, J., Gundersen, H.J.G. and Melsen, F. (1987) Unbiased stereologic estimation of surface density in bone using vertical sections. *Bone* **8**, 13–17.
5. Kragstrup, J., Melsen, F. and Mosekilde, L. (1982) Reduced wall thickness of completed remodelling sites in iliac trabecular bone following anticonvulsant therapy. *Metabolic Bone Disease and Related Research* **4**, 181–185.
6. Kragstrup, J., Gundersen, H.J.G., Mosekilde, L. and Melsen, F. (1982) Estimation of the three dimensional wall thickness of completed remodelling sites in iliac trabecular bone. *Metabolic Bone Disease and Related Research* **4**, 113–119.
7. Garrahan, N.J., Mellish, R.W.E. and Compston, J.E. (1986) A new method for the analysis of two-dimensional trabecular bone structure in human iliac crest biopsies. *Journal of Microscopy* **142**, 341–349.
8. Compston, J.E., Garrahan, N.J., Croucher, P.I. and Yamaguchi, K. (1993) Quantitative analysis of trabecular bone structure. *Bone* **14**, 187–192.
9. Parfitt, A.M., Drezner, M.K., Glorieux, F.H. *et al.* (1987) Bone histomorphometry. Standardization of nomenclature, symbols and units. *Journal of Bone and Mineral Research* **2**, 595–610.
10. Schwartz, M.P. and Recker, R.R. (1981) Comparison of surface density and volume of human iliac trabecular bone measured directly and by applied sterology. *Calcified Tissue International* **33**, 561–565.
11. Frost, H.M. (1963) Mean formation time of human osteons. *Canadian Journal of Biochemistry and Physiology* **41**, 1307–1310.
12. Agerbaek, M.O., Eriksen, E.F., Kragstrup, J. *et al.* (1991) A reconstruction of the remodelling cycle in normal human cortical iliac bone. *Bone and Mineral* **12**, 101–112.
13. de Vernejoul, M.C., Kuntz, D., Miravet, L. *et al.* (1981) Histomorphometric reproducibility in normal patients. *Calcified Tissue International* **33**, 369–374.
14. Chavassieux, P.M., Arlot, M.E. and Meunier, P.J. (1985) Intermethod variation in bone histomorphometry: comparison between manual and computerised methods applied to iliac bone biopsies. *Bone* **6**, 211–219.
15. Wright, C.D.P., Vedi, S., Garrahan, N.J. *et al.* (1992) Combined inter-observer and inter-method variation in bone histomorphometry. *Bone* **13**, 205–208.
16. Compston, J.E., Crowe, J.P., Wells, I.P. *et al.* (1980) Vitamin D prophylaxis and osteomalacia in chronic cholestatic liver disease. *Digestive Diseases* **25**, 28–32.
17. Eventov, I., Frisch, B., Cohen, Z. and Hammel, I. (1991) Osteopenia, hematopoiesis, and bone remodelling in iliac crest and femoral biopsies: a prospective study of 102 cases of femoral neck fractures. *Bone* **12**, 1–6.

18. Compston, J.E. and Croucher, P.I. (1991) Histomorphometric assessment of trabecular bone remodelling in osteoporosis. *Bone and Mineral* **14**, 91–102.

19. Frost, H.M. (1969) Tetracycline-based histological analysis of bone remodelling. *Calcified Tissue International* **3**, 211–237.

20. Parfitt, A.M., Foldes, J., Villanueva, A.R. and Shih, M.S. (1991) Difference in length between demethylchlortetracycline and oxytetracycline: implications for the interpretation of bone histomorphometric data. *Calcified Tissue International* **48**, 74–77.

21. Compston, J.E., Vedi, S. and Webb, A. (1985) Relationship between toluidine blue-stained calcification fronts and tetracycline labelled surfaces in normal human iliac crest biopsies. *Calcified Tissue International* **37**, 32–35.

22. Vedi, S. and Compston, J.E. (1984) Direct and indirect measurements of osteoid seam width in human iliac crest biopsies. *Metabolic Bone Disease and Related Research* **5**, 269–274.

23. Parfitt, A.M. (1990) Osteomalacia and related disorders, in *Metabolic Bone Disease*, 2nd edn, (ed. S.M. Krane), Grune and Stratton, New York.

24. Frost, H.M. (1983) Bone histomorphometry: analysis of trabecular bone dynamics, in *Bone Histomorphometry: Techniques and Interpretations*, (ed. R. Recker), CRC Press, Boca Raton, Florida, pp. 109–131.

25. Foldes, J., Shih, M.-S. and Parfitt, A.M. (1990) Frequency distributions of tetracycline-based measurements: implications for the interpretation of bone formation indices in the absence of double-labelled surfaces. *Journal of Bone and Mineral Research* **5**, 1063–1067.

26. Frost, H.M. (1983) Bone histomorphometry: choice of marking agent and labelling schedule, in *Bone Histomorphometry: Techniques and Interpretations*, (ed. R. Recker), CRC Press, Boca Raton, Florida, pp. 37–51.

27. Arlot, M., Edouard, C., Meunier, P.J. *et al.* (1984) Impaired osteoblast function in osteoporosis: comparison between calcium balance and dynamic histomorphometry. *British Medical Journal* **289**, 517–520.

28. Frost, H.M. (1985) The pathomechanics of osteoporosis. *Clinical Orthopaedics and Related Research* **200**, 198–225.

29. Lips, P., Courpron, P. and Meunier, P.J. (1978) Mean wall thickness of trabecular bone packets in the human iliac crest: changes with age. *Calcified Tissue Research* **26**, 13–17.

30. Derkz, P. and Birkenhäger-Frenkel, D.H. (1995) A thionin stain for visualizing bone cells, mineralizing fronts and cement lines in undecalcified bone sections. *Biotechnic and Histochemistry* **70**, 70–74.

31. Steiniche, T., Eriksen, E.F., Kudsk, H. *et al.* (1992) Reconstruction of the formative site in trabecular bone by a new, quick, and easy method. *Bone* **13**, 147–152.

32. Vedi, S., Tighe, J.R. and Compston, J.E. (1984) Measurement of total resorption surface in iliac crest trabecular bone in man. *Metabolic Bone Disease and Related Research* **5**, 275–280.

33. Burstone, M.S. (1959) Histochemical demonstration of acid phosphatase activity in osteoclasts. *Journal of Histochemistry and Cytochemistry* **7**, 39–41.

34. Evans, R.A., Dunstan, C.R. and Baylink, D.J. (1979) Histochemical identification of osteoclasts in undecalcified sections of human bone. *Mineral and Electrolyte Metabolism* **2**, 179–185.

35. Eriksen, E.F., Gundersen, H.J.G., Melsen, F. and Mosekilde, L. (1984) Reconstruction of the resorptive site in iliac trabecular bone; a kinetic model for bone resorption in 20 normal individuals. *Metabolic Bone Disease and Related Research* **5**, 235–242.

36. Palle, S., Chappard, D., Vico, L. *et al.* (1989) Evaluation of the osteoclastic population in iliac crest biopsies from 36 normal subjects: a histoenzymologic and histomorphometric study. *Journal of Bone and Mineral Research* **4**, 501–506.

37. Garrahan, N.J., Croucher, P.I. and Compston, J.E. (1990) A computerised technique for the quantitative assessment of resorption cavities in trabecular bone. *Bone* **11**, 241–246.

38. Cohen-Solal, M.E., Shih, M.-S., Lundy, M.W. and Parfitt, A.M. (1991) A new method for measuring cancellous bone erosion depth: application to the cellular mechanisms of bone loss in postmenopausal osteoporosis. *Journal of Bone and Mineral Research* **6**, 1331–1338.

39. Weinstein, R.S. and Hutson, M.S. (1987) Decreased trabecular width and increased trabecular spacing contribute to bone loss with ageing. *Bone* **8**, 137–142.

40. Mellish, R.W.E., Garrahan, N.J. and Compston, J.E. (1989) Age-related changes in trabecular width and spacing in human iliac crest biopsies. *Bone and Mineral* **6**, 331–338.

41. Parfitt, A.M. (1991) Bone remodelling in Type 1 osteoporosis (letter). *Journal of Bone and Mineral Research* **6**, 95–97.

42. Wakamatsu, E. and Sissons, H.A. (1969) The cancellous bone of the iliac crest. *Calcified Tissue Research* **4**, 147–161.

43. Whitehouse, W.J. (1974) The quantitative morphology of anisotropic trabecular bone. *Journal of Microscopy* **101**, 153–168.

44. Aaron, J.E., Makins, N.B. and Sagreiya, K. (1987) The microanatomy of trabecular bone loss in normal aging men and women. *Clinical Orthopaedics and Related Research* **215**, 260–271.

45. Compston, J.E. (1994) Connectivity of cancellous bone: assessment and mechanical implications. *Bone* **15**, 463–466.

46. Feldkamp, L.A., Goldstein, S.A., Parfitt, A.M. *et al.* (1989) The direct examination of bone architecture in vitro by computed tomography. *Bone* **4**, 3–11.

47. Odgäard, A. and Gundersen, H.J.G. (1993) Quantification of connectivity in cancellous bone with special emphasis on 3-D reconstruction. *Bone* **14**, 173–182.

48. Clermonts, E.C.G.M. and Birkenhäger-Frenkel, D.H. (1985) Software for bone histomorphometry by means of a digitizer. *Comput. Math. Prog. Biomed.* **21**, 185–194.

49. Garrahan, N.J., Mellish, R.W.E., Vedi, S. and Compston, J.E. (1987) Measurement of mean trabecular plate thickness by a new computerized method. *Bone* **8**, 227–230.

50. Parfitt, A.M., Mathews, C.H.E., Villanueva, A.R. *et al.* (1983) Relationship between surface, volume and thickness of iliac trabecular bone in aging and in osteoporosis: implications for the microanatomic and cellular mechanism of bone loss. *Journal of Clinical Investigation* **72**, 1396–1409.

51. Mellish, R.W.E., Ferguson-Pell, M.W., Cochran, G.V.B. *et al.* (1991) A new manual method for assessing two-dimensional cancellous bone structure: comparison between iliac crest and lumbar vertebra. *Journal of Bone and Mineral Research* **6**, 689–696.

52. Serra, J. (1982) *Image Analysis and Mathematical Morphology*, Academic Press, London.

53. Vesterby, A. (1990) Star volume of marrow space and trabeculae in iliac crest: sampling procedure and correlation to star volume of first lumbar vertebra. *Bone* **11**, 149–155.

54. Croucher, P.I., Garrahan, N.J. and Compston, J.E. (1996) Assessment of cancellous bone structure: comparison of strut analysis, trabecular bone pattern factor and marrow space star volume. *Journal of Bone and Mineral Research* **11**, 955–961.

55. Hahn, M., Vogel, M., Pompesius-Kempa, M. and Delling, G. (1992) Trabecular bone pattern factor – a new parameter for simple quantification of bone microarchitecture. *Bone* **13**, 327–330.

56. Mandelbrot, B.B. (1977) *Fractals: Form, Chance and Dimension*, W.H. Freeman, San Francisco.

57. Weinstein, R.S., Majumdar, S. and Genant, H.K. (1992) Fractal geometry applied to the architecture of cancellous bone biopsy specimens. *Bone* **13**, A38.

58. Jacquet, G., Ohley, W.J., Mont, M.A. *et al.* (1990) Measurement of bone structure by fractal dimension. *Proceedings of the Annual Conference of IEEE/EMBS* **12**, 1402–1403.

59. Majumdar, S. and Genant, H.K. (1995) A review of the recent advances in magnetic resonance imaging in the assessment of osteoporosis. *Osteoporosis International* **5**, 79–92.

60. Genant, H.K., Engelke, K., Fuerst, T. *et al.* (1996) Noninvasive assessment of bone mineral and structure: state of the art. *Journal of Bone and Mineral Research* **11**, 707–730.

61. Goldstein, S.A., Goulet, R. and McCubbrey, D. (1993) Measurement and significance of three-dimensional architecture to the mechanical integrity of trabecular bone. *Calcified Tissue International* **53** (Suppl. 1), S127–133.

62. Dempster, D.W., Ferguson-Pell, M.W., Mellish, R.W.E. *et al.* (1993) Relationship between bone structure in the iliac crest and bone structure and strength in the lumbar spine. *Osteoporosis International* **3**, 90–96.

63. Vesterby, A., Mosekilde, L., Gundersen, H.J.G. *et al.* (1991) Biologically meaningful determinants of the in vitro strength of lumbar vertebrae. *Bone* **12**, 219–224.

64. Gundersen, H.J.G., Boyce, R.W., Nyengaard, J.R. and Odgäard, A. (1993) The Conneulor: unbiased estimation of connectivity using physical disectors under projection. *Bone* **14**, 217–222.

CHAPTER EIGHT

The application of immunocytochemistry and in situ hybridization to cryostat sections of undecalcified bone

Robert A. Dodds, Janice R. Connor and Ian E. James

8.1 INTRODUCTION

This chapter presents the *in situ* techniques for determining the temporal and spatial expression patterns of both mRNA (*in situ* hybridization) and protein (immunocytochemistry) in individual cells within sections of bone.

In situ hybridization techniques involve the annealing of a specifically labelled nucleic acid probe to the complementary sequences of cellular RNA. Genes are expressed at highly variable levels (potentially undetectable); within one tissue, genes may be in a minority of the cells or temporally expressed during the differentiation of a certain cell type. Although *in situ* hybridization identifies which cells synthesize mRNA transcripts for a given gene, it cannot determine if the mRNAs are translated into protein. Furthermore, the relative amount of mRNA may not reflect similar amounts of synthesized protein. However, this technique is especially powerful when used in conjunction with immunocytochemistry – namely, the high affinity binding of antibodies to specific antigens within tissue sections. Similarly, immunolocalization of secreted proteins may give rise to 'false' expression patterns if the proteins are taken up by other cells or bind to extracellular matrix immediately adjacent to neighbouring cells; mRNA expression of the corresponding protein would clarify the expression pattern.

8.2 *IN SITU* ANALYSIS OF BONE SECTIONS

The use of osteoblast-like and osteoclast-like cells in culture coupled to the use of molecular and immunobiological techniques has enabled rapid advances in our understanding of the expression, regulation and role of proteins involved in bone formation and resorption (reviewed in Chapters 1 and 3). *In situ* hybridization and

Methods in Bone Biology. Edited by Timothy R. Arnett and Brian Henderson.
Published in 1997 by Chapman & Hall, London. ISBN 0 412 75770 2.

immunocytochemical studies on sections of bone would allow observations of bone cells to be interpreted in relation to the morphology and the disposition of other cell types present in bone, including adipocytes, macrophages, mast cells and endothelial cells. Although *in situ* studies of adult bone remodelling are generally impeded by the nature of the tissue (consisting of a hard mineralized matrix with relatively few bone cells per unit volume), the ability to section unfixed, undecalcified bone sections by suitably modified cryostat microtomy provides an ideal substrate for both *in situ* hybridization and immunolocalization [1]. Used in combination with histomorphometric and cytochemical methodology, analysis could be made at the level of the basic multicellular unit of bone (BMU) and related to the metabolic characteristics of the cell at definitive stages of bone remodelling (discussed by Bradbeer [2]). The difficulty of applying *in situ* techniques to adult remodelling bone *in situ* has also led bone biologists to rely on the use of developing fetal, neonatal or growing animal bones. Despite the softer nature of these tissues, researchers generally fix, demineralize and embed the bone prior to sectioning. Although these harsh procedures preserve adequate morphology, they would ultimately be deleterious if the various *in situ* techniques are to be performed individually on serial sections. This chapter focuses on the use of cryostat sections of fresh, unfixed, undecalcified developing and remodelling bone. When applicable, fixation and demineralization procedures are done subsequently on the undecalcified sections, leaving the original bone sample untreated.

In situ hybridization and immunolocalization of bone sections requires high quality sectioning. In fact many of the problems that arise when using these techniques are a consequence of poor tissue procurement, freezing or sectioning. To approach the quality of sections obtained by pre-embedding techniques, a number of crucial steps must be followed. The first part of this chapter will deal with the basic processing and sectioning of frozen bone. A detailed account of cryotomy [3] is not within the scope of this chapter so this report will only outline the various bone-related problems associated with this difficult technique. The bulk of the chapter will describe the methods, pitfalls (and troubleshooting) of immunocytochemistry and *in situ* hybridization as applied to undecalcified bone sections, as used routinely in our laboratory. These are rapidly advancing fields and a number of novel approaches (often in well characterized kit forms) will also be briefly discussed (e.g *in situ* PCR, non-radioactive *in situ* hybridization, and new amplification systems for immunolocalization). Finally, we will review some of the recent work from our laboratory utilizing these techniques on human bone and human bone-derived cells.

8.3 BASIC TISSUE PROCESSING

Although the protocols detailed in this chapter specify cryostat sections, modified methods can be used successfully with dewaxed and hydrated paraffin sections and plastic embedded (undecalcified) sections [4–7]. We have not discussed or outlined ultrastructural methodologies. This section will outline the procedures (including troubleshooting) for bone procurement, freezing and microtomy.

8.3.1 Bone procurement and freezing

A critical account of freezing procedures (and associated ice damage) is reviewed by Chayen and Bitensky [3]. Briefly, bones should be trimmed to an appropriate size (<5 mm^3) and frozen directly after they are removed from the animal or human. To prevent cracking during freezing, bone samples should be dipped in a 5% aqueous solution of polyvinyl alcohol (PVA; Sigma). The layer of PVA that coats the block also helps to bond the specimen to the microtome chuck and prevents thawing during mounting (see later). The two most widely used forms of freezing are the hexane (or isopentane) chilling bath and immersion in liquid nitrogen. The recommended procedure is the former in which each specimen is individually cooled by precipitate immersion in hexane at −70°C (see procedure below). In principle, this method causes the water in the tissue to be supercooled, thus avoiding the phase change that leads to ice damage. Freezing in liquid nitrogen actually forms an insulating layer of gaseous nitrogen around the warm specimen, resulting in ice damage. A number of other freezing procedures cause severe deterioration of tissue and must be avoided: (i) freezing bone slowly on a block of solid carbon dioxide or cryostat freezing bar; (ii) freeze-spraying within the cryostat chamber; or (iii) placing bone directly in a freezer (−20°C or −70°C) until frozen. These procedures cause severe damage within cells characterized by denaturation of the cytoplasm, rupturing of lipid–protein complexes, loss of membrane integrity and ultimately mechanically shattered cells.

Another important consideration is tissue storage after freezing: bone samples (or any tissue) should be stored at −70°C (in airtight plastic or glass tubes), immediately after freezing. A small piece of tissue paper at the bottom of the tube soaks up the excess hexane (hexane will cause severe solvent damage, even at temperatures as low as −30°C). Tissue gradually dessicates in storage at low temperatures; this can be minimized by including some ice powder (frost from inside the cryostat) or pre-frozen wetted tissue paper in the storage tube.

Materials and equipment

- **Hexane** (BDH 'low in aromatic hydrocarbons' grade, boiling range 67–70°C)
- **Ethanol** (Baxter Scientific Products)
- **Chilling bath** (see procedures)
- **Dry ice** (crushed)
- **Polyvinyl alcohol** (5% PVA; Sigma)
- **Forceps** (room temperature and −70°C).

Chilling bath procedure

1. Small chips of solid carbon dioxide (crushed dry ice) are added to alcohol in an insulated glass container (pyrex) until a saturated slurry is obtained.
2. A pyrex beaker containing approximately 30 ml hexane is inserted into the bath and allowed to stand for 20 minutes. (An alternative to this procedure is to fill a small 100 ml bottle with hexane, surround with dry ice and leave for 15 minutes prior to adding to the beaker; this provides a supply of fresh chilled hexane.)
3. The temperature of the hexane should be tested and be at least −64°C. (This should be tested periodically when chilling numerous blocks of bone.)
4. Immerse the bone samples (< 5 mm^3 in size) into 5% PVA solution.
5. Drop PVA-coated bone (using room temperature forceps) into the −70°C hexane and chill for 30 seconds.

6. Transfer the frozen bone (with –70°C forceps) to a dry tube prechilled to –70°C, containing a small piece of tissue paper.
7. Store frozen tissue at –70°C.

8.3.2 Mounting the bone

While mounting the frozen bone onto the chuck, care must be taken not to allow the tissue to warm. The procedure described below is performed within a Bright/Hacker cryostat (model OTF/AS/MR), fitted with a freezing bar. Alternative methods are described in Chayen and Bitensky [3].

Materials and equipment
- **Brass chucks** (25 mm long/22 mm diameter; Chuck Banski, Machine Shop Service.)
- **5% PVA**
- **Plastic pipette**
- **Cryostat freezing bar** (Bright/Hacker cryostat, OTF/AS/MR)
- **–70°C forceps**.

Procedure
1. Place a solid brass chuck onto the freezing bar within the cryostat (–50°C) and allow to stand for 10 seconds. Pipette a few drops (5–10) of 5% PVA (not water) on top of the chuck and allow to freeze partially (PVA turns white).
2. Orientate tissue appropriately, place it in the remaining liquid PVA; the residual PVA will freeze the tissue to the already frozen PVA. Additional PVA can then be added to the sides of the tissue block to ensure that it is secure. (The original coating of PVA prevents any thawing at this stage.)

8.3.3 Microtomy

For detailed procedures of cryostat microtomy refer to Chayen and Bitensky [3]. This section will highlight some of the important features associated with sectioning bone, together with many of the problems and remedies. As in the previous section, the information presented refers to the use of a Bright/Hacker cryostat (model OTF/AS/MR); we consider this cryostat to be particularly suitable for bone microtomy.

- To section bone it is essential to use a sharpened, finely polished tungsten steel knife [8].
- The angle of the microtome knife is critical (and differs between different manufacturers and the size/profile of knife). With the large 'C' profile ARP tungsten steel knife (angle 42–52° blade) the knife angle should be at its maximum (25°) when using the Bright/Hacker cryostat; for smaller 'C' profile knives (22°) the angle should be set to 17–19°.
- The chuck/bone sample should be orientated appropriately in the microtome specimen holder. The smallest edge should face downwards; for example, animal long bones (e.g. tibia) must be orientated at a 45° angle (approximately) to the knife, such that the growth plate is sectioned first. In this way the collagen fibres are orientated appropriately. After trimming the bone block the knife should be moved to an unused zone.

- In general, sections 5–10 µm thick are suitable for both *in situ* hybridization and immunocytochemistry.
- To ensure that serial sections are of equal thickness (important for comparative studies) it is advisable to use a cryostat fitted with an automatic drive.
- Cryostat sections of soft tissues are generally flash-dried onto glass slides. However, for bone (and cartilage) it may be necessary to touch the section gently (for less than 2 seconds) with the slide; do not let the section refreeze.
- The cryostat cabinet temperature should be –25°C; if sections appear brittle or powdery, then the cabinet temperature may be too low. Alternatively the block may have dehydrated due to excessive storage time in the freezer (> 1 month). Storing tissue in the presence of ice powder significantly reduces tissue dehydration for up to 1 year.
- It is important to let frozen bone (–70°C) equilibrate to the temperature of the cryostat prior to sectioning.
- The temperature of the knife can be further cooled by packing the knife handle with dry ice (important for dense cortical bone).
- The specialized slides (TESPA; enhanced adhesion) used for *in situ* hybridization are described later in the relevant section.
- Collect serial sections on three or four well slides; include sections for immunolocalization, cytochemistry and basic histology.
- See Bradbeer *et al.* [9] for alternative methods for sectioning dense adult bone.

Materials and equipment

- **Cryostat and microtome equipped with antiroll plate** (e.g. Bright/Hacker model OTF/AS/MR)
- **Tungsten tipped steel knife** (Bright/Hacker Instruments, Autoradiographic [ARP] products)
- **Brush and scalpel**
- **Low temperature oil** (Bright/Hacker Instruments)
- **Glass microscope slides** (4-well; Cel-Line Associates)
- **Coverslips**.

(a) Microtomy troubleshooting

Cryostat sections should show excellent retention of morphology and provide an ideal substrate for both immunocytochemistry and *in situ* hybridization. Poor section quality (e.g. crumpled, smeared, shattered, torn, folded) may result from one or many of the causes shown in Table 8.1.

8.4 IMMUNOCYTOCHEMISTRY

8.4.1 Background

Immunocytochemistry was first described in the early 1940s when Coons [10] coupled anti-pneumococcal antibodies to β-anthracene to detect bacterial antigens in tissue sections. However, because of the blue autofluorescence that is inherent in many tissues, this method was soon superseded by a fluorescein-conjugated antibody system. This fluorochrome gave a strong specific signal which drastically reduced problems with autofluorescence and has become the most popular

Table 8.1 Problems with poor section quality

Problem and causes	Solution
Scoring; a blunt or chipped knife	Sharpen routinely on a Shandon knife sharpener
Crumpled; tissue remnants/ice	Clean knife with low temperature oil on knife after each section; or clean with a fine brush. This also prevents section sticking to the knife (i.e. not flash drying)
Thin/thick sections; loose tissue block	Secure by adding additional PVA, or remount block
Shattering	
Incorrect knife angle	Optimize knife angle
Inappropriate tissue orientation	Optimize orientation
Incorrectly positioned anti-roll plate	Finite adjustment of anti-roll plate
Poor morphology	
Excessive delay between dissection and freezing (lysis, dehydration)	Discard bone sample
Freezing artefacts	Discard bone sample
Static (especially prevalent with TESPA coated slides)	Use antistatic gun on slide/section

Notes: Good quality sections may fold underneath the anti-roll plate. If repositioning the anti-roll plate, cleaning the knife, etc. does not improve matters, it may mean that the block is too large. Trim appropriately (0.5 cm^3) with a cold scalpel (–70°C); it is advisable either to remove the knife or to place a guard over the knife if this procedure is done within the cryostat.

fluorochrome for many of the standard immunofluorescence techniques used today. The use of enzyme-labelled antibodies in immunocytochemistry was first described by Nakane and Pierce [11]. In addition to their great sensitivity, these enzyme-based techniques have a distinct advantage over immunofluorescence in that the reactivity in tissues and cells can be observed with an ordinary light microscope.

8.4.2 General considerations

(a) Detection options
Sensitivity, resolution and cost are all important criteria to be considered when choosing an immunocytochemical detection method. Fluorescence techniques, for example, provide excellent resolution, but are relatively insensitive; they require expensive and specialized equipment to be visualized and the preparations are not permanent. Enzyme-based techniques, on the other hand, while lacking the resolution of fluorescence techniques, have superior sensitivity, require only a simple light microscope for visualization and provide a permanent record. In addition, these enzyme techniques are compatible with histological counterstains, such as haematoxylin, which provide a means of examining the overall histology of the tissue.

(b) Amplification techniques
Immunocytochemical techniques are now available that include steps to amplify the initial signal provided by the binding of the primary antibody. A very popular

detection method is the biotin/streptavidin system, which provides increased sensitivity for both fluorescence and enzyme-based techniques. Purified primary antibody is conjugated to biotin and this is subsequently detected in tissue preparations using streptavidin labelled with either a fluorochrome or an enzyme.

Another amplification technique involves using murine monoclonal antibodies that have been raised to an enzyme, e.g. alkaline phosphatase or peroxidase. The antibodies are incubated with excess enzyme to form an antibody/enzyme complex to form the basis of the alkaline phosphatase anti-alkaline phosphatase (APAAP) (Plate 14) and peroxidase anti-peroxidase (PAP) techniques (excellent staining kits based on these methods are available commercially, e.g. Dako, Carpinteria, CA). These methods involve incubating the primary murine monoclonal antibody with the tissue/cell substrate, followed by the addition of an excess of rabbit anti-mouse secondary antibody. This secondary antibody not only recognizes the bound primary antibody but also binds the final reagent, the murine antibody/enzyme complex (APAAP or PAP). Finally, the reactivity is visualized by the addition of the appropriate substrate (see below).

The tyramide amplification system is relatively new and is claimed to be up to 1000-fold more sensitive than standard fluorescence and enzymatic detection methods. In this technique the primary antibody step is followed sequentially by the addition of a horseradish peroxidase conjugated secondary antibody and then by the tyramide reagent. The signal amplification is achieved because the horseradish peroxidase catalyses the deposition of activated tyramide molecules at the enzyme site with minimal loss in resolution. The tyramide can be linked to fluorescein for direct observation by fluorescence microscopy, or to biotin, which permits its detection with either streptavidin/enzyme or streptavidin/fluorescein conjugates. This is a particularly useful system if only small volumes of primary antibody are available. We have had success with this technique using the DuPont Renaissance kit.

8.4.3 Choice of primary antibody

Either polyclonal or monoclonal antibodies can be used as sensitive tools to detect antigens in tissue or cell preparations. When choosing an immunocytochemical technique the following points should be taken into consideration.

- The antibody should bind specifically to the antigen of interest.
- Antibodies that work well in techniques such as immunoprecipation or Western blotting may not work at all well immunocytochemically.

(a) Polyclonal antibodies

- The antisera will contain multiple antibodies that react with different epitopes on the same antigen, giving rise to a strong staining signal.
- Polyclonal antibodies will often work well on formalin-fixed samples in which denaturation has taken place.
- Background staining may be present due to the unknown reactivities of the irrelevent antibodies that constitute the antibody repertoire of the immunized animal. These non-specific reactivities may be removed or reduced by: careful titration (see below); preadsorption with protein that does not contain the antigen of interest; or affinity purification.

(b) Monoclonal antibodies

- These antibodies often demonstrate high specificity with low background staining, particularly on tissues and cells that have been fixed in organic solvents (e.g. acetone) or paraformaldehyde.
- Denaturation of tissues by harsh fixation (e.g. prolonged fixation in formalin) often ablates the reactivity of these antibodies. This problem can be overcome by: immunization with denatured antigen; unmasking epitopes with a protease such as trypsin; or unmasking antigen by microwave irradiation.

8.4.4. Unmasking hidden epitopes

Harsh fixation with formalin or glutaraldehyde can often lead to loss of antigenicity due to epitopes becoming hidden or denatured. These processes can sometimes be overcome by re-exposing the epitopes with mild protease treatment or microwave treatment. Although these techniques apply primarily to formalin fixed/paraffin-embedded tissues, we have had occasion to apply them to cryostat sections. Both of the following methods should be performed on sections that have been cut onto slides that have been coated with a strong adhesive, e.g. TESPA (section 8.5.4).

(a) Protease treatment

Reagents

- **Protease solution**: 0.1% trypsin/0.1% calcium chloride (w/w) in 20 mM Tris (pH 7.8).

Proccdure

1. Incubate the samples for approximately 10 minutes at room temperature (the optimum time should be determined by the researcher) in the protease solution.
2. Terminate the enzyme digestion by rinsing the samples in distilled water for 5 minutes.
3. Proceed with the immunolocalization procedure.

(b) Microwave treatment

Reagents

- **Buffer**: 10 mM citrate (pH 6.0).

Procedure

1. Rehydrate the sections and place the slides in a coplin jar (two/jar) containing citrate buffer.
2. Cover the coplin jar with cling-film (punch holes to allow the steam to escape) and microwave (a domestic microwave with an output of 650 W and a rotating turntable is ideal) at full power for 5–30 minutes (the optimum time for each antigen should be determined by the researcher).
3. Fill the coplin jar regularly (every 5 minutes) to replace solution lost by evaporation.

4. Remove the coplin jar from the microwave and allow the buffer to cool to room temperature (15–20 minutes).
5. Rinse the slides in the appropriate buffer (e.g. TBS or PBS) and proceed with the immunolocalization procedure.

8.4.5 Immunocytochemical methods

We have not attempted to give an exhaustive review of all the immunocytochemical techniques that are currently available (see Watkins [4]). The methods outlined below (indirect immunofluorescence, alkaline phosphatase and horseradish peroxidase) are those that have proved successful in localizing antigens on bone-related tissue and cell substrates. Where possible, we have tried to pre-empt some of the problems that might be experienced and have described some of their possible solutions. Unless otherwise stated, the reagents and chemicals used in the following assays were purchased from Sigma Chemicals Co.

General equipment

- **Staining dish** (e.g. coplin jar)
- **Slide racks**
- **Humidified chamber** (essential to prevent evaporation of small volume samples)
- **Coverslips** (size depends on size of tissue sections)
- **Mounting medium**: the type (i.e. aqueous or non-aqueous) depends on the immunocytochemical method chosen (see below)
- **Microscope**: a standard light microscope for enzymatic techniques and a microscope equipped with epifluorescence for fluorescence techniques. The latter microscope should be equipped with filters for fluorescein and rhodamine and objectives with a high numerical aperture.

(a) Indirect immunofluorescence
The following protocol describes a fluorescence technique in which the primary antibody is localized with a fluorescein conjugated secondary antibody (for example, Plate 14g).

- Cryostat sections can be stored at –20°C until required for screening but the antigen of interest may degrade with time and give a false negative result. The stability of a particular antigen should therefore be determined by the individual researcher.
- The optimal dilution of primary and secondary antibodies should be determined, since high concentrations will produce high backgrounds and low concentrations will yield weak signals. Careful titration of these antibodies will help to decrease the signal obtained from non-specific interactions and the checkerboard titration that is required to determine the optimal concentrations of these is described elsewhere [12].
- Relevant controls should always be screened and these should include:
    - Tissue or cell controls in which the antigen of interest is known to reside.
    - Primary antibody controls generated against an irrelevant antigen that match the isotype (for monoclonals) and the form (serum, tissue culture supernatant, ascites, purified antibody, etc.) of the specific antibody that is being

tested. This will detect any non-specific binding of the primary antibody which can be removed by either preincubating the sections in 1–5% serum (usually the same species as the secondary antibody) or by directly diluting the primary antibodies in 1–10% of this serum.

- A conjugate control to detect any non-specific binding of the secondary antibody (substitute the primary antibody with antibody diluent). Non-specific Fc binding can be a problem although excellent conjugates are now available, consisting of antibodies that have had their Fc moieties removed by enzyme digestion.

- To determine the extent of autofluorescence in tissues or cell preparations, run a sample that is not overlaid with primary or secondary antibody. (Remember: specific fluorescein fluorescence is a bright apple green colour, while autofluorescence is yellow.) If the autofluorescence is really troublesome, try incubating the sections in freshly prepared 0.5% sodium borohydride in water (because of bubble generation, this treatment may cause the section to lift off the slide). If this is unsuccessful another staining technique may be the best option.

● Although tetramethyl rhodamine is less susceptible to fading, photobleaching can be a major problem when photography of a fluorescein-labelled specimen is required. This problem can be minimized by the addition of an anti-bleaching agent to the mounting medium, e.g. 1,4-diazobicyclo-[2,2,2]-octane (DABCO, 2.5% w/v), *p*-phenylenediamine or *n*-propyl gallate.

● The fluorescence techniques require that aqueous mountants are used. These include 90% glycerol in PBS (v/v), which has the disadvantage of being very messy unless the coverslips are secured with nail varnish. Suitable alternatives include Gelvatol (Monsanto Chemicals) and Mowiol 4-88 (Hoechst), which will set overnight.

● To prevent fading of the labelled preparations before screening under the microscope, store the slides in the dark at 4°C, or for extended periods at –20°C.

Reagents

● **Wash buffer**: phosphate buffered saline (PBS, pH 7.6). containing 2% foetal calf serum (FCS).

● **Blocking solution**: 1–10% serum (usually derived from the same species as the secondary antibody) in wash buffer.

● **Secondary (fluorochrome-labelled) antibody**: optimally diluted (usually fluorescein or rhodamine-labelled) in wash buffer. Excellent secondary antibodies can be obtained from many companies but the authors routinely use reagents from Sigma. Commercially available antibodies are often manipulated by the manufacturers to reduce the problems of non-specific binding, e.g. Fab_2 fragments to reduce Fc binding and/or preabsorbed with serum to reduce non-specific binding to the tissue/cell substrate.

● **Nuclear counterstain (optional)**: 0.002% (w/v) ethidium bromide in PBS. (**CAUTION**: ethidium bromide is a carcinogen and should be handled with extreme care.)

● **Mounting medium**: 90% glycerol in PBS (v/v). For alternative mounting reagents and the addition of anti-bleaching agents, see the general points above.

Procedure
Allow all reagents and tissue sections to equilibrate to room temperature. Perform all incubations at room temperature and do not allow the tissue sections to dry during the staining procedure.

1. Place the slides in a moist chamber. (A 230 × 230 mm tissue culture plate, with wet paper towels at the base, will suffice. Keep the slides off the towels with plastic 1 ml pipettes placed approximately 40 mm apart. This will comfortably hold 12 slides without them touching each other.)
2. Rehydrate the sections in wash buffer for 5 minutes.
3. Preincubate the sections for 30 minutes in blocking solution.
4. Aspirate the blocking solution and incubate for 30 minutes with optimally diluted primary antibody (dilute in wash buffer).
5. Wash the sections twice for 15 minutes in wash buffer.
6. Centrifuge the optimally diluted secondary antibody for 10 minutes at 1500 g.
7. Overlay the sections with the secondary antibody and incubate for 30 minutes.
8. Aspirate excess antibody and wash the sections twice for 15 minutes in wash buffer and once for 15 minutes in PBS alone.
9. Counterstain the nuclei for 30 seconds by immersion in ethidium bromide (optional step).
10. Mount the slides by placing a drop of the mountant in the middle of the appropriate number of coverslips. Invert a slide and touch it onto the mountant until it spreads beneath the coverslip. To prevent damage to the sections, do not try to remove air bubbles by applying pressure to the coverslip. For PBS/glycerol mountant, seal the coverslips with nail varnish to prevent leakage of this aqueous mountant onto the microscope stage.
11. Immediately screen the slides under the microscope or store at 4°C (–20°C for longer periods) in a sealed box.

(b) Alkaline phosphatase-based immunoenzymatic techniques
We have tried a number of commercially available kits from a number of sources that incorporate alkaline phosphatase as the enzyme detection system. However, we routinely use the Dako LSAB kit (code K674) which consistently gives us excellent results (Plate 14a–e). Alkaline phosphatase activity is detected by the naphthol phosphate method. In this procedure the released naphthol group of the substrate (e.g. naphthol AS-MX phosphate) is trapped by coupling with a diazonium salt (e.g. Fast Red TR) to give the final coloured precipitate (e.g. red); the coupling can either be performed simultaneously or by a post-coupling method.

- Endogenous osteoblast and marrow cell-derived alkaline phosphatase activity must be blocked when using this technique. Addition of either levamisole or bromotetramisole (0.1 mM) to the substrate solution can effectively block endogenous alkaline phosphatase activity in bone; however, the alkaline phosphatase isozyme (intestinal) used to detect the primary antibody binding is unaffected. This blocking procedure is considerably less detrimental to the technique than the blocking steps required for the peroxidase methods (see below).
- When using the Fast Red simultaneous coupling procedure, the concentration of bromotetramisole (Aldrich Chemical Co.) or levamisole required to block endogenous alkaline phosphatase activity in bone (> 10^{-4} M) can cause the

substrate to precipitate out of solution, resulting in a diminished level of specific reactivity (unpublished observation). To prevent this precipitation, we have replaced the conventional simultaneous coupling procedure with a post-coupling method; the conjugated alkaline phosphatase, acting at its optimal pH of 9.2, produces an insoluble naphthol AS-MX, which can be coupled subsequently in a second buffer containing Fast Red TR (at pH 7.4). This procedure enables us to block endogenous alkaline phosphatase activity totally while retaining the specific antibody binding (Plate 14c).

- It is important that the substrate system used is compatible with the mountant that has been chosen, since positive reactivity can be lost if the incorrect mountant is used. The Fast Red system, for example, requires an aqueous mountant (e.g. PBS glycerol, Gelvatol or Mowiol) while the New Fuchsin system is compatible with both aqueous and non-aqueous mountants (e.g. DPX).

Reagents

- **Wash buffer**: TRIS buffered saline (TBS, 0.05 M TRIS/HCl, pH 7.6, 0.15 M sodium chloride).
- **Blocking solution**: 1–10% serum (usually derived from the same species as the secondary antibody) in wash buffer.
- **Substrate solutions**: prepare immediately before use.
  Fast Red TR post coupling solution (step 1): Dissolve 2 mg of naphthol AS-MX phosphate in 0.2 ml of dimethylformamide in a glass tube. Make up to 10 ml with 9. 7ml 0.1 M TRIS (pH 9.2) and 0.1 ml of 0.1 M L-*p*-bromotetramisole (10^{-4} M). **(step 2)**: Dissolve 10 mg of Fast Red TR in 10 ml of cold 0.1 M TRIS buffer (pH 7.4).
  New Fuchsin substrate solution: Use this method when solvent-based mountants (e.g. DPX) are preferred.
  Solution I: mix 18 ml of propandiol buffer (0.2 M 2-amino-2-methyl-1,3-propanediol) with 50 ml of 0.05 M TRIS buffer (pH 9.7) containing 600 mg sodium chloride. To this solution add 28 mg of levamisole.
  Solution II: dissolve 35 mg of naphthol AS-BI phosphate in 0.42 ml of *N,N*-dimethylformamide.
  Solution III: mix 5% New Fuchsin (5 g in 100 ml 2M HCl) with 4% sodium nitrite (4 mg in 1 ml of distilled water – prepare fresh) in a fume hood and mix for 1 minute.
- **Counterstain**: Mayer's haematoxylin
- **Mounting medium**: The type of mounting medium is very important and is dependent upon the substrate system used (see general points above).

Ready-prepared Fast Red (Dako code K597) and New Fuchsin (Dako code K596) substrate kits can be purchased commercially.

Procedure

Allow all reagents and tissue sections to equilibrate to room temperature. Perform all incubations at room temperature and do not allow the tissue sections to dry during the staining procedure. This procedure is based on the Dako LSAB kit which contains a secondary antibody that contains both goat anti-mouse and anti-rabbit biotinylated antibodies. Therefore, it can be applied to murine monoclonal and rabbit polyclonal antibodies.

1. Place the slides in a moist chamber.
2. Rehydrate the sections in TBS.
3. Preincubate the sections for 30 minutes in 1–10% whole goat serum blocking solution. (This is used to block the tissue to prevent non-specific binding.)
4. Aspirate the diluted serum and incubate for 30 minutes with optimally diluted primary antibody.
5. Aspirate excess antibody and wash twice for 15 minutes in TBS.
6. Carefully wipe around the tissue/cells, overlay with optimally diluted biotinylated secondary antibody and incubate for 30 minutes.
7. Aspirate excess secondary antibody and wash twice in TBS for 15 minutes.
8. Carefully wipe around the preparations and incubate for 30 minutes with the biotin/enzyme complex.
9. Aspirate the excess complex and wash twice in TBS for 15 minutes.
10. Add the appropriate substrate solution (Fast Red or New Fuchsin alkaline phosphatase solutions):
    - If the post-coupling Fast Red system is being used, overlay the sections with the step 1 reagents (see above) and incubate for 10 minutes. Wash in cold 0.1 M TRIS buffer (pH 7.4), and overlay the sections with the step 2 post-coupling solution for 10 minutes at 4°C.
    - For the New Fuchsin method, mix solutions 1 and 2, and then add solution 3. Adjust the pH to 8.7 using HCl and then filter the mixture onto the slides (Whatman no. 1 filter paper). Incubate at room temperature for 10–15 minutes.
11. For both of the substrate methods described above, terminate the reaction by immersing the sections in distilled water.
12. Counterstain the nuclei in Mayer's haematoxylin for 45 seconds. (This is an optional step since weak staining could be masked by the haematoxylin. We suggest mounting the slides in the glycerol-based mountant and then viewing under the microscope. If sufficiently strong staining is present, remove the coverslips in distilled water, allow the sections to dry and mount in DPX.)
13. Remove the excess stain and wash the sections in tap water for 2 minutes.
14. Rinse the sections in distilled water and, if using an alcohol soluble substrate, mount the slides in 90% glycerol/10% PBS (v/v).
15. If an alcohol-soluble substrate system is being used, allow the sections to dry completely and mount in DPX. (An alternative is to dehydrate the sections in graded ethanol, 2–3 minutes each in 30%, 60%, 75%, 90% and absolute ethanol, and then into xylene before mounting in DPX.)

(c) Horseradish peroxidase-based immunoenzymatic techniques

The protocol below is for use with mouse monoclonal antibodies (anti-mouse peroxidase conjugated secondary antibody) since this is the system that we use routinely. However, because of the availability of high grade peroxidase-labelled secondary antibodies to a number of other species, this technique can also be applied to polyclonal primary antibodies.

Endogenous peroxidase activity can vary greatly from tissue to tissue and is often less of a problem in frozen sections of unfixed tissues, e.g. red blood cell peroxidase is lost due to cell lysis. However, some tissues, such as bone marrow, may express high endogenous activity and this may mask specific activity due to antibody binding. Therefore, a number of methods have been developed to block endogenous peroxidase activity. Many of these are quite harsh and the effect they have on individual antigens must be determined before running a full assay.

Reagents

- **Wash buffer**: TBS (pH 7.6).
- **Blocking solution**: 1–10% serum (usually derived from the same species as the secondary antibody) in wash buffer.
- **Substrate solutions**: 3,3′-diaminobenzidine (DAB) is a commonly used horseradish peroxidase substrate, which yields a dark brown precipitate that is insoluble in both aqueous and alcohol-based mountants. Dissolve 6 mg of DAB hydrochloride in 10 ml of 0.05 M TRIS buffer (pH 7.6). Add 10 ml of 30% hydrogen peroxidase immediately before use (the 30% hydrogen peroxide should be kept at 4°C and should be discarded after 1 month).

Procedure

1. Endogenous peroxidase activity can be blocked by overlaying the sections with either (i) 3% hydrogen peroxide for 3 minutes (the vigorous generation of bubbles with this procedure may cause the sections to lift off the slides) or (ii) a mixture of 4 ml of methanol and 1 ml of 3% hydrogen peroxide for 20 minutes.
2. Wash the slides for 5 minutes in TBS.
3. Overlay the sections for 30 minutes with the blocking solution.
4. Aspirate the blocking solution and overlay the sections for 30 minutes with the primary murine monoclonal antibody.
5. Aspirate excess antibody and wash twice for 15 minutes in TBS.
6. Carefully wipe around the tissue/cells and overlay for 30 minutes with optimally diluted goat/anti-mouse immunoglobulin conjugated to peroxidase.
7. Aspirate excess secondary antibody and wash twice in TBS for 15 minutes.
8. Using Whatman no. 1 filter paper filter freshly prepared DAB substrate directly onto the slides and allow the colour reaction to develop for 5–10 minutes.
9. Stop the reaction by immersing the slides in distilled water.
10. Lightly counterstain the preparations in Mayer's haematoxylin for 45 seconds. (This is an optional step – see the comments in the alkaline phosphatase procedure.)
11. Remove the excess stain with tap water and then rinse in distilled water.
12. Allow the sections to dry and mount in DPX. (See the comments in the alkaline phosphatase procedure, above, regarding the dehydration of tissue sections.)

8.5 *IN SITU* HYBRIDIZATION

8.5.1 Background

In situ hybridization was initially used to localize chromosomal DNA sequences [13] using double- or single-stranded tritiated DNA probes [14, 15]. The sensitivity of this methodology has advanced rapidly with the introduction of single-stranded RNA probes (riboprobes) [16], and the use of ^{35}S-labelled [17] and ^{33}P-labelled probes

[18]. The rapidly expanding use of *in situ* hybridization in clinical and research applications has prompted the development of alternative non-radioactive labelling techniques for cDNA and riboprobes. The most commonly used methods involve biotin, digoxigenin or fluorescein-substituted nucleotides to label probes which are then detected by either a fluorescent or enzymatic system [19, 20]. The inability to detect very low levels of mRNA expression (below 20–30 copies per cell) by *in situ* hybridization is a major limitation of the method. This is confounded by the level of non-specific binding of probe to sections. New methods for amplifying the signal by reverse transcriptase *in situ* PCR [21] will address these concerns and significantly enhance the application of *in situ* hybridization.

8.5.2 General considerations

(a) Probe and detection options
A number of different types of nucleic acid probes can be prepared for use in *in situ* hybridization. These include double- and single-stranded DNA probes, oligonucleotides (20–30 bases in length) and single-stranded RNA probes. The different types of probes available, their preparation and synthesis, and the different isotopes available, are fully discussed by Wilkinson [22] and Angerer and Angerer [23]. Briefly, ^{35}S-labelled probes offer both higher sensitivity and autoradiographic efficiency (1–2 weeks) when compared with ^{3}H- and ^{32}P-labelled probes. However, due to the relatively short half-life, the ^{35}S probes can only be stored for short time periods (see later). Non-radioactive probes have reduced sensitivity but offer a number of advantages, including a more rapid development time (<24 hours), safer handling and the ability to prepare and store the probes for several months.

This section will outline *in situ* hybridization procedures using both ^{35}S-labelled and fluorescein-labelled (RNA colour kit, Amersham) single-stranded RNA probes (riboprobes) on cryostat sections of bone (Plate 15), describing the fixation and storage of bone sections, the selection and preparation of the labelled riboprobes and, finally, the hybridization and autoradiography/enzyme detection. The advantages of using riboprobes include high sensitivity and specificity, and the ability to synthesize complementary sense strands that serve as negative controls. For detailed reviews of the theory and methodology of *in situ* hybridization see Wilkinson [22], Angerer and Angerer [23] and Polak and McGee [24].

(b) Optimization of conditions
- It is essential to have positive and negative control tissue sections for evaluating the method development (Plate 15).
- To establish the methods, select a target mRNA that is highly and selectively expressed in bone tissue (e.g. TRAP, collagen type 1, osteocalcin, etc.).
- Check the condition of the probe by Northern blot analysis to ensure that the appropriate size band is recognized and that no background or ribosomal bands are evident. If there is background, consider subcloning the probe.
- Initially, experiments should be organized conservatively when estimating the number of slides to be processed simultaneously.
- Optimization of fixation and hybridization time is critical to maximize the signal-to-noise ratio; overfixation will result in low or negligible signals and extended hybridization time may increase the non-specific background with ^{35}S-labelled riboprobes.

- As a control to detect non-specific background signal, hybridize serial sections with the corresponding sense strand riboprobe under identical conditions as the antisense riboprobe.

8.5.3 Preparation of RNase-free reagents

(a) RNase contamination

All reagents and glassware must be treated to remove RNases, which will cause degradation of both the target mRNA and the RNA probes. Ribonucleases are inactivated by diethylpyrocarbonate (DEPC) treatment (see below).

- Gloves must be worn during all procedures to avoid contamination from RNases on the hands.
- It is advisable to purchase glassware that can withstand repeated autoclaving and which is reserved solely for the *in situ* hybridization reagents (many suppliers now offer molecular biology-grade reagents which are nuclease-free).
- Disposable plastic items are generally free of RNases and are recommended.

(b) Preparation of RNase-free water and solutions

- All work with diethylpyrocarbonate (DEPC, D-5758, Sigma) must be done in a chemical fume hood.
- DEPC is not stable and should only be purchased in small aliquots and used within one month after opening.
- To prepare RNase-free water, add DEPC at a final concentration of 0.1% to distilled, deionized water. Cap and shake the solution well. Loosen the cap and allow the solution to stand overnight. Autoclave to remove the residual DEPC.
- Because TRIS buffers cannot be DEPC treated, molecular biology-grade reagent stocks should be purchased.
- Glassware can be decontaminated by baking at 180°C for 8 hours or by treating with DEPC, as described, and then draining and air-drying to remove residual DEPC.

8.5.4 Preparation of serial cryostat sections for *in situ* hybridization

This section outlines the procedures for TESPA (amino-propyl triethoxy silane) coating slides to enhance tissue adhesion, fixation and dehydration.

Reagents and materials

- **Slide staining dishes** (Baxter Scientific)
- **Teflon-coated multi-well slides** (Cel-Line Associates)
- **Coverslips**
- **Contra 70 soap** (no. 1003, Decon Laboratories)
- **Drying oven** (80°C)
- **Acetone** (9002-03, J.T. Baker)
- **TESPA** (amino-propyl triethoxy silane; A-3648, Sigma)
- **Slide storage boxes**

- **Silica gel dessicant packets** (S-8394, Sigma)
- **10X Dulbecco's phosphate-buffered saline** (D-PBS; 21600-069, Gibco)
- **4% Paraformaldehyde** (P-6148, Sigma) in 1X D-PBS; see below
- **Ethanol (30, 60, 80, 95 and 100%) in DEPC-treated water/glass jars**.

(a) Preparation of TESPA-coated slides

Procedure

1. Place glass slides in staining dishes and soak overnight in a 5% solution of Contra 70 soap. Rinse the slides in running tap water for 2 hours. Finally, rinse three times in deionized water and air-dry.
2. Bake the slides at 80°C for 1 hour, then allow to cool.
3. Fill a slide staining dish with a 2% solution of TESPA diluted with acetone. Fill a second staining dish with acetone. Soak the clean, dry slides in the TESPA/acetone solution for 2 minutes at room temperature, then transfer to the acetone for 2 minutes. Drain and air-dry the slides.
4. Store the coated slides at 4°C in boxes containing dessicant. Equilibrate to room temperature before opening the slide box. The coated slides should be used within 3 weeks.

(b) Preparation of 4% paraformaldehyde
This work must be performed under a chemical fume hood.

Procedure

1. Heat 300 ml DEPC-treated water to 60°C on a stirring hot plate.
2. Add 20 g paraformaldehyde with stirring. Stir for 5 minutes.
3. Add 100 µl 2 M NaOH.
4. Add 50 ml 10X D-PBS.
5. Adjust pH to 7.2 with HCl.
6. Add 1.25 ml 2 M $MgCl_2$.
7. Adjust final volume to 500 ml with DEPC-treated water. Cool to room temperature and filter through a 0.2 µm filter. Store at room temperature.

(c) Fixation and dehydration
The extent to which probes penetrate cryostat sections will depend on the size of the probe, the choice and length of fixation, and whether pretreatments (e.g. proteinase K) have been performed [22, 23]. For micron-thin sections cut from unfixed undecalcified bone blocks, we advocate a short fixation in the cross-linking fixative paraformaldehyde and no proteinase K treatment. This procedure preserves good histological detail and can also be used successfully on sections of soft tissues.

Procedure

1. Cryostat sections (5 µm) on TESPA-coated slides should be air-dried briefly (15 minutes) before fixing (wear gloves at all times).
2. Fix sections in freshly prepared 4% paraformaldehyde for 5 minutes at room temperature.
3. Rinse in 1X D-PBS for 4 minutes at room temperature; repeat 3 times.

4. Dehydrate through 30, 60, 80, 95 and 100% ethanol at room temperature (2 minutes each step). Air-dry at room temperature.
5. Store the slides in a dessicated box at –20°C (equilibrate to room temperature before opening the slide box).

8.5.5 Preparation of ^{35}S-labelled riboprobes

Preparation of riboprobes is accomplished in three basic steps: (a) selection of a clone containing the sequence of interest; (b) linearization of the cDNA template; and (c) an *in vitro* transcription reaction to generate the radiolabelled probe. Although the preparation of ^{35}S-labelled RNA probes can be time consuming, they are considered to be the most sensitive and specific. The advantages include:

- the synthesis of complementary single-stranded probes (the sense strand serves as a negative control);
- strong, stable hybridization with mRNA, which allows for high stringency washes;
- the ability to digest non-hybridized probe with RNase during the washes;
- the ability to synthesize high specific activity probes by the *in vitro* transcription reaction.

(a) Selection of a clone
Good microbiological practices (detailed in Sambrook *et al.* [25]) should be followed in the selection and preparation of clones for use in *in situ* hybridization. Clones should be plated on Luria Bertani, or other suitable agar with the appropriate antibiotic, and single colonies selected for mini-preparation and restriction endonuclease analysis to confirm the presence of the insert cDNA. Selected single colonies are then scaled up for cDNA template preparation.

The *in vitro* transcription reaction requires a cDNA template, which is prepared by either gradient density centrifugation [25] or the Qiagen method (Qiagen Inc.). We routinely use the Qiagen method for cDNA template preparation, following the manufacturer's instructions supplied with the kit.

(b) Linearization of the cDNA template
A detailed description of restriction endonuclease digestion can be found in Bloch and Bartos [26]. RNA transcripts are generated from cDNA templates that have been inserted into vectors which contain the promoter sites for the RNA polymerases SP6 or T3 and T7. Numerous commercial vectors, such as the pBluescript vectors (Stratagene) and the pGEM series (Promega), are available. If the insert cDNA is linearized at either the 5′ end or the 3′ end, the transcription reaction will generate both antisense and sense run-off transcripts of the insert sequence. The size of the riboprobe can be controlled by subcloning a specific size insert into the vector or by linearizing the insert at a site that will generate the desired size run-off transcript. Inclusion of a radiolabelled nucleotide in the *in vitro* transcription reaction will generate radiolabelled transcript, which is used to detect a positive hybridization signal in the tissue section through autoradiographic methods.

To linearize the cDNA template, select a unique restriction endonuclease site in the insert sequence which will yield the desired length riboprobe. Although the size of RNA probes is not as critical as for DNA probes, we generally use probes around 600 base pairs or less in length. Probes greater than 1 kb may have reduced accessibility to the target mRNA. It is generally advisable to use probes to the coding sequence and to avoid 3′ untranslated regions [24]. Analyse the linearized cDNA by agarose gel elctrophoresis to ensure complete cutting. Purify the linearized template by phenol/chloroform extraction and ethanol precipitation [25]. Resuspend the template in nuclease-free water at a concentration of 1 μg/μl. It is also advisable to run a 'cold' *in vitro* transcription reaction (without radiolabelled nucleotide) and to check the size of the run-off transcripts by gel electrophoresis. This step is not necessary for every *in situ* hybridization, but should be performed on newly received probes or clones that have been subcloned.

Occasionally, certain sequences will give non-specific background hybridization. In these cases, it may be necessary to select another region of the gene for hybridization or to subclone the insert. Recent reports on the preparation of riboprobe cDNA templates by PCR technology [27] will reduce the time-consuming subcloning procedure.

(c) In vitro transcription reaction

This section describes the preparation of the radiolabelled riboprobes from the linearized cDNA template. We routinely use a commercially available *in vitro* transcription kit (*in vitro* Transcription Systems; Promega). All reagents are supplied with the kit. By following the detailed instructions provided by the supplier, high specific activity antisense and sense RNA probes can be routinely synthesized. Special attention should be given to the following.

- Linearized cDNA templates for the reactions should be at a concentration of 1 μg/μl. ^{35}S-labelled UTP or CTP at high specific activity (800 Ci/mmol, Amersham) is required for the reaction. To protect the sulphur from oxidation, all solutions containing the ^{35}S probe should contain a reducing agent such as dithiothreitol.
- Following the 1-hour transcription reaction, the cDNA template is digested by the addition of 1 μl of RQ1 RNase-free DNase incubated at 37°C for 15 minutes.
- Finally, add 30 μl of nuclease-free water to the reaction tube, cap tightly, vortex and briefly centrifuge. Remove a 1 μl aliquot to determine the total radioactivity in the reaction.
- Separation of unincorporated nucleotides from the labelled transcripts is efficiently accomplished using Sephadex G50 Quickspin columns for RNA (Boehringer Mannheim). Following the manufacturer's procedure, the columns are mixed gently to suspend the Sephadex, allowed to drain by gravity, then centrifuged at $1100 \times g$ for 2 minutes. The RNA sample (50 μl total volume) is loaded and the column is centrifuged at the same speed (the eluate contains the purified RNA transcripts). It may be necessary to adjust the times of the centrifugation steps to obtain consistent volumes of eluate.
- Add 2 μl of 100 mM dithiothreitol to the eluate, mix thoroughly and remove a 1 μl aliquot to determine the radioactivity incorporated into the RNA transcript.
- The purified RNA transcripts are stored at −20°C and used within 1 week.

8.5.6 *In situ* hybridization methodology

The hybridization procedure, originally published by Zeller and Rogers [7], has been modified for bone tissue. There are several notable changes from standard protocols, which include the following.

- The sections are not digested with proteinase K. (As described earlier, cryostat sections require only brief fixation in paraformaldehyde. For this reason digestion is unnecessary and may in fact cause loss of mRNA [20].)
- Pretreatment with HCl decalcifies the bone. (This procedure also reduces non-specific background and permeabilizes the section.)
- Hybridization time is reduced to 4 hours. (This reduced hybridization time significantly reduces non-specific background.)

Reagents for hybridization
All reagents are prepared with DEPC-treated water.

- **Glycine buffer**, 2 mg/ml in 1X D-PBS (autoclave and store at room temperature)
- **1X D-PBS**
- **0.2 M HCl**
- **0.1 M triethanolamine (TEA)** (T9534, Sigma) buffer, pH 8.0, prepared same day
- **Acetic anhydride** (A6404, Sigma)
- **2X SSC** (20X SSC = 0.3 M sodium citrate, 3.0 M NaCl, pH 7, 15557-044, Gibco)
- **Prehybridization dehydration solutions**: 30, 60, 80, 95 and 100% ethanol
- **Hybridization mixture**: 1.2 M NaCl; 20 mM TrisHCl, pH 7.5; 4 mM EDTA; 2X Denhardt's (D2532, Sigma); 1 mg yeast tRNA/mg (R 9001, Sigma); 200 mg poly (A)/ml (27-7836-02, Pharmacia); in DEPC-treated water, pH 7.5. Aliquot and store frozen at –20°C
- **50% dextran sulphate** (D8906, Sigma) prepared in DEPC-treated water (aliquot and store at 4°C)
- **Deionized formamide** (47671, Fluka), aliquot stock and store at –80°C
- **3.3 M dithiothreitol** (DTT, D9779, Sigma) (this must be freshly prepared)
- **Humidified chamber solution**: 50% formamide (47670, Fluka), 0.6 M NaCl, 10 mM TRIS HCl, pH 7.5, and 2 mM EDTA, prepared as sterile stock and stored at room temperature
- **RNA wash solutions**:
  - Solution I: 2X SSC, 50% formamide (47670, Fluka), 0.1% β-mercaptoethanol (63689, Fluka), preheated to 50°C
  - Solution II: prepare 10 mg ribonuclease A/ml (R6513, Sigma) in 10 mM TRIS, pH 7.5, 15 mM NaCl, aliquot and store at –20°C; at time of use, dilute stock to 20 μg ribonuclease A/ml in 0.5 M NaCl, 10 mM TRIS HCl, pH 8.0, preheated to 37°C
  - Solution III: 0.1X SSC, 1% β-mercaptoethanol (63689, Fluka), preheated to 50°C
- **Dehydration solutions**: 0.6 M NaCl in 30% and 60% ethanol; 80, 95 and 100% ethanol
- **Shandon Lipshaw humidity chamber** (no. 197 Shandon Lipshaw)
- **42°C incubator**
- **37°C and 50°C water baths**

(a) Prehybridization steps
All prehybridization steps are performed at room temperature. Use slide dishes that will accommodate 10 ml of buffer for each slide.

Procedure

1. Remove slide boxes from freezer and allow the box to equilibrate to room temperature before opening.

2. Hydrate the sections in glycine buffer for 2 minutes, then rinse twice in 1X D-PBS for 2 minutes each.
3. Decalcify sections in 0.2 M HCl for 20 minutes.
4. Treat in TEA buffer for 5 minutes, then acetylate for 10 minutes in TEA buffer to which acetic anhydride has just been added (0.25% final concentration).
5. Rinse twice in 2X SSC for 5 minutes each.
6. Dehydrate in 30, 60, 80, 95 and 100% ethanol for 2 minutes each; air-dry. Place the slides in the racks that are supplied with the humidity chamber.
7. Prepare the Shandon humidity chamber by soaking a paper towel with the chamber solution and placing it in the bottom of the chamber.
8. Proceed immediately to the hybridization steps.

(b) Hybridization steps

Procedure

1. Prepare the hybridization solution by combining the hybridization mixture, deionized formamide (47671, Fluka) and 50% dextran sulphate in a $2:2:1$ volume ratio. Prepare sufficient quantity for approximately 30 µl per 25 mm^2 tissue section.
2. Thaw the previously prepared ^{35}S-labelled RNA probes and add to the hybridization solution at a concentration of 2×10^5 cpm/µl. Add freshly prepared 3.3 M DTT to a final concentration of 50 mM. Mix well.
3. Apply the hybridization solution to the tissue sections. Use the pipette tip to distribute the solution evenly, taking care not to scratch the sections.
4. Place the slides in the humidity chamber, seal and incubate at 42°C for 4 hours.

(c) Wash steps
The wash steps should be performed in a shaking water bath. Dispose of radioactive waste appropriately.

Procedure

1. Wash the slides in solution I at 50°C for 15 minutes. Repeat at least once.
2. Wash the slides in solution II at 37°C for 30 minutes.
3. Wash the slides in solution I at 50°C for 15 minutes. Repeat at least once.
4. Wash the slide in solution III at 50°C for 15 minutes. Repeat once.
5. Dehydrate the slides through 0.6 M NaCl, 30% ethanol; 0.6 M NaCl, 60% ethanol; 80% ethanol; 95% ethanol and 100% ethanol at room temperature for 3 minutes each. Air-dry slides. The slides may be kept at room temperature overnight prior to autoradiography.

8.5.7 Emulsion autoradiography

This section outlines the detection of radioactively labelled probes by autoradiography. The exposure of slides must be performed in a darkroom and in the complete absence of moisture, which reduces latent grains in the emulsion. Emulsions are very sensitive to pressure and stretching artefacts (section 8.5.9); problems include grain accumulation at the edge of bone and within internal marrow spaces, and embedded cell lacunae. Detection of mRNA with the ^{35}S-labelled probe requires 2–4 weeks.

Reagents and equipment for autoradiography, counterstaining and microscopy
- **LM-1 autoradiographic emulsion** (RPN.40, Amersham)
- **Developer, D19** (146 4593, Kodak)
- **Fixative** (197 1746, Kodak)
- **Stop solution** (0.5% glacial acetic acid)
- **Plastic microscope slide mailers** (362-863, Curtin Matheson Scientific)
- **43°C water bath** (in the darkroom)
- **Stainless steel tray inverted over an ice-filled pan**
- **Open-faced cardboard slide trays**
- **Lightproof box**
- **Slide boxes with dessicant**
- **Darkroom with safelights**
- **Diff Quik**: methylene blue (Solution II; B4132-10,11,12, Baxter Scientific)
- **DPX mountant** (36029 2F, BDH Lab Supplies)
- **Photomicroscope** (optional: dark field optics)

(a) Initial preparations and slide dipping

Procedure

1. The air-dried slides should be arranged in a slide holder such that they are easily handled during the coating step (performed in the darkroom).
2. Preheat the water bath to 43°C. Arrange the iced stainless steel tray so that slides which have been coated with the emulsion can be chilled prior to the drying step. This chilling step helps to solidify the emulsion and allows for a more consistent coating.
3. Under safelights, remove the emulsion from the protective package and place in the water bath for approximately 10 minutes. With gentle tilting, mix the emulsion as it liquifies. The emulsion may be diluted slightly with distilled water.
4. Pour the melted emulsion into a plastic slide mailer (a convenient container for dipping the slides).
5. Slowly immerse each slide for 5 seconds, remove slowly and drain for 5 seconds.
6. Carefully wipe (once) the back of the slide with a paper towel. Place the slide horizontally on the iced tray for about 10 minutes.
7. The slides are placed on open-faced cardboard slide holders and placed in a lightproof box to dry (at room temperature) for at least 2 hours.
8. Transfer the dried slides to a plastic slide box containing a dessicant pack, wrap the box in aluminium foil, label and expose at 4°C.

(b) Development

Procedure
The time for exposure varies (and must be optimized), but we recommend an initial 2-week exposure.

1. Remove the wrapped slide boxes from the refrigerator and equilibrate to room temperature.
2. Prepare a series of slide staining dishes containing developer, stop and fixer solutions at 15–20°C (this is accomplished by placing the dishes in a shallow pan containing chilled water).

3. In the darkroom (under safelights), place the slides in the staining racks.
4. Develop the slides for 5 minutes, followed by 1 minute in the stop solution, and then fix for 8 minutes (all steps performed under safelights).
5. Wash in gently running tapwater (in daylight) for 20 minutes. Finally, rinse twice in deionized water.
6. Counterstain the slides as desired (e.g. three times 3-second dips in methylene blue) and mount (DPX) under coverslip.
7. The extent of the hybridization signal is assessed by the autoradiographic silver grain density over the cell (Plate 15a–e).

8.5.8 Non-isotopic *in situ* hybridization

There are several commercial kits available for the preparation of non-radioactive RNA and DNA probes. The method of choice will depend on the type of probes and vectors which are available to the investigator. Our experience has been with riboprobes synthesized by *in vitro* transcription where the ^{35}S is replaced by digoxigenin- or fluorescein-labelled UTP. The positive hybridization signal is then detected by anti-digoxigenin or anti-fluorescein antibody conjugated to alkaline phosphatase. The detection of the alkaline phosphatase is via the colour-substrate solution, nitroblue tetrozolium/bromochloroindolyl phosphate (NBT/BCIP) which forms a dark purple formazan precipitate (Plate 15f). It is important to assess the sensitivity of this method compared with the standard isotope method for each specific application.

Procedure

We have experience with the anti-fluorescein labelled RNA kit ('RNA colour kit' supplied by Amersham; RPN330). The '*In situ* Hybridization Detection System' for fluorescein labelled probes (Dako; k0607) gives very good signal and low background when used in combination with the RNA kit. The instructions provided with the kit are designed to ensure optimum results with the reagents supplied. The following considerations are noted for *in situ* hybridization on cryostat sections of bone.

1. Probe preparation: the *in vitro* transcription reaction is completed with the RNase-free DNase digestion of the cDNA template and the precipitation of the labelled RNA, as described. We do not reduce the size of the probe by alkaline hydrolysis. RNA probes can be stored in the long term at –70°C.
2. Pretreatment of cryostat sections: the manufacturer's instructions are followed with the exception of the proteinase K digestion, which is omitted. The slides are dehydrated through an ethanol series, then air dried prior to hybridization.
3. Hybridization and washes: use the hybridization buffer supplied with the kit according to the instructions. Hybridize with a probe concentration of 600 ng/ml at 55°C for 3 hours. Wash in 1X SSC, 0.1% SDS twice for 5 minutes at room temperature, then 0.1% SDS at 55°C twice for 10 minutes. Rinse in 2X SSC for 2 minutes. Digest with 10 µg/ml RNase A in 2X SSC at 37°C for 20 minutes. Rinse in 2X SSC.
4. Blocking, antibody incubation and washes: wash, block and rinse the slides as directed by the manufacturer. Dry the area around the tissue sections. The optimum antibody dilution should be determined for each application. Dilutions of 1 : 500 and 1 : 1000 are recommended to start.

5. Detection: wash slides in detection buffer, drain, and dry the areas around the sections. Place slides in a rack in a humidity chamber, apply the colour detection reagent and incubate the slides in the dark for 4–24 hours at room temperature. Rinse in distilled water twice for 2 minutes.
6. Counterstain with 0.5% Fast Green and mount in DPX.

8.5.9 Troubleshooting

Many of the problems associated with *in situ* hybridization can be solved by appropriate tissue procurement/freezing and good quality bone sections (section 8.3). In general, fixation and hybridization times are critical and must be optimized for each probe.

(a) Loss of sections during procedure

- Inadequate TESPA coating.
- Bone section not flattened onto slide.

(b) Low signal, high overall background
If the size and concentration of the probe are correct, check the following.

- Old tissue blocks.
- Outdated reagents; ^{35}S oxidized (avoid extensive freeze-thawing; store in presence of DTT); DTT not added.
- Gene sequence associated with high background; use a different portion of the gene.
- Insufficient washes; increase stringency.
- Check that the post-hybridization RNase treatment is working.

(c) Low signal, clean background

- Section overfixed.
- Poor tissue procurement (see above); avoid nuclease contamination.
- Low abundance of the target (e.g. Plate 15b).
- Suggest double-labelled (i.e. UTP and CTP) probe.
- Check probe size: 400–600 bp balances between sufficient signal and lack of assessibility.

(d) High signal, high background

- Reduce hybridization time.
- Reduce concentration of probe.

(e) Non-specific localization patterns

- ^{35}S-labelled riboprobes bind non-specifically to certain osteoblast populations, osteocytes, chondrocytes and macrophages (especially populations containing oxidized lipids or azurophilic granules); use of sense strand controls and negative tissue controls is of paramount importance to ascertain non-specific staining.
- Digest non-hybridized probe with RNase during washes.

(f) Dense non-specific signal

For dense non-specific signal trapped under bone, within cellular lacunae of osteocytes and chondrocytes in calcified areas, and general edge effects, check the following.

- Dehydrated bone block; discard.
- Poor tissue procurement/sectioning (section 8.3).
- Thick, unflattened section; remount block (section 8.3).

(g) Dense background over entire glass slide and section

- Mistreatment of the emulsion; check the condition of the emulsion by dipping a blank slide. Never re-use emulsion.
- Emulsion drying artefact; increase the time to dry the emulsion.

8.6 UTILITY OF *IN SITU* HYBRIDIZATION AND IMMUNOLOCALIZATION ON CRYOSTAT SECTIONS OF BONE

This section briefly reviews some of the recent work from our laboratory using the techniques of *in situ* hybridization and immunolocalization studies on human bone and human bone-derived cells.

8.6.1 Identification of novel genes and their products

Future development of drugs to combat diseases such as osteoporosis may rely on the identification of novel genes specific to osteoblast or osteoclast function. Random high throughput sequencing of human osteoblast and osteoclast cDNA libraries [28] could potentially identify such genes. The specificity or selectivity of these genes and their products to bone cells relies on tissue distribution studies using the techniques of *in situ* hybridization and immunolocalization. A novel osteoclast-specific cysteine protease, termed cathepsin K, has been identified using this regimen [29] (Plate 15).

8.6.2 Isolation and characterization of a human preosteoclast

Not only could cathepsin K represent the essential proteolytic enzyme central to the resorption process and an obvious target for inhibition strategy, it also provides a specific marker for cells of the osteoclast lineage. To identify novel factors/targets that are involved in osteoclast differentiation, it is essential to develop a human cell line of this phenotype. To address this we have used immunocytochemical and *in situ* hybridization techniques to isolate and characterize a mononuclear cell population, derived from osteoclastomas, that in the presence of osteoclastoma-derived stromal cells fused to form multinucleated osteoclasts that resorbed dentine slices [30]. The mononuclear cell population expressed a number of osteoclast 'selective' markers, including the vitronectin receptor, cathepsin K and osteopontin.

8.6.3 Osteoclasts and bone matrix proteins

We first demonstrated, by these combined *in situ* techniques, that the bone matrix protein, osteopontin, is expressed by cells of the osteoclast lineage [31, 32, 33].

Immunolocalization of this secreted protein demonstrated dense osteopontin staining at bone resorption surfaces that coincided with the intense expression of osteopontin mRNA and protein in resident and resorbing osteoclasts [1]. At adjacent formation sites, osteopontin expression in osteoblasts was rare, implying that during bone remodelling (as opposed to modelling [32]) the osteoclast may be the cell responsible for depositing osteopontin at resorption sites prior to osteoblast bone formation. Although the function of this bone matrix protein remains unknown (osteoclast adhesion, mineralization, osteoblast recruitment, cell signalling?), this controversial result highlights the importance of using these two techniques in tandem to identify the cell types responsible for secreting connective tissue proteins. It also clearly demonstrates that the osteoclast is not purely a resorptive cell but may play a key role in matrix deposition that is potentially coupled to osteoblast activity.

8.6.4 Severe combined immunodeficient (SCID) mouse as a surrogate in study of human bone cell differentiation

It has been demonstrated that the intramuscular injection of osteoclastoma-derived stromal cells into SCID mice results in ectopic bone formation [34]; immunolocalization studies using human and mouse-specific antibodies confirmed the human origin of this bone. Similarly, at these sites it was demonstrated that the bone formation was preceded by mouse osteoclast infiltration and differentiation [35]. This model system, used in conjunction with *in situ* hybridization and immunolocalization, represents a powerful tool for studying both osteoblast and osteoclast differentiation.

8.7 CONCLUSION

In situ expression studies will clarify the bone cell types expressing a certain mRNA and/or protein, and potentially relate this expression to the stage of differentiation at definitive stages of bone remodelling [1, 36]. Ideally, information gained may clarify or refute established dogmas and elucidate the mechanism of action or role of bone-active proteins in normal versus abnormal bone remodelling, rather than merely confirming that a certain protein 'plays a role'. The use of frozen, untreated, undecalcified bone tissue allows optimization for each technique at the level of the section. In view of the urgent need for *in situ* expression studies in adult bone pathology, future methodology may rely on the use of novel resin embedding techniques (see, for example, Onetti-Muda *et al.* [5]) that can be used universally for both techniques in combination with enzyme cytochemistry and histomorphometry.

ACKNOWLEDGEMENTS

We thank Dr Michael Lark and Dr Jeremy Bradbeer for critical review of this chapter.

APPENDIX 8.A LIST OF SUPPLIERS

Aldrich Chemical Co.,
1001 West St Paul,
PO Box 355,
Milwaukee,
WI 53201, USA.
414-273-3850.

Amersham Life Science,
2636 Clearbrook Drive,
Arlington Heights,
IL 60005, USA.
708-593-6300.

J.T. Baker, Phillipsburg,
NJ 08865, USA.
908-859-2151.

Baxter Scientific Products,
1430 Waukegan Road,
McGaw Park,
IL 69985, USA.
708-689-8410.

BDH Lab Supplies,
Poole,
BH15 1TD, UK.
01202-669700.

Boehringer Mannheim Corporation,
Indianapolis,
IN 46250, USA.
317-849-9350.

Bright Instrument Co. Ltd,
Huntingdon,
Cambs PE18 6EB, UK.
01480-454528.

Cel-Line Associates,
PO Box 35,
Newfield,
NJ 08344, USA.
609-697-4590.

Curtin Matheson Scientific, Inc.,
500 American Road,
Morris Plains,
NJ 07950, USA.
201-644-9500.

Dako Corporation,
6392 Via Real,
Carpinteria,
CA 93013, USA.
805-566-6655.

Decon Laboratories, Inc.,
Bryn Mawr,
PA 19010,
USA.
610-520 0610

DuPont-NEN Research Products,
549 Albany Street,
Boston,
MA 02118, USA.
617-482-9595.

Fluka Chemical Corp.,
980 South Second Street,
Ronkonkoma,
New York 11779,
USA.
516-467-0980.

GibcoBRL,
PO Box 6009,
Graithersburg,
MD 20884-9980, USA.
301-840-8000.

Hacker Instruments, Inc.,
Box 10033,
Fairhill,
New Jersey, USA.
201-226-8450.

Machine Shop Service,
Chuck Banski,
28 Vaux Lane,
Phoenixville,
PA 19426, USA.
610-935-1340.

Pharmacia Biotech, Inc.,
800 Centennial Avenue,
Piscataway,
NJ 08854, USA.
908-457-8000.

Promega Corporation,
2800 Woods Hollow Road,
Madison,
WI 53711-5399, USA.
608-274-4330.

Qiagen Inc.,
9600 DeSoto Avenue,
Chatsworth,
CA 91311,
USA.
800-426 8157

Shandon Lipshaw Inc.,
171 Industry Drive,
Pittsburgh,
PA 15275, USA.
412-788-1133.

Sigma Chemical Company,
PO Box 14508,
Saint Louis,
MO 63178, USA.
314-771-5750.

REFERENCES

1. Dodds, R.A., Connor, J.R., James, I.E. *et al.* (1995) Human osteoclasts, not osteoblasts, deposit osteopontin onto resorption surfaces: an *in vitro* and *ex vivo* study of remodeling bone. *Journal of Bone and Mineral Research* **10**, 1666–1680.
2. Bradbeer, J. (1992) The cell biology of bone remodelling, in *Recent Advances in Clinical Endocrinology and Metabolism*, Vol. 4, (eds D.W.Lincoln and C.R.W. Edwards), Churchill Livingstone, London, pp. 95–114.
3. Chayen, J. and Bitensky, L. (eds) (1991) *Practical Histochemistry*, 2nd edn, John Wiley and Sons, London.
4. Watkins, S. (1989) Immunohistochemistry, in *Current Protocols in Molecular Biology*, (eds F.M. Ausubel, R. Brent, R.E. Kinston *et al.*), John Wiley and Sons, Inc., Greene Publishing Associates, New York, pp. 14.6.1–14.6.13.
5. Onetti-Muda, A., Riminucci, P. and Bianco, P. (1992) Freeze-drying of bone tissue: immunocytochemistry and enzyme histochemistry on paraffin embedded and low-temperature resin embedded specimens. *Histochemistry* **98**, 283–288.
6. Ikeda, T., Nomura, S., Yamaguchi, A. *et al.* (1992) *In situ* hybridisation of bone matrix proteins in undecalcified adult rat bone sections. *Journal of Histochemistry and Cytochemistry* **40**, 1079–1088.
7. Zeller, R. and Rogers, M. (1989) *In situ* hybridisation to cellular RNA, in *Current Protocols in Molecular Biology*, (eds F.M. Ausubel, R. Brent, R.E. Kinston *et al.*), John Wiley and Sons, Inc., Greene Publishing Associates, New York, pp. 14.2.1–14.3.11.
8. Johnstone, J.J.A. (1979) The routine sectioning of undecalcified bone for cytochemical studies. *Histochemistry Journal* **11**, 359–365.

9. Bradbeer, J.N., Lindsay, P.C. and Reeve, J. (1994) Fluctuation of mineral apposition rate at individual bone-remodelling sites in human iliac cancellous bone: independent correlations with osteoid width and osteoblastic alkaline phosphatase activity. *Journal of Bone and Mineral Research* **9**, 1679–1686.

10. Coons, A.H., Creech, H.J. and Jones, R.N. (1941) Immunological properties of an antibody containing a fluorescent group. *Proceedings of the Society of Experimental Biology and Medicine* **47**, 200–205.

11. Nakane, P.K. and Pierce G.B. (1967) Enzyme-labeled antibodies: preparations and applications for the localization of antigens. *Journal of Histochemistry and Cytochemistry* **14**, 929–930.

12. Johnson, G.D. and Dorling, J. (1981) Immunofluorescence and peroxidase techniques, in *Techniques in Clinical Immunology*, (ed. R.A. Thompson), Blackwell Scientific Publications, London, pp. 106–137.

13. Gall, J.G. and Pardue, M.L. (1971) Nucleic acid hybridization in cytological preparations. *Methods in Enzymology* **38**, 470–480.

14. Hafen, E., Levine, M., Garber, R.L. and Gehring, W.J. (1983) An improved *in situ* hybridisation method for the detection of cellular RNA in drosophila tissue sections and its application for localizing transcripts of the homeotic antennapedia complex. *EMBO Journal* **2**, 617–623.

15. Akam, M.E. (1983) The location of ultrabithorax transcripts in *Drosophila* tissue sections. *EMBO Journal* **2**, 2075–2084.

16. Cox, K.H., DeLeon, D.V., Angerer, L.M. and Angerer, R.C. (1984) Detection of mRNAs in sea urchin embryos by *in situ* hybridisation using asymmetric RNA probes. *Developmental Biology* **101**, 485–502.

17. Awgulewitsch A., Utset, M.F., Hart, C.P. *et al.* (1986) Spatial restriction in expression of a mouse homeobox locus within the central nervous system. *Nature* **320**, 328–335.

18. McLaughlin, S.K. and Margolskee, R.F. (1993) ^{33}P is preferable to ^{35}S for labeling probes used in *in situ* hybridisation. *BioTechniques* **15**, 506–511.

19. Singer, R.H. and Ward, D.C. (1982) Actin gene expression visualized in a chicken muscle tissue culture by using *in situ* hybridisation with a biotinated nucleotide analog. *Proceedings of the National Academy of Sciences USA* **79**, 7331–7335.

20. Singer, R.H., Lawrence, J.B. and Villnave, C. (1986) Optimization of *in situ* hybridisation using isotopic and nonisotopic methods. *BioTechniques* **4**, 230–250.

21. Nuovo, G. (1994) *PCR In Situ Hybridisation*, 2nd edn, Raven Press, New York.

22. Wilkinson, D.G. (1992) The theory and practice of *in situ* hybridisation; in *In Situ Hybridisation: a Practical Approach* (ed. D.G. Wilkinson), Oxford University Press, New York, pp. 1–13.

23. Angerer, L.M. and Angerer, R.C. (1992) *In situ* hybridisation to cellular RNA with radiolabelled RNA probes, in *In Situ Hybridisation: a Practical Approach* (ed. D.G. Wilkinson), Oxford University Press, New York, pp. 15–32.

24. Polak, J.M. and McGee, J.O'D. (eds) (1990) *In Situ Hybridisation, Principles and Practice*, Oxford University Press, New York.

25. Sambrook, J., Fritsch, E.F. and Maniatis, T. (1989) *Molecular Cloning: a Laboratory Manual*, 2nd edn, Cold Spring Harbor Laboratory Press, Cold Spring Harbor, NY.

26. Bloch, K.D., and Bartos, B. (1992) Digestion of DNA with restriction endonucleases, in *Current Protocols in Molecular Biology*, (eds F.M. Ausubel, R. Brent, R.E. Kinston *et al.*), John Wiley & Sons, Inc., Greene Publishing Associates, New York, pp. 3.1.1–3.1.9.

27. Lu, L.H. and Gillett, N. (1994) An optimized protocol for *in situ* hybridisation using PCR-generated ^{33}P-labeled riboprobes. *Cell Vision* **1**, 169–176.

28. Drake, F.H., Dodds, R.A., James, I.E. *et al.* (1996) Large scale sequencing of ESTs from human osteoclast cDNA library: 'Electronic northern blot'. *Bone* **17**, 579.

29. Drake, F.H., Dodds, R.A., James, I.E. *et al.* (1996) Cathepsin K, but not cathepsins B, L, or S is abundantly expressed in human osteoclasts. *Journal of Biological Chemistry* **271**, 12511–12516.

30. James, I.E., Dodds, R.A., Rykaczewski, L.E. *et al.* (1996) Purification and characterization of fully functional human osteoclast precursors. *Journal of Bone and Mineral Research,* **11**, 1608–1618.
31. Merry, K.H., Dodds, R.A., Littlewood, A.J. and Gowen, M. (1992) Osteoclasts express osteopontin mRNA in sections of human bone as detected by *in situ* hybridisation. *Calcified Tissue International* **50** (S1), A40.
32. Merry, K., Dodds, R.A., Littlewood, A. and Gowen, M. (1993) Expression of osteopontin mRNA by osteoclasts and osteoblasts in modelling adult human bone. *Journal of Cell Science* **104**, 1013–1020.
33. Connor, J.R., Dodds, R.A., James, I.E. and Gowen, M. (1995) Human osteoclast and giant cell differentiation: the apparent switch from nonspecific esterase to tartrate resistant acid phosphatase activity coincides with the *in situ* expression of osteopontin mRNA. *Journal of Histochemistry and Cytochemistry* **43**, 1193–1201.
34. James, I.E., Dodds, R.A., Olivera, D.L. *et al.* (1996) Human osteoclastoma-derived stromal cells : correlation of the ability to form mineralized nodules *in vitro* with formation of bone *in vivo*. *Journal of Bone and Mineral Research* **11**, 1453–1460.
35. Dodds, R.A., James, I.E., Olivera, D.L. *et al.* (1996) Formation of bone by human osteoclastoma-derived stromal cells in the SCID mouse model is preceded by mouse osteoclast differentiation. *Bone* **17**, 578.
36. Dodds, R.A., Merry. K., Littlewood, A. and Gowen, M. (1994) Expression of mRNA for IL1b, IL6, and TGFb in developing bone and cartilage. *Journal of Histochemistry and Cytochemistry* **42**, 733–744.

CHAPTER NINE

Biochemical markers of bone turnover

Simon P. Robins

9.1 INTRODUCTION

Much of the research effort in bone biology is directed towards understanding the mechanisms of bone modelling and remodelling. This includes attempts to elucidate the influences of a large number of biochemical and biophysical factors on cellular processes in bone. The development of techniques to quantify specific facets of these processes in bone provides important tools to facilitate research in this area. Markers of bone turnover also have more general applications clinically in the detection and management of disorders involving the skeleton. Indeed, the increasing interest in osteoporosis and related disorders in recent years has probably contributed some impetus to the major advances in developing new markers that have occurred recently.

Generally, the approaches adopted for developing bone markers are to utilize specific enzyme activities or cellular products of osteoblasts or osteoclasts, which are taken to be indicative of the rates of bone formation and resorption, respectively. The aim in this chapter is not to present a catalogue of all the methods available, but initially to review briefly the biochemical basis and relative merits of the more recently developed techniques, with particular emphasis on the often overlooked assumptions that are involved in their application. This will be followed by a more detailed description of the methods of choice. The methods will be considered primarily with reference to their application to measuring whole-body bone metabolism by measuring bone components in blood or urine.

9.2 FORMATION MARKERS

9.2.1 Bone-specific alkaline phosphatase

Measurement of total alkaline phosphatase represents a well established technique in clinical chemistry that has long been applied to the assessment of conditions involving increased bone turnover, such as Paget's disease. For detecting more subtle changes in bone metabolism such as those occurring in osteoporosis,

Methods in Bone Biology. Edited by Timothy R. Arnett and Brian Henderson.
Published in 1997 by Chapman & Hall, London. ISBN 0 412 75770 2.

however, measurements of total enzyme activity in serum have proved less useful. Attention has therefore focused on the measurement of the bone isoform of alkaline phosphatase but it is only in recent years that simple immunoassays have been developed.

The four genes that code for alkaline phosphatase give rise to four true isoenzymes. The tissue-nonspecific form expressed in many tissues, including liver, bone and kidney, constitutes the major form of this enzyme in blood. Tissue-specific gene products are expressed in the intestine, germ cells (e.g. testis, thymus) and in the placenta at later stages of development. In practice, the quantification of bone-specific alkaline phosphatase (BAP) requires the ability to differentiate between liver and bone isoforms, the principal constituents of the enzyme activity in blood. These isoforms of tissue-nonspecific alkaline phosphatase differ in their carbohydrate attachments and the degree of sialation. Until recently, the main methods available for separating and quantifying different isoforms of alkaline phosphatase were selective heat inactivation (often combined with chemical inhibition) and electrophoretic techniques [1]. These techniques are laborious to perform and lack precision, particularly when one isoform of the enzyme is in large excess. Lectin precipitation methods provided a simpler technique in which alkaline phosphatase activity is measured before and after precipitation of the bone isoform by wheat germ agglutinin. This method also forms the basis of a commercial assay (section 9.5.2).

Major advances in assay technology came with the development of antibodies specific for the bone isoform. An immunoradiometric assay based on two monoclonal antibodies provides a quantitative measurement of the amount of the enzyme protein in serum [2]. An alternative approach has been to develop a microtitre plate assay based on a bone-specific monoclonal antibody attached to the plate to capture the alkaline phosphatase, the activity of which is measured colorimetrically in the plate [3]. The latter method has some advantage in expressing the results in activity units that can be compared directly with the results for total alkaline phosphatase measurements. Both of the immunoassays provide reasonable specificity for the bone isoform with only about 10% cross-reactivity with the liver isoform.

Clearly, these immunoassay kits are now the methods of choice for determining bone-specific alkaline phosphatase. Whether this assay is also the method of choice for assessing bone formation rate is less clear since the function of this enzyme is poorly understood. The concentration of BAP in serum is taken to indicate the number and activity of osteoblasts, though this also involves enzymatic release of the enzyme from the plasma membrane – a process for which the control is unclear. Whether the production of BAP is indicative of mineralization rather than new bone formation is also uncertain but these two processes are likely to be closely linked. The role of bone-specific alkaline phosphatase in bone cell function is reviewed in Chapter 1.

9.2.2 Osteocalcin

Synthesized primarily by osteoblasts, osteocalcin (or bone Gla protein, BGP) appears to provide a good candidate as a bone marker and, following development of the first immunoassay in 1980 [4], various forms of assays for serum osteocalcin have been used extensively (Chapter 1). This 49-residue protein is characterized by the presence of up to three γ-carboxyglutamic acid residues, formed post-ribosomally by a vitamin K-dependent enzymatic carboxylation of glutamate residues at positions 17, 21 and 24. These residues accentuate the calcium-binding properties and most of the newly synthesized protein is immediately bound to the bone mineral, with only a proportion being secreted directly into the blood. Most osteocalcin assays are now designed to measure this fraction of the intact protein in blood, and it is assumed that the proportion of total osteocalcin synthesis that this represents is consistent, though there appears to be no direct evidence to support this assumption.

The aim of measuring primarily intact osteocalcin led to the development more recently of two-site immunoassays or the use of conformation-dependent antibodies that would eliminate possible cross-reaction with fragments of osteocalcin emanating from bone matrix degradation. The latter probably contributed significantly to the wide variations in osteocalcin measurements that have been observed with different assays. There are other confounding factors that contribute to assay variations and there have been many reports of immunochemical heterogeneity of osteocalcin (e.g. [5, 6]). A general finding is that different combinations of monoclonal antibodies used in sandwich assays give different results when applied to the same serum samples and with the same standard. Since different assays cannot be compared directly, therefore, all measurements of osteocalcin should be assessed with respect to a reference population determined using the same assay in the same laboratory [7].

Incomplete carboxylation of the susceptible glutamate residues presents another potential complication although most assays react equally well with both carboxylated and non-carboxylated forms of osteocalcin. Deficiencies in vitamin K, particularly in the elderly, would result in a larger proportion of under-carboxylated osteocalcin which has a lower affinity for the mineral and might contribute to the increase with age in osteocalcin concentrations [8]. Methods to distinguish between fully and under-carboxylated forms of osteocalcin have been developed using hydroxyapatite *in vitro* [8, 9], and the concentration of under-carboxylated form may be related to the risk of hip fracture [10]. Monoclonal antibody-based assays for determining under-carboxylated osteocalcin are being developed but additional data are required to confirm the utility of these assays.

The susceptibility of many osteocalcin assays to problems associated with the collection and storage of the blood sample has been well recognized [11], but recent evidence characterizing the forms of osteocalcin present in blood (Fig. 9.1) has suggested ways in which these problems can be overcome [12]. Rapid cleavage in serum between residues 43 and 44 is common, leading to the conclusion that assays

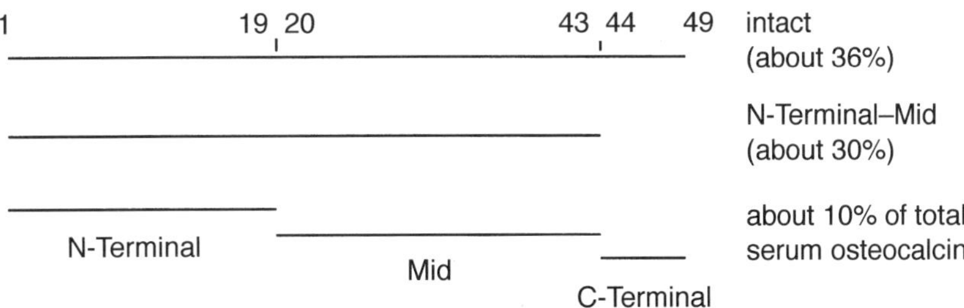

Figure 9.1 Major forms of osteocalcin in serum determined by immunoassay with antibodies specific for different parts of molecule. (Data from [12].)

which recognize the N-terminal mid-fragment will be less susceptible to storage problems. Recently, tests of the stability of different assays have confirmed that those with specificity for the N-terminal mid-fragment are less susceptible to loss of epitopes on storage of serum at room temperature [13].

In addition to the vitamin K dependence, osteocalcin synthesis depends on 1,25 dihydroxyvitamin D_3, having a vitamin D receptor element in the promoter region of the gene. This property also contributes to the unique susceptibility of osteocalcin measurements to corticosteroid treatment, which acts through interaction with the vitamin D receptor element. Despite the many assumptions involved in applying osteocalcin measurements, the assay has been applied in a large number of clinical studies. In common with bone alkaline phosphatase, the function of osteocalcin is unclear and its measurement seems to reflect mainly the activity and number of osteoblasts rather than being directly related to bone mineralization. Recent evidence from gene knockout studies [14] indicates that osteocalcin may play some inhibitory role in bone formation, since mice lacking the protein had larger and stronger bones than their osteocalcin-replete counterparts.

9.2.3 Procollagen propeptides

As collagen type I constitutes about 90% of the protein matrix in bone, it is reasonable that many of the biochemical markers of bone metabolism should be based on this collagen type. Procollagen type I, the initially synthesized product, is about 50% larger than the collagen molecule in fibrils, having large extension peptides at both the N- and C-terminal ends. These propeptides are removed *en bloc* by separate proteases at or near the cell surface during secretion of the molecule. The intact C-terminal propeptide (PICP) containing intermolecular disulphide bonds can be detected in the blood as a 100 kDa component [15] and several commercial assays are now available. The assay has been used successfully to assess growth [16] but its sensitivity to relatively small changes in bone formation, such as those accompanying the menopause, has been rather limited.

Immunoassays for the N-terminal propeptide of procollagen I (PINP) have recently received renewed interest. A component isolated from amniotic fluid referred to as fetal antigen 2 was shown to be the *N*-propeptide [17]. In serum, there are components related to the *N*-propeptide having apparent molecular masses of about 100 kDa and 30 kDa, but different assays react differently with these components [18, 19]. It has been suggested [20] that the smaller component is a degradation product related to the short helical domain, but there is currently some doubt about this interpretation [21]. Although the preliminary clinical data using PINP assays are encouraging, the fact that the PINP and PICP assays can in certain clinical applications give different results emphasizes the importance of gaining further knowledge about the degradative pathways and clearance of these molecules.

9.3 RESORPTION MARKERS

In contrast to the formation markers which are all measured in serum, most assays designed to monitor bone resorption are measured in urine. An exception is the measurement of tartrate-resistant acid phosphatase (TRAP) in serum, for which a number of assays have been developed [22]. Although TRAP gives an indication of osteoclastic activity, there are several isoenzymes and TRAP is not entirely specific for the osteoclast. Most of the urinary bone resorption assays are based on the detection of fragments of bone collagen. Archetypal amongst these is the determination of urinary hydroxyproline, but this assay was shown to have major drawbacks and has now been largely replaced by collagen crosslink-related assays. The main disadvantages of urinary hydroxyproline are the lack of specificity for bone, the dependence on adequate dietary control to avoid the contributions of exogenous hydroxyproline, the high level (about 90%) of metabolism of hydroxyproline in the liver and the contributions of hydroxyproline from the degradation of newly synthesized collagen. Each of the newer assays developed should therefore be assessed with regard to how well all of the above factors have been overcome.

9.3.1 Galactosyl hydroxylysine

The formation of both galactosyl hydroxylysine (GHL) and glucosyl-galactosyl hydroxylysine are intracellular events during procollagen synthesis. Early analytical studies showed that the monosaccharide predominated in bone collagen whereas the disaccharide was more prevalent in skin collagen as well as in collagen-like proteins such as C1q [23]. The monosaccharide, GHL, therefore appears to be relatively specific for bone and, for human studies, the renal conversion of the disaccharide to GHL that was demonstrated in rats does not occur [24]. The initial studies of the glycosides showed that these compounds were essentially not metabolized in the body and contributions from dietary sources were much less than for hydroxyproline [25]. As GHL is the product of an intracellular modification, it is possible that degradation of newly synthesized collagen may contribute to the rate of excretion. For many cell types, there appears to be a basal level of lysosome-mediated intracellular procollagen degradation of about 15%, but this may rise to as much as 40% in certain circumstances [26]. At present, however, there is no direct evidence that excretion of GHL is affected significantly by this phenomenon.

Much of the recent interest in GHL has been engendered by the work of Moro and colleagues in developing an HPLC method of analysis for urine [27]. The method requires no preliminary fractionation of the sample, but involves pre-column dansylation followed by a reversed phase separation. Wider utilization of this technique will depend on the development of direct immunoassays and, although preliminary reports have recently appeared from two companies, no commercial assay is yet available.

9.3.2 Pyridinium crosslinks

Following the initial characterization of pyridinoline [28], the potential value of this crosslink as a marker was suggested by the discovery that the compound could be detected and quantified in urine [29, 30]. The development of a reversed phase HPLC separation [31] facilitated analysis of tissue hydrolysates for both pyridinoline (Pyd), also referred to as hydroxylysyl pyridinoline (HP), and deoxypyridinoline (Dpd), an analogue that was shown to be derived from a lysyl residue in the collagen helix [32] giving rise to the alternative nomenclature of lysyl pyridinoline (LP). Inclusion of a pre-fractionation step using cellulose partition chromatography facilitated analysis by HPLC of both crosslinks in hydrolysed urine samples [33] and, as discussed later, the technique can be fully automated [34].

Considering the parameters discussed earlier as being necessary to provide improvements over urinary hydroxyproline as a marker, the pyridinium crosslinks do not themselves appear to be metabolized, though there is no direct evidence for this; there is no dietary contribution to their excretion [35] and, because the crosslinks are formed only at the final stages of fibril formation, the crosslinks indicate degradation only of mature, functional tissue and are unaffected by degradation of newly synthesized collagen. Another major issue to be considered is that of tissue specificity. Clearly, the pyridinium crosslinks have advantages over hydroxyproline since these compounds are essentially absent in skin. Analyses of tissues has indicated that Pyd has a much wider tissue distribution than Dpd [31, 36]. Initially it appeared that Dpd was present only in mineralized tissues, but it is now known to be present in cardiovascular tissue, intramuscular collagen and some ligaments. These findings do not negate the original suggestion that Dpd should be considered as a bone-specific marker, since the concentration in tissue is only one of the factors that has to be taken into account. When the pool size of the tissue and its turnover rate are also considered (Fig. 9.2), it is clear that the low metabolic turnover of the main soft tissues in which Dpd has been detected make their contribution to urinary Dpd negligible. Thus, the major advantage of using Dpd as a urinary marker is to gain specificity for bone.

All of the initial analyses of crosslinks were performed using acid hydrolysates to liberate all peptide-bound and other conjugated forms of the crosslink. Although preliminary analyses using an ELISA for Pyd indicated that no crosslinks were present in free form in urine, the HPLC chromatography system combined with

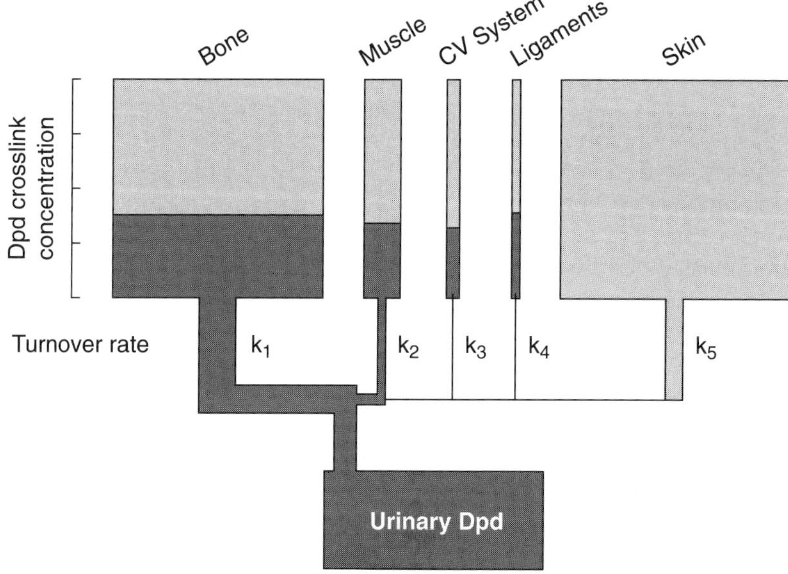

Figure 9.2 Contributions of body pools of collagen type I to deoxypyridinoline (Dpd) excretion. Dpd concentrations in different tissues, shown by the dark-shaded area of each tissue box, make varying contributions to urinary excretion of the crosslink, depending on pool size (denoted by size of box) and turnover rate. Bone is the major contributor to urinary Dpd owing to its large pool size and high turnover rate.

pre-fractionation on cellulose indicated that 40–50% of the crosslinks were free and could be analysed directly without hydrolysis [37, 38]. Further analysis of urine from patients with a series of different disorders associated with increased bone resorption revealed that a similar proportion of crosslink was free in each case, thus indicating that similar information about bone resorption could be obtained by direct analysis of urine without the time-consuming hydrolysis step. These results also suggested the possible application of direct immunoassays for the crosslinks and a number of ELISA systems have recently been reported [39–41], which will be described in more detail in section 9.5.4.

9.3.3 Peptide assays related to crosslinks

A number of assays have been described recently that are based on measurement of peptides associated with the crosslink regions in collagen rather than measuring the crosslinks themselves. It is known that the crosslinks are derived from (hydroxy)lysine residues present in both the N- and C-terminal telopeptides which are converted to aldehydes by lysyl oxidase. The peptide assays are therefore based on measuring peptides from either end of the collagen type I molecule.

The ICTP assay refers to measurement of a component from collagen type I C-terminal telopeptide [42]. Antibodies were raised against a crosslink-containing 8.5 kDa peptide partially purified from a bacterial collagenase digest of human bone collagen. Although the peptide contained pyridinium crosslinks, other forms of crosslink were also detected, and antibody recognition does not depend on the form of crosslink present. In contrast to other crosslink or peptide resorption assays, ICTP is a serum assay.

An assay based on a crosslinked peptide from the N-terminal telopeptide, referred to as NTx, was developed using a monoclonal antibody raised against a peptide isolated from urine of a patient with Paget's disease of bone [43]. The ELISA was shown to react with several pyridinium crosslink-containing peptides in urine, though the original description of the assay states that the pyridinium crosslink is not essential for reactivity. Sequences from the α-2 chain telopeptide were an essential part of the epitope and the presence of this telopeptide chain crosslink was claimed to confer specificity for bone, as soft tissue collagen type I crosslinking was said to involve primarily the α-1 chain telopeptides [43]. However, the fact that a digest of skin collagen peptides showed the same molar reactivity with the NTx assay as those from bone [44] casts some doubt on this contention. The NTx assay therefore appears to be measuring collagen type I degradation products from any source. Since the antibody does not react with the linear telopeptide sequences, this assay is unlikely to detect degradation of newly formed collagen.

By raising antibodies against a synthetic, linear octapeptide which contained the crosslinking site of the C-terminal telopeptide of collagen I, Bonde and colleagues [45] intended to develop an assay (referred to as CTx) that would recognize degradation products of all crosslink forms of this collagen. Subsequent analysis has shown that the assay (termed 'Crosslaps') in fact recognized only a form of the peptide containing an isoaspartyl (or β-aspartyl) peptide bond [46]. The transformation of aspartyl to isoaspartyl residues in proteins is well known and, for extracellular proteins, the process is thought to be time-dependent as well as a function of the particular environment of the susceptible bond. The interpretation of results from this assay is therefore complex without knowledge of the rate of isomerization that occurs in collagen type I of bone and other tissues. The Crosslaps assay also reacts with peptides from a human digest of skin, suggesting that conversion to the isoaspartyl peptide form is common in other tissues containing collagen type I.

9.4 WHICH RESORPTION MARKER?

Before describing methods in detail, it is pertinent to review briefly the choices available. The foregoing discussion has been included in order to provide a background to the biochemical basis for the different assays, and it seems clear that some form of crosslink assay is to be preferred. The main decisions to be made, therefore, are whether to measure pyridinium crosslinks by HPLC or by an ELISA kit, whether to measure free or total crosslinks and finally, if it is decided to opt for immunoassay kits, whether this should be the crosslinks themselves or the peptide assays. This section will consider some additional points that are relevant to these decisions.

Whether to use HPLC or immunoassay kits for pyridinium crosslink determination is relatively straightforward and will depend to a large extent on the expertise and equipment available. For those laboratories lacking suitable HPLC facilities or where the systems are fully occupied in running other assays, the advantages of using immunoassay kits far outweighs the time, equipment costs and potential technical problems in setting up an HPLC system. Several automated systems have been described (for review, see [47]) but their use adds appreciably to the equipment cost.

The decisions on whether to measure free or total crosslinks and on whether the crosslinks themselves or the crosslinked peptides should be assayed are in fact linked, and depend on some consideration of the degradative metabolism of collagen. The measurement of free crosslinks, and particularly free Dpd, as a valid marker of bone resorption depends on these values representing a consistent proportion of the total Dpd. Implicit in this statement is an understanding that measuring total Dpd is the most accurate parameter available for determining the true rate of bone resorption – a fact supported experimentally by comparing this method with radioisotopic exchange [48]. The proportion of free crosslink was shown to be consistent in a series of studies including a range of patients with osteoporosis and other metabolic bone diseases [37]. Some disparity in urinary free/bound crosslink ratios has been noted, however, particularly in patients treated with bisphosphonates, such that the proportion of free Dpd was shown to increase during the 4-week treatment period [49]. In other studies of short-term intravenous administration of bisphosphonate, the change over 3 days in urinary free crosslink concentration was less than that for the total crosslinks [50]. The same study showed much larger percentage changes during treatment for both the NTx and CTx telopeptide assays. These results have been widely interpreted to indicate that the telopeptide assays are 'more specific' for bone than the free crosslink assays. This is unlikely to be the case and it is relevant that generally the magnitude of the changes in telopeptide assays is greater than that for the total crosslinks. The concept has therefore been put forward that some treatments for high bone resorption states may affect not only true bone resorption but also the patterns of free crosslinks and peptides released into urine [44]. These effects may be at the osteoclast level in bone or at other sites of the body, such as liver or kidney. The most reasonable conclusion in the present state of knowledge is that, with respect to bisphosphonate treatments, the free crosslink assays underestimate the true changes in bone resorption whereas the telopeptide assays probably overestimate these effects. The latter point cannot be demonstrated directly as it cannot be determined whether the urinary peptides measured by these assays constitute a consistent proportion of the bone collagen degraded.

It has been suggested that changes in free crosslink proportions are simply a function of the rate of bone collagen resorption and not related to any particular form of treatment. Some evidence for this was provided by an observed inverse relationship between the proportion of free Dpd crosslink and the total amount of Pyd excreted [50]. Our own experience, examining a wide range of results from both healthy volunteers and patients with metabolic bone disease not receiving treatment, is that the relationship between the proportion of free Dpd and the total Dpd excreted is present (Fig. 9.3) but is unlikely to contribute significantly to the observed effects of treatment.

Issues related to the specificity for bone of the different assays have been discussed in section 9.3.3. The use of Dpd provides a more bone-specific assay but the fact that the telopeptide assays reflect only collagen type I degradation may not seriously affect their application in most situations, because bone collagen is

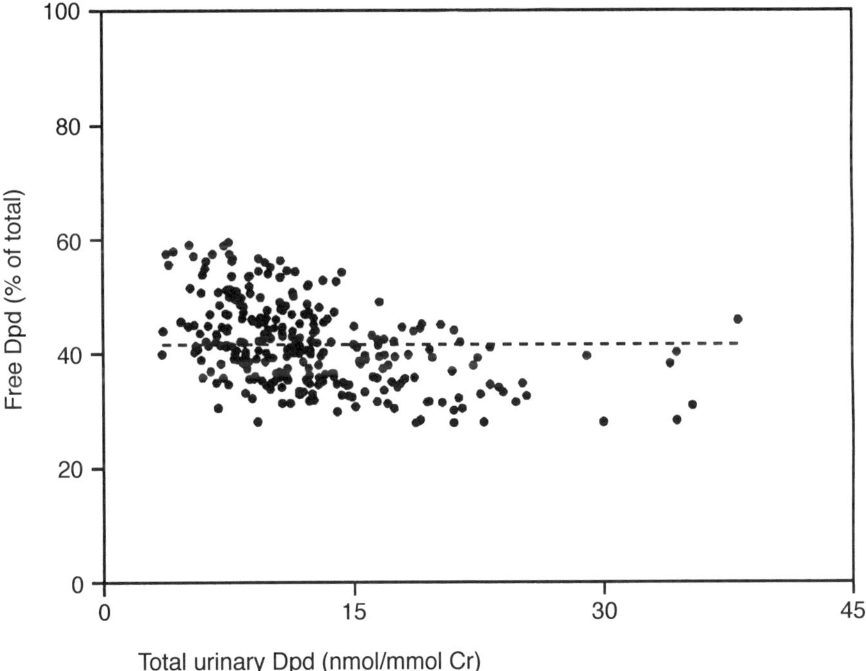

Figure 9.3 Changes in proportion of free Dpd in urine in relation to total excretion expressed as a creatinine ratio. Data (n = 318) are from both healthy volunteers and patients with metabolic bone diseases, to give a wide range in total excretion, with each point representing one individual. Broken line indicates mean proportion of free Dpd (41.5%). Linear regression analysis gave a line with the equation $y = -0.56x + 48.2$, with $r^2 = 0.18$. There is little variation in proportion of free Dpd within normal range of total excretion (up to 15 nmol/mmol Cr).

usually the major contributor to urinary collagen I degradation products. In most cases, therefore, results for the peptide assays and the crosslinks are closely correlated.

A further consideration in this section should be the clinical variability of the assays used. The purpose of the assays in the context being considered is to provide the clinician with information about the rates of bone resorption. As with most biochemical markers, more useful data are provided for individual patients using longitudinal samples where the effects of any intervention are being monitored. The magnitude of change for any particular marker must be viewed not only in terms of whether the effect is a true reflection of bone resorption, but also with regard to the significance of the change with respect to the variability of the assay.

A decision on which resorption assay to use is not a simple one and there is currently a need for more information on the degradative metabolism of collagen, including any hepatic or renal influences. Currently, there are fewest assumptions involved in measuring total Dpd, which can be done either by HPLC or by immunoassay.

9.5 METHODOLOGY

9.5.1 Sample collection and storage

One of the main considerations for collecting serum or urine samples for measuring bone markers is to take account of the circadian or diurnal variations that occur. Virtually all of the markers are higher at night compared with the day and, although estimates of the magnitude of this effect vary, it is undoubtedly sufficient to affect the results significantly. Provided that samples are collected at similar times of day, these difficulties can be adequately overcome.

(a) Serum
The conditions of blood collection and storage are particularly important for studies where osteocalcin is to be measured. Any haemolysis during blood collection or processing will result in proteolysis of osteocalcin and such samples should be discarded. The relative instability of osteocalcin in serum also requires that the time from blood collection to freezing the serum should be as short as possible and no longer than 2 hours. For long-term storage, the serum should be maintained at below –70°C but periods of a few weeks at –20°C cause no significant losses. Some commercial assays measure the primary fragments produced on storage in serum (section 9.2.2). Alkaline phosphatase and the procollagen peptides are more robust in terms of their collection and storage requirements but, for all assays, freeze–thawing of serum samples should be avoided and storage should be in multiple aliquots of 0.2–0.5 ml serum.

(b) Urine
Wherever possible, urinary bone markers should be measured in 24-hour collections, so that excretion rates can be expressed as absolute amounts per day rather than the more usual ratio with urinary creatinine which corrects for urine dilution and muscle mass. In practice, full 24-hour collections present many logistical problems and the collection of first morning urine is the best alternative. This sample

relates to urine produced during the night hours and the markers will therefore be present at higher concentrations than in the 24-hour samples. Where long-term storage is contemplated, it is preferable to adjust the samples to pH 2–3 by the addition of acid (0.1 ml 6 M HCl/10 ml urine). Both the creatinine and crosslinks are slightly more stable at this pH, but also the pH adjustment minimizes precipitate formation in the urine, particularly after freeze–thawing, which can lead to removal of some peptide-bound crosslink components. Within about 12 hours of collection, the samples should be stored at –20°C and, for the crosslink assays, the samples are stable at this temperature for many years. Our studies indicate that the crosslink assays are not susceptible to multiple freeze–thaw cycles.

9.5.2 Assay of bone formation markers

Several commercial assays are available for bone-specific alkaline phosphatase, osteocalcin, PICP and, more recently, PINP and no further details of the methodology are necessary here. Some of the factors relevant to which assay to adopt have been considered earlier and the choice will depend on the specific information required in particular studies. The characteristics of some of these assays are included in Table 9.1.

9.5.3 Assay of urinary pyridinium crosslinks by HPLC

(a) Hydrolysis
Additions to urine of an equal volume of concentrated HCl followed by heating at 107°C for 20 hours is an effective means of hydrolysis. This can be done in screw-cap glass vials or in disposable polypropylene tubes (Appendix 9.A). Hydrolysis of urine by heating in 3 M HCl (3 vol. urine to 1 vol. concentrated HCl) has been shown to release free crosslinks as efficiently as the more concentrated acid. The more dilute acid has some practical advantages in some systems (section 9.4.3d) and can provide increased sensitivity.

(b) Pre-fractionation
In order to remove interfering compounds, hydrolysed urine must be partially purified before application to the HPLC column. Partition chromatography on cellulose (Appendix 9.A) is a robust and reliable technique based on the adsorption of pyridinium compounds to this support in a mobile phase consisting of butanol : acetic acid : water (4 : 1 : 1 by volume). A fully automated procedure has been described incorporating both pre-fractionation and HPLC [34]. Although such a technique is the method of choice in terms of precision and analysis time, the method requires a considerable amount of specialized equipment that may not be generally available. A manual, batchwise procedure which can readily deal with 40 samples concurrently will therefore be described, indicating where improvements in precision can be made. For the manual method, columns are prepared containing about 100 mg CF1 cellulose, added as a slurry of 5% (w/v) CF1 in mobile phase. The columns should either have extensions fitted or be of a size to accommodate at least 5 ml of fluid above the cellulose to facilitate subsequent washing steps. The freshly prepared columns are washed with 5 ml of mobile phase. To the sample (0.5 ml hydrolysate) is added acetic acid (0.5 ml), the CF1 slurry (0.5 ml) and butanol (2 ml), and the mixture is shaken to allow adsorption of the crosslinks

Table 9.1 Characteristics of serum assays for bone formation

Marker	Measured analyte	Assay types	Examples of commercially available kits
Osteocalcin	Human intact osteocalcin	IRMA, double antibody	ELSA-OST-NAT, CIS-bio International, Gif-sur-Yvette, France
		ELISA, double antibody	Biomedical Technologies Inc., Stoughton, MA, USA
	Human intact and N-terminal mid molecule	IRMA, double antibody	ELSA-OSTEO, CIS-bio International, Gif-sur-Yvette, France
		ELISA, double antibody	Biomedical Technologies Inc., Stoughton, MA, USA
	Bovine osteocalcin[a]	ELISA, single monoclonal antibody	NovoCalcin, Metra Biosystems Inc., Mountain View, CA, USA
	Rat osteocalcin[b]		Biomedical Technologies Inc., Stoughton, MA, USA
Procollagen propeptide	PICP	RIA	Orion Diagnostica, Espoo, Finland
		ELISA, double antibody	Prolagen-C, Metra Biosystems Inc., Mountain View, CA, USA
Alkaline phosphatase	Bone-specific isoform	Lectin precipitation	Iso-ALP, Boehringer Mannheim, Mannheim, Germany
		IRMA, double antibody	Tandem-R Ostase, Hybritech Inc., San Diego, CA, USA
		ELISA, single monoclonal	Alkphase-B, Metra Biosystems Inc., Mountain View, CA, USA

[a] Cross-reacts with human osteocalcin
[b] Antibodies to human or bovine osteocalcin show little cross-reactivity with the rat protein

to the cellulose. The contents are then added to the CF1 columns, effecting quantitative transfer by rinsing the sample tube twice with mobile phase and adding each washing to the CF column. The column is washed with more mobile phase so that the total volume of washing is at least 15 ml. There is no limit to the volume of washings as the crosslinks remain firmly attached to the cellulose: generally, increasing the volume of washing will result in fewer interfering compounds in the HPLC chromatograms. The crosslinks are eluted from the CF column with water (5 ml) into conical tubes, which are then centrifuged briefly to enhance separation of the butanolic and aqueous phases. The latter is aspirated carefully with a Pasteur pipette and lyophilized.

(c) Chromatography

For HPLC separation of the pyridinium crosslinks, the precise conditions depend on the system used and on other factors such as the condition of the chromatography column. The conditions for reversed-phase chromatography shown in Table 9.2, using heptafluorobutyric acid (HFBA) as ion-pair with an acetonitrile gradient, may therefore require slight modification for individual systems, and this section will concentrate on the effects of changing various parameters.

Although isocratic systems with faster throughput and requiring less instrumentation have been used, the lack of a high acetonitrile step for each sample application can lead to a build-up of contaminants and can shorten column life. Also, the time-limiting step is generally sample preparation. With respect to the instrumentation, the only critical issue is the type of fluorimeter used to monitor the natural fluorescence of the crosslinks. As an excitation wavelength of 295 nm is required, an instrument with a xenon lamp is essential to provide the necessary power and appropriate band-width; instruments with a mercury lamp and wavelength filters are unsatisfactory for this application.

The purity of the ion-pairing reagent is critical and HFBA has been found to be the most reliable (Appendix 9.A). Separation of Pyd and Dpd can generally be achieved using 10 mM HFBA elution buffers by small adjustments to the gradient. Higher HFBA concentrations of 15 mM or even 20 mM give progressively increased separation between the two crosslink analogues, though the lower pH tends to shorten column life.

(d) Internal and external standards

The use of an appropriate internal standard has been problematic and although fluorescent pyridinium compounds such as pyridoxine and pyridoxamine have

Table 9.2 Conditions for reversed phase HPLC separation of pyridinium crosslinks

Sample buffer:	0.1 M-heptafluorobutyric acid (HFBA)
Buffer A:	10 mM-HFBA in distilled deionized water
Buffer B:	10 mM-HFBA in 75% acetonitrile
Column:	Microsorb C_{18} (10 cm × 4.6 mm)
Flow rate:	1 ml/min
Gradient[a] (time, %B):	0, 15; 7, 17; 12, 20; 15, 23; 15.1, 70; 21, 70; 21.1, 15; 26, 15.
Monitor:	Fluorescence (excitation 295 nm, emission 400 nm)

[a] Chromatography conditions will vary with different systems and may also need modification with increased usage of the separating column. Product specifications are given in Appendix 9.A.

been used to standardize the HPLC, these compounds are not recovered from the pre-fractionation step. Acetylation of the aliphatic hydroxyl group of Pyd provided a compound which eluted about 1 minute later than Dpd (Fig. 9.4) and corrected both for the recovery from the cellulose pre-fractionation step and the HPLC. This internal standard was designed for use with an automated system, where timing of particular steps is strictly controlled. Acetyl-Pyd is susceptible to hydrolysis when subjected to prolonged incubation in acid, and this standard is not suitable for manual operation unless special precautions are taken. Hydrolysis of urine in 3 M HCl (section 9.5.3a) reduces considerably the rate of hydrolytic removal of the acetyl group. A second precaution is to initiate processing of samples individually rather than as a batch. To each hydrolysate (0.5 ml) is added butanol (2 ml) and 90% acetic acid containing about 50 pmol of acetyl-Pyd. After rapid mixing, the solution is immediately applied to the CF column together with the preliminary washings and the procedure is repeated for the next sample. Once each sample has been processed in this way, the additional washing and elution steps are completed batchwise, as described in section 9.5.3b. Because of the short, consistent time period in the butanolic, acid environment, it is unnecessary to add CF1 slurry to the sample.

An alternative internal standard, isodesmosine, has been proposed [35] which has the advantage of being fully stable to the acid hydrolysis conditions. The main disadvantage is that the fluorescence characteristics of isodesmosine are not the same as those of the 3-hydroxypyridinium compounds, which may necessitate some compromise in monitoring conditions, though the use of a UV monitor to determine isodesmosine in tandem with the fluorimeter can overcome this problem [51].

External standardization of the crosslinks, Pyd and Dpd, raises a number of issues which have recently been addressed [52] and will not be dealt with here in

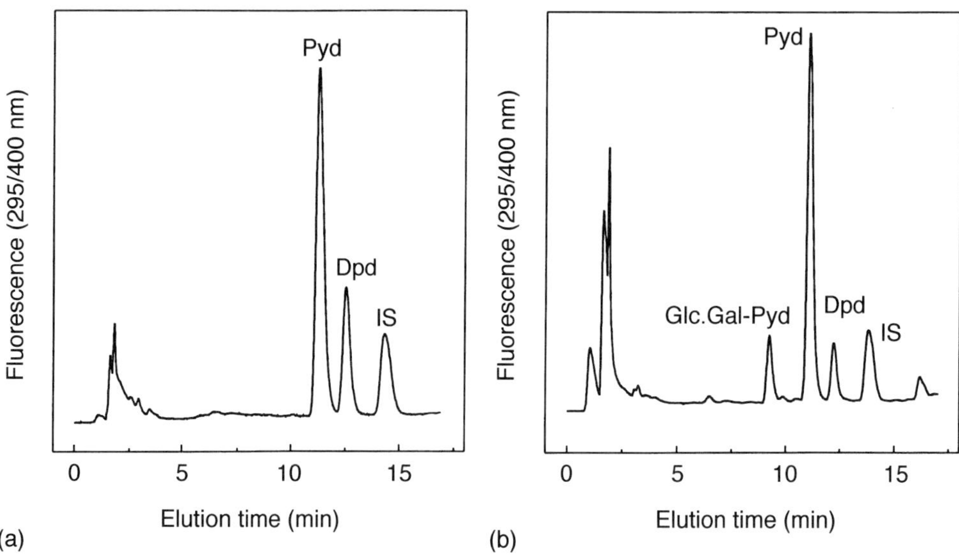

Figure 9.4 HPLC profiles of pyridinium crosslinks in urine after pre-fractionation of (a) hydrolysed or (b) unhydrolysed urine using cellulose columns. An internal standard (IS; *O*-acetyl-pyridinoline) was added to the samples before pre-fractionation. Additional peak in unhydrolysed chromatogram was shown to be the *O*-linked pyridinoline glycoside, glucosyl-galactosyl-pyridinoline (Glc.Gal-Pyd).

detail. Irrespective of the system being run, it is important to ensure that authentic standards are submitted to the full pre-fractionation procedure in order to validate sample quantification. Both internal and external standards are available commercially (Appendix 9.A).

(e) Free and total crosslinks

Free crosslinks can be measured without hydrolysis of urine, using the same methodology as for hydrolysates, though some precautions should be taken. The main difficulty is that the chromatograms of free crosslinks are much more susceptible to interference from unrelated fluorescent compounds in urine. These can generally be removed at the pre-fractionation stage by additional washing with mobile phase: the volume of washing for measuring free crosslinks should be increased by at least 50% compared with the protocol for total crosslinks. The chromatograms of unhydrolysed urine have an additional component, glucosylgalactosyl hydroxylysine which elutes about 1 minute ahead of Pyd (Fig. 9.4) and usually comprises 10–15% of the total urinary Pyd. One advantage in analysing free crosslinks by HPLC using a manual system is that acetyl-Pyd internal standard can be used without any special procedures (section 9.5.3d).

(f) Potential problems

Although setting up the HPLC analysis system is generally straightforward, problems can arise in maintaining a high standard of analysis and the question of what is likely to go wrong may need to be addressed. Where internal standard is not being used, the most common problem is variable recovery from the pre-fractionation step. Because this step involves conversion of the sample to a butanol–acetic acid–aqueous (4 : 1 : 1) mixture, the pyridinium compounds are rendered insoluble and will adhere to vessel surfaces, particularly if cellulose slurry is not added to the sample. A consistency in timing of the pre-fractionation steps is therefore essential.

For the HPLC analysis, the most common problems are loss of peak resolution or peak splitting. The latter can arise, for example, when insufficient ion-pair is present in the sample: the concentration of ion-pair is 10-fold higher than the running buffer in order to ensure full ion-pairing of the sample. Peak-splitting can also be caused by dead-space above the HPLC column but, for many columns, this can be eliminated by adjustment. The loss of resolution for crosslink separations that can sometimes occur after many months' running can be cured by passivating the system. This involves pumping 30% nitric acid through the whole system up to the HPLC column for about 1 hour at 1 ml/minute. Greater use of chemically inert PEEK (polyetheretherketone) tubing rather than stainless steel can lessen the need for passivation. Loss of peak resolution can also be an indication for re-packing or replacing the guard column (Appendix 9.A). A guard column is essential to improve the life of the separating column; depending on the types of samples being analysed, renewing the guard column should be done after 200–300 analyses.

9.5.4 Immunoassay of bone resorption markers

Several assays for measuring bone resorption have become available commercially within the past few years, and the characteristics of these urinary assays are given in Table 9.3. Protocols for sample collection are essentially those described in section 9.5.1b.

Table 9.3 Characteristics of commercially available urinary assays for bone resorption

Analyte (Assay kit[a])	Type	Specificity	Source reference
Free crosslinks (Pyrilinks®)	ELISA, monoclonal antibody	Equal reactivity with free Pyd and Dpd; < 2.5% cross-reaction with peptides	[41]
Free Dpd (Pyrilinks-D®)	ELISA or RIA[b], monoclonal antibody	Specific for free Dpd; < 1% reaction with Pyd; no reaction with peptides	[40]
NTx peptides (Osteomark®)	ELISA, monoclonal antibody	Collagen type I N-terminal crosslinked peptides which may contain pyridinium or other crosslinking compounds	[43]
CTx peptides (CrossLaps®)	ELISA, polyclonal antibody	Collagen type I C-terminal peptides containing isoaspartyl bonds; peptides may be crosslinked with pyridinium or other types of bond	[45]

[a] Pyrilinks and Pyrilinks-D are registered trademarks of Metra Biosystems Inc, Mountain View, CA, USA; Osteomark is the trademark of Ostex International Inc, Seattle, WA, USA; CrossLaps is the trademark of Osteometer A/S, Copenhagen, Denmark.
[b] Developed by IDS Ltd, Boldon, NE35 9PD, UK.

(a) Pyridinium crosslinks
Assays for free Dpd and for combined Pyd and Dpd have been described. Both of these assays are also capable of measuring total crosslinks in urine hydrolysates neutralized with 1 M NaOH.

(b) Peptide assays
Except for the serum ICTP assay, all of the peptide assays, in common with urinary crosslink measurements, are usually expressed relative to urinary creatinine. The urinary peptide assay kits are not directly comparable either with each other or with the crosslink assays, for the reasons outlined in section 9.4. In particular, the assays are standardized differently, with the NTx kit utilizing a bacterial collagenase digest of bone, whereas the CTx standards are the linear peptide antigens. It should be borne in mind that in neither case are these the true standards, since the analytes being measured in urine are not the same as the materials used for quantification and may not have equivalent antibody reactivities.

9.6 APPLICATIONS OF BONE TURNOVER MARKERS

9.6.1 Clinical studies

The main areas of application for biochemical markers are in the detection and diagnosis of disorders affecting bone metabolism and in monitoring therapy for these disorders. With the current focus on osteoporosis, a major area of interest is in the potential of these markers to help to identify those at risk of hip fracture. There has been a large number of publications in recent years on the application of bone markers and this summary will necessarily be selective and incomplete.

This research is underpinned by a series of careful studies of the physiological and experimental variables, particularly for serum osteocalcin [7, 13] and the crosslinks [35, 53–55]. The correlations with bone histomorphometry obtained for several markers [56, 57] also provide a sound basis for the application of these techniques. Chapter 7 gives a review of histomorphometry.

Bone markers have been shown to be elevated in a wide range of clinical disorders involving altered bone metabolism (reviewed in [36, 58]), including hyperparathyroidism, Paget's disease, osteomalacia and thyroid dysfunction. Alone, the markers rarely provide diagnostic information and monitoring the efficacy of treatment is a much more important application. In patients with breast or prostate cancer, both bone formation and resorption markers have been used to identify those where tumours have metastasized to bone.

The potential of biochemical markers of bone metabolism to predict the risk of future hip fracture has been highlighted recently with publications of the results of large prospective population studies. Increased resorption markers were associated with increased risk of hip fracture [59, 60], whereas bone formation markers were not linked to significant changes in risk [60]. An important feature of these results was that the increased risk of hip fracture associated with high bone turnover was independent of bone density. The predictive value of increased bone resorption markers for hip fracture appears to be of the same order as that for low BMD, but the results of the EPIDOS study showed that using a combination of low BMD and high turnover resulted in an improved assessment of risk [60]. Although significant in large group studies, these tests will not in themselves provide unequivocal information for individual patients. For prediction of osteoporotic fracture, the biochemical markers provide the physician with an additional tool to be used in conjunction with a panel of other techniques to give an improved assessment of risk.

9.6.2 *In vitro* studies

For cell or organ culture systems, many of the problems associated with the metabolism of bone-derived components *in vivo* are obviated. Thus, simple alkaline phosphatase assays provide an excellent early phenotypic marker for osteogenic activity with osteocalcin providing a marker of a later stage of osteoblast differentiation. The newer resorption markers provide simpler assessments of osteoclastic activity than using radiocalcium prelabelling or resorption pit assays. For the *in vitro* assay, the question of metabolism of bone collagen fragments at other body locations is irrelevant. Assays for the N- and C-terminal telopeptides have been applied successfully to organ cultures [61, 62] and, in our laboratory, an assay for the N-terminal telopeptide $\alpha 2(I)$ chain has given good correlations with a resorption pit assay [63].

9.7 CONCLUDING REMARKS

In a rapidly developing field, the number of assays for biochemical markers for bone metabolism has increased dramatically. The biochemical basis for the different assays has been emphasized in this chapter as this is crucial for the interpretation of clinical data: assumptions based on clinical studies alone should not be the primary 'validation' of any particular assay. Further advances continue to be made in many areas, particularly with respect to the development of serum assays for measuring bone resorption. Potentially useful assays for both pyridinium crosslinks and telopeptides have been reported, though their clinical utility is not yet fully established. The biochemical markers are increasingly being viewed not only as a useful research technique but also as a valuable adjunct for clinical studies to monitor bone metabolism.

ACKNOWLEDGEMENTS

I am grateful to the Scottish Office Agriculture, Environment and Fisheries Department for support and to Sandy Duncan for the benefit of his wealth of experience in HPLC techniques.

APPENDIX 9.A MATERIALS FOR PYRIDINIUM CROSSLINK ANALYSIS BY HPLC

Hydrolysis vessels	'Multi-Max Seal Tubes' (1.7 or 2.0 ml capacity) from Bioquote Ltd, York YO1 3DW, UK; or polypropylene 'Laboratory Microtubes' (1.5 or 2.0 ml capacity) from Alpha Laboratories, Eastleigh, Hants SO50 4NU, UK. Provided 'Lid Lock Clips' (Bioquote Ltd) are fitted, the evaporation losses on hydrolysis are less than 2% by weight. The tubes are designed for use in microcentrifuges and are disposable.
Prefractionation	CF1 Cellulose from Whatman International Ltd, Maidstone, Kent ME16 0LS, UK.
Chromatography	Heptafluorobutyric acid (Sequenal/HPLC grade) from Pierce Chemical Co., Rockford, IL 61105, USA.
Fluorescence monitor	RF-551 from Shimadzu Corp, Chiyoda-ku, Tokyo 101, Japan; or FP-920 from Jasco UK Ltd, Great Dunmow, Essex CM6 1XG, UK.
HPLC column	Microsorb C_{18} 3μ bead size (10 cm × 4.6 mm; cat. no. 80-200-C3) from Rainin Instrument Co. Inc., Woburn, MA 01888-4026, USA.
HPLC guard column	'Uptight precolumn kit' model C-135B from Upchurch Scientific Inc., Oak Harbor, WA 98277, USA. The column is dry packed with any type of 10μ C_{18} packing.
External standards	Pyd/Dpd HPLC calibrators from Metra Biosystems Inc., Mountain View, CA 94043-3911, USA.

Internal standards   *O*-Acetyl-Pyd 'Int-Pyd' from Metra Biosystems Inc., Mountain View, CA   94043-3911, USA; or Isodesmosine from ICN Pharmaceuticals Inc, Costa Mesa, CA 92626, USA.

REFERENCES

1. Price, C. (1993) Multiple forms of human serum alkaline phosphatase: detection and quantitation. *Annals of Clinical Biochemistry* **30**, 355–372.
2. Garnero, P. and Delmas, P. (1993) Assessment of the serum levels of bone alkaline phosphatase with a new immunoradiometric assay in patients with metabolic bone disease. *Journal of Clinical Endocrinology and Metabolism* **77**, 1046–1053.
3. Gomez, B., Ardakani, S., Ju, J. *et al.* (1995) Monoclonal antibody assay for measuring bone-specific alkaline phosphatase activity in serum. *Clinical Chemistry* **41**, 1560–1566.
4. Price, P.A. and Nishimoto, S.K. (1980) Radioimmunoassay for the vitamin K-dependent protein of bone and its discovery in plasma. *Proceedings of the National Academy of Sciences USA* **77**, 2234–2238.
5. Deftos, L.J., Wolfert, R.L., Hill, C.S. and Burton, D.W. (1992) Two-site assays of bone gla protein (osteocalcin) demonstrate immunochemical heterogeneity of the intact molecule. *Clinical Chemistry* **38**, 2318–2321.
6. Hellman, J., Kakonen, S., Matikainen, M. *et al.* (1996) Epitope mapping of nine monoclonal antibodies against osteocalcin: Combinations into two-site assays affect both assay specificity and sample stability. *Journal of Bone and Mineral Research* **11**, 1165–1175.
7. Delmas, P., Christiansen, C., Mann, K. and Price, P. (1990) Bone Gla protein (osteocalcin); assay standardization report. *Journal of Bone and Mineral Research* **5**, 5–11.
8. Knapen, M., Hamulyak, K. and Vermeer, C. (1989) The effect of vitamin K supplementation on circulating osteocalcin (bone gla protein) and urinary calcium excretion. *Annals of Internal Medicine* **111**, 1001–1005.
9. Merle, B. and Delmas, P.D. (1990) Normal carboxylation of circulating osteocalcin (bone gla protein) in Paget's disease of bone. *Bone and Mineral* **11**, 237–245.
10. Szulc, P., Chapuy, M., Meunier, P. and Delmas, P. (1996) Serum undercarboxylated osteocalcin is a marker of the risk of hip fracture: a three year follow-up study. *Bone* **18**, 487–488.
11. Tracy, R., Andrianorivo, A., Riggs, B.L. and Mann, K. (1990) Comparison of monoclonal and polyclonal antibody-based immunoassays for osteocalcin: a study of sources of variation in assay results. *Journal of Bone and Mineral Research* **5**, 451–461.
12. Garnero, P., Grimaux, M., Seguin, P. and Delmas, P. (1994) Characterization of immunoreactive forms of human osteocalcin generated in vivo and in vitro. *Journal of Bone and Mineral Research* **9**, 255–264.
13. Blumsohn, A., Hannon, R. and Eastell, R. (1995) Apparent instability of osteocalcin in serum as measured with different commercially available immunoassays. *Clinical Chemistry* **41**, 318–319.
14. Ducy, P., Desbois, C., Boyce, B. *et al.* (1996) Increased bone formation in osteocalcin-deficient mice. *Nature* **382**, 448–452.
15. Melkko, J., Niemi, S., Risteli, L. and Risteli, J. (1990) Radioimmunoassay of the carboxyterminal propeptide of human type I procollagen. *Clinical Chemistry* **36**, 1328–1332.
16. Trivedi, P., Risteli, J., Risteli, L. *et al.* (1991) Serum concentrations of the type I and III procollagen propeptides as biochemical markers of growth velocity in healthy infants and children and in children with growth disorders. *Pediatric Research* **30**, 276–280.
17. Teisner, B., Rasmussen, H., Hojrup, P. *et al.* (1992) Fetal antigen 2: an amniotic protein identified as the aminopropeptide of the alpha1 chain of human procollagen type I. *Acta Pathologica, Microbiologica et Immunologica Scandinavica* **100**, 1106–1114.

248 *Biochemical markers of bone turnover*

18. Orum, O., Hansen, M., Jensen, C. *et al.* (1996) Procollagen type I N-terminal propeptide (PINP) as an indicator of type I collagen metabolism: ELISA development, reference interval, and hypovitaminosis D induced hyperparathyroidism. *Bone* **19**, 157–163.

19. Melkko, J., Kauppila, S., Niemi, S. *et al.* (1996) Immunoassay for intact amino-terminal propeptide of human type I procollagen. *Clinical Chemistry* **42**, 947–954.

20. Risteli, J., Niemi, S., Kauppila, S. *et al.* (1995) Collagen propeptides as indicators of collagen assembly. *Acta Orthopaedica Scandinavica* **Suppl. 266**, 183–188.

21. Jensen, C., Hansen, M., Brandt, J. *et al.* (1997) Quantification of the N-terminal propeptide of human procollagen type I (PINP): comparison of ELISA and RIA with respect to different molecular forms. *Clinical Chemistry* (in press).

22. Cheung, C., Panesar, N., Haines, C. *et al.* (1995) Immunoassay of a tartrate-resistant acid phosphatase in serum. *Clinical Chemistry* **41**, 679–686.

23. Krane, S., Kantrowitz, F., Byrne, M. *et al.* (1977) Urinary excretion of hydroxylysine and its glycosides as an index of collagen degradation. *Journal of Clinical Investigation* **59**, 819–827.

24. Moro, L., Noris-Suarez, K., Michalsky, M. *et al.* (1993) The glycosides of hydroxylysine are final products of collagen degradation in humans. *Biochimica Biophysica Acta* **1156**, 288–290.

25. Segrest, J.P. and Cunningham, L.W. (1970) Variations in human urinary *O*-hydroxylysyl glycoside levels and their relationship to collagen metabolism. *Journal of Clinical Investigation* **49**, 1497–1509.

26. Bienkowski, R. (1984) Intracellular degradation of newly synthesized collagen. *Collagen and Related Research* **4**, 399–412.

27. Moro, L., Modricky, C., Rovis, L. and de Bernard, B. (1988) Determination of galactosyl hydroxylysine in urine as a means for the identification of osteoporotic women. *Bone and Mineral* **3**, 271–276.

28. Fujimoto, D., Moriguchi, T., Ishida, T. and Hayashi, H. (1978) The structure of pyridinoline, a collagen crosslink. *Biochemical and Biophysical Research Communications* **84**, 52–57.

29. Gunja-Smith, Z. and Boucek, R.J. (1981) Collagen crosslink components in human urine. *Biochemical Journal* **197**, 759–762.

30. Robins, S.P. (1982) An enzyme-linked immunoassay for the collagen crosslink, pyridinoline. *Biochemical Journal* **207**, 617–620.

31. Eyre, D.R., Koob, T.J. and Van Ness, K.P. (1984) Quantitation of hydroxypyridinium crosslinks in collagen by high-performance liquid chromatography. *Analytical Biochemistry* **137**, 380–388.

32. Ogawa, T., Ono, T., Tsuda, M. and Kawanashi, Y. (1982) A novel fluor in insoluble collagen: a crosslinking molecule in collagen molecule. *Biochemical and Biophysical Research Communications* **107**, 1252–1257.

33. Black, D., Duncan, A. and Robins, S.P. (1988) Quantitative analysis of the pyridinium crosslinks of collagen in urine using ion-paired reversed-phase high-performance liquid chromatography. *Analytical Biochemistry* **169**, 197–203.

34. Pratt, D.A., Daniloff, Y., Duncan, A. and Robins, S.P. (1992) Automated analysis of the pyridinium crosslinks of collagen in tissue and urine using solid-phase extraction and reversed-phase high-performance liquid chromatography. *Analytical Biochemistry* **207**, 168–175.

35. Colwell, A., Russell, R. and Eastell, R. (1993) Factors affecting the assay of urinary 3-hydroxy pyridinium crosslinks of collagen as markers of bone resorption. *European Journal of Clinical Investigation* **23**, 341–349.

36. Seibel, M.J., Robins, S.P. and Bilezikian, J.P. (1992) Urinary pyridinium crosslinks of collagen: specific markers of bone resorption in metabolic bone disease. *Trends in Endocrinology and Metabolism* **3**, 263–270.

37. Robins, S.P., Duncan, A. and Riggs, B.L. (1990) Direct measurement of free hydroxy-pyridinium crosslinks of collagen in urine as new markers of bone resorption in osteoporosis. *Osteoporosis 1990* (eds C. Christiansen and K. Overgaard), Osteopress ApS, Copenhagen, pp. 465–468.

38. Abbiati, G., Bartucci, F., Longoni, A. *et al.* (1993) Monitoring of free and total urinary pyridinoline and deoxypyridinoline in healthy volunteers: sample relationships between 24-h and fasting early morning urine concentrations. *Bone and Mineral* **21**, 9–19.

39. Seyedin, S., Kung, V., Daniloff, Y. *et al.* (1993) Immunoassay for urinary pyridinoline: the new marker of bone resorption. *Journal of Bone and Mineral Research* **8**, 635–641.

40. Robins, S.P., Woitge, H., Hesley, R. *et al.* (1994) Direct, enzyme-linked immunoassay for urinary deoxypyridinoline as a specific marker for measuring bone resorption. *Journal of Bone and Mineral Research* **9**, 1643–1649.

41. Gomez, B., Ardakani, S., Evans, B. *et al.* (1996) Monoclonal antibody assay for free urinary pyridinium cross-links. *Clinical Chemistry* **42**, 1168–1175.

42. Risteli, J., Elomaa, I., Niemi, S. *et al.* (1993) Radioimmunoassay for the pyridinoline cross-linked carboxy-terminal telopeptide of type I collagen: a new serum marker of bone collagen degradation. *Clinical Chemistry* **39**, 635–640.

43. Hanson, D., Weis, M., Bollen, A. *et al.* (1992) A specific immunoassay for monitoring human bone resorption: quantitation of type I collagen cross-linked N-telopeptides in urine. *Journal of Bone and Mineral Research* **7**, 1251–1258.

44. Robins, S.P. (1995) Collagen crosslinks in metabolic bone disease. *Acta Orthopaedica Scandinavica* **Suppl. 266**, 171–175.

45. Bonde, M., Qvist, P., Fidelius, C. *et al.* (1994) Immunoassay for quantifying type I degradation products in urine evaluated. *Clinical Chemistry* **40**, 2022–2025.

46. Bonde, M., Fledelius, C., Qvist, P. and Christiansen, C. (1996) Coated-tube radioimmunoassay for C-telopeptides of type I collagen to assess bone resorption. *Clinical Chemistry* **42**, 1639–1644.

47. James, I.T., Walne, A.J. and Perrett, D. (1996) The measurement of pyridinium crosslinks: a methodological overview. *Annals of Clinical Biochemistry* **33**, 397–420.

48. Eastell, R., Colwell, A., Hampton, L. and Reeve, J. (1997) Biochemical markers of bone resorption compared with estimates of bone resorption from radiotracer kinetic studies in osteoporosis. *Journal of Bone and Mineral Research* **12**, 59–65.

49. Tobias, J., Laversuch, C., Wilson, N. and Robins, S.P. (1996) Neridronate preferentially suppresses the urinary excretion of peptide-bound deoxypyridinoline in post-menopausal women. *Calcified Tissue International* **59**, 407–409.

50. Garnero, P., Gineyts, E., Arbault, P. *et al.* (1995) Different effects of bisphosphonate and estrogen therapy on free and peptide-bound bone cross-links excretion. *Journal of Bone and Mineral Research* **10**, 641–649.

51. Randall, A., Kent, G., Garcia Webb, P. *et al.* (1996) Comparison of biochemical markers of bone turnover in Paget's disease treated with pamidronate and a proposed model for the relationships between measurements of the different forms of pyridinoline cross-links. *Journal of Bone and Mineral Research* **11**, 1176–1184.

52. Robins, S.P., Duncan, A., Wilson, N.J. and Evans, B.J. (1996) Standardization of the pyridinium crosslinks, pyridinoline and deoxypyridinoline, for use as biochemical markers of collagen degradation. *Clinical Chemistry* **42**, 1621–1626.

53. McLaren, A., Isdale, A., Whiting, P. *et al.* (1993) Physiological variations in the urinary excretion of pyridinium crosslinks of collagen. *British Journal of Rheumatology* **32**, 307–312.

54. Schlemmer, A., Hassager, C., Pedersen, B. and Christiansen, C. (1994) Posture, age, menopause, and osteopenia do not influence the circadian variation in the urinary excretion of pyridinium crosslinks. *Journal of Bone and Mineral Research* **9**, 1883–1888.

55. Gerrits, M., Thijssen, J.H. and van Rijn, H.J. (1995) Determination of pyridinoline and deoxypyridinoline in urine, with special attention to retaining their stability. *Clinical Chemistry* **41**, 571–574.

56. Eriksen, E., Charles, P., Melsen, F. *et al.* (1993) Serum markers of type I collagen formation and degradation in metabolic bone disease: correlation with bone histomorphometry. *Journal of Bone and Mineral Research* **8**, 127–132.

57. Delmas, P., Schlemmer, A., Gineyts, E. *et al.* (1991) Urinary excretion of pyridinoline crosslinks correlates with bone turnover measured on iliac crest biopsy in patients with vertebral osteoporosis. *Journal of Bone and Mineral Research* **6**, 639–644.

58. Calvo, M., Eyre, D. and Gundberg, C. (1996) Molecular basis and clinical application of biological markers of bone turnover. *Endocrinology Reviews* **17**, 333–368.

59. van Daele, P., Seibel, M.J., Burger, H. *et al.* (1996) Case-control analysis of bone resorption markers, disability and hip fracture risk: the Rotterdam study. *British Medical Journal* **312**, 482–483.

60. Garnero, P., Hausherr, E., Chapuy, M. *et al.* (1996) Markers of bone resorption predict hip fracture in elderly women: the EPIDOS prospective study. *Journal of Bone and Mineral Research* **11**, 1531–1538.

61. Apone, S., Lee, M. and Eyre, D.R. (1996) Osteoclasts generate NTX, the collagen telopeptide marker of bone resorption, but not free pyridinolines. *Journal of Bone and Mineral Research* **11** (Suppl. 1), S189.

62. Foged, N., Delaisse, J-M., Hou, P. *et al.* (1996) Quantification of the collagenolytic activity of isolated osteoclasts by enzyme-linked immunosorbent assay. *Journal of Bone and Mineral Research* **11**, 226–237.

63. Scheven, B., Milne, J. and Robins, S.P. (1997) A novel culture system to generate osteoclasts and bone resorption using porcine bone marrow cells: role of M-CSF. *Biochemical and Biophysical Research Communications*, **231**, 231–235.

CHAPTER TEN

Radiographic measurement of bone turnover: microfocal radiography

J. Chris Buckland-Wright and John A. Lynch

10.1 INTRODUCTION

The radiograph is a shadow image of the X-ray beam as it passes through a bone. Variations in mineral content or subtle changes in the shape or contour of the tissue can indicate either an increase or loss of mineral consistent with alterations in bone turnover. The ability of standard X-ray units to detect such changes is limited by the poor image quality due to the large size of their X-ray source (0.3–1 mm), whereas microfocal X-ray units are characterized by a micron-sized X-ray source (5–100 μm) in which the object being examined is placed close to the X-ray tube, with the film at 1.5–2 m away, resulting in magnification (macroradiographs) of the image with high spatial resolution (Fig. 10.1). Such techniques have been used widely in the United States [1] and Japan [2]. The work carried out by these groups has failed largely due to limitations in the design of their X-ray machines, in which the minute size of the X-ray source could not be maintained. The macroradiographs they obtained, of either the hand or fingers, tended to be restricted in radiographic magnification (usually ×4, rarely ×6) and spatial resolution [1, 3]. The design of the British tube [3, 4], based on a different approach, ensures that the micron-sized X-ray source is retained throughout the life of the machine. The present equipment is capable of producing radiographs of small specimens (post-mortem samples, animals) to most parts of the human body at higher magnifications (from ×5 to ×20) and at a spatial resolution approximating to that of histology [3]. This is clearly better than those of other non-invasive medical imaging techniques.

The advantages of point-source X-ray machines are:

- high magnification – up to ×20 but more usually ×4 to ×10;
- high spatial resolution – the smallest structures recorded in macroradiographs of patients are 25–50 μm and > 5 μm for post-mortem material;
- minimum penumbral blurring – the radiographic margins of features are sharply defined and all planes of the object are in focus;
- because of the above, direct measurements of radiographic features are possible.

A limitation is the longer exposure time than normally used with standard X-ray machines. The length of the exposure is due to the low current loading on the

Methods in Bone Biology. Edited by Timothy R. Arnett and Brian Henderson.
Published in 1997 by Chapman & Hall, London. ISBN 0 412 75770 2.

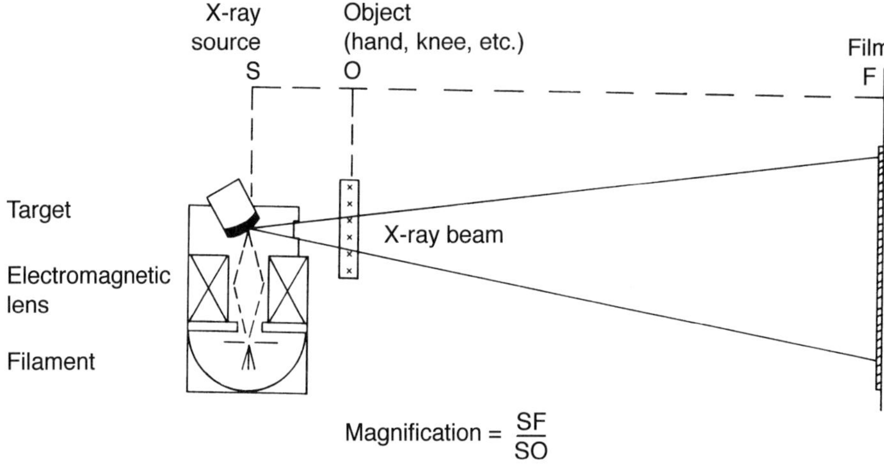

Figure 10.1 Microfocal X-ray unit. The electrons emitted from the filament are focused by the electromagnetic lens onto the target to form a point-source of X-rays. The object is placed close to the source and the projected shadow image is recorded on a film placed some distance away.

target, necessary to maintain the micron-sized X-ray source. The development of rare-earth film screen combinations has reduced the exposure time to less than a second for macroradiographs of most parts of the human body.

Clinical applications for magnification radiography have been in the study of the arthritides and metabolic bone diseases [1], where early investigators scored the extent of bone changes in rheumatoid arthritis [5], or undertook direct measurement of changes in cortical thickness or cortical striations in patients with metabolic bone disease [6, 7]. Developments in X-ray technology and computing now permit the assessment of cancellous bone organization. Three-dimensional computed tomography detects alterations in the structural organization of cancellous bone samples obtained from animals and patients [8–10]. The complex pattern of cancellous bone recorded by macroradiography can be analysed using textural methods of analysis. These can either be based upon a single fractal dimension [11, 12], a multi-fractal dimension estimation [13] or fractal signature analysis [14–19], which separately quantifies changes in the horizontal and vertical trabecular organization. Techniques for assessing changes in cancellous bone are distinct from those used in examining whole bones, and this chapter has been divided into two to reflect this difference.

10.2 QUANTITATIVE MICROFOCAL RADIOGRAPHY

Microfocal radiography, with its advantages of magnification and high spatial resolution, permits accurate and reproducible measurements to be made of its radiographic features. This approach depends on good image quality, which is determined by the X-ray source size, method of image acquisition and the radiographic procedure. X-ray source size is determined by the type of X-ray unit employed and is outside the scope of this chapter, whereas radiographic procedure and image acquisition are determined by the specimen and the nature of the investigation. Much has already been written about the different methods of image acquisition [20, 21] and less about the standardization of the radiographic procedure necessary for quantitative methods of analysis [22]. These procedures require precision in the radio-anatomical position of the specimen, clearly defined anatomical boundaries for measurement, correction for radiographic magnification, and an assessment of the reproducibility of the method of measurement [23–27].

10.2.1 Radiographic procedure

It is a principle of radiography that optimal alignment of the object is obtained by centring that part of the object under examination with respect to the central ray of the X-ray beam. Thus, where measurements of radiographic features are to be carried out, it is mandatory that the specimen be positioned in precisely the same plane both within and between samples of the same kind. In order to avoid variation between examination, protocols should be prepared that define:

- the radio-anatomical plane of the sample;
- the alignment of the central ray of the X-ray beam relative to a specified anatomical plane;
- correction for radiographic magnification, usually by means of a radio-opaque device such as a metal grid or bead of known dimensions.

The use of fluoroscopy or other method of real-time imaging can facilitate positioning and overcomes individual variation between specimens.

(a) Stereotaxic or positioning apparatus
In microfocal X-ray units, the X-ray source is fixed and rigidly mounted to a cement floor or shock-absorbing table, since the slightest vibration from other equipment such as centrifuges (even on adjacent floors), swinging doors, etc. will degrade the image quality. The X-ray beam is usually horizontal, parallel to the floor or table. With this arrangement, the specimen is brought as close as possible to the source and is aligned within the X-ray beam to obtain the desired radio-anatomical position. This is achieved using a positioning or stereotaxic apparatus which also keeps the specimen motionless during the exposure [4,22]. For small specimens, simple retaining devices made from perspex can be mounted on the sample manipulator, which moves in the X–Y plane, rotates, and tilts towards or away from the X-ray source (X-Tek Systems Ltd, Hertfordshire, UK). Larger systems may need to be made to meet the specific requirements of the investigator [4] (Fig. 10.2).

Stereotaxic devices allow the specimen either to be tilted between X-ray exposures, through an angle of 5° on either side of the plane perpendicular to the centre of the beam, or to be displaced horizontally by 5–10 mm (depending on the degree of radiographic magnification) (Fig. 10.3). The stereo-pair macroradiographs thus obtained are examined under a stereo-viewer, for use with radiographs of either small or large specimens (Figs 10.4 and 10.5) (Ross Instruments, Salisbury, Wiltshire). In these units, the stereo-viewer is attached to a moveable arm, allowing the viewing assembly to traverse the entire image. The virtually unlimited depth of focus obtained with point-source radiography and the perspective, as in normal vision, produced when the radiographs are examined under the stereoscope permit lesions within bone to be readily identified. This approach has been used to determine the interface between regions of osteopaenic bone and the endosteal margin

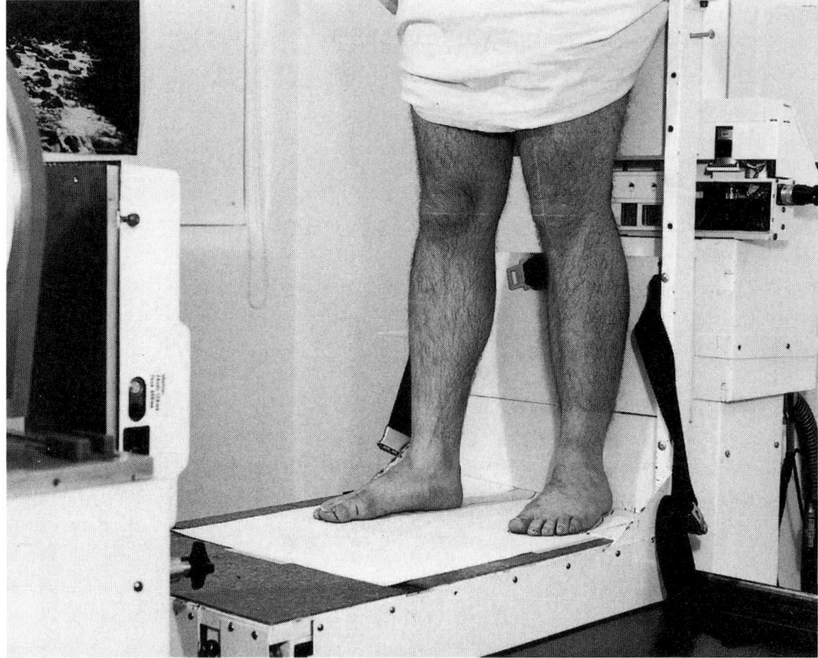

Figure 10.2 General arrangement of the microfocal X-ray unit with the patient standing close to the X-ray tube on the right. The X-ray film and cassette are placed in front of the image intensifier on the far left. The platform on which the patient stands can be moved in three planes and permits the patient to be positioned in relation to the X-ray beam. In front of the image intensifier is a laser light-source; its cross-beam is used to align the region of the patient to be radiographed with the centre of the X-ray beam. Prior to taking the radiograph, the laser is dropped out of view.

of erosions in the hands of rheumatoid patients [3, 28, 29]. A more detailed account of the advantages and limitations of stereo-microfocal radiography has been given elsewhere [30, 31].

10.2.2 Image acquisition

It is essential to select the most appropriate recording medium at the start of an investigation, since this will determine whether the features to be assessed are optimally defined for the analysis. The medium used to record the shadow images produced by the X-ray beam passing through the object depends on two factors: their size and whether the bone lies within a living or post mortem specimen, either animal or human.

(a) Specimen size
The size of the object, and hence the mass of tissue through which the X-ray beam passes, determines the dimensions of the bony features recorded in the image, i.e. in small samples fine detail can be recorded on fine-grained X-ray film [20, 22] (Table 10.1). With increased object size there is a reduction in fine structural detail. In large specimens, small changes in mineral density are attenuated by the greater mass of the tissue, leading to the use of coarser grained, fast-speed X-ray films or film-screen combinations [4, 22] (Table 10.1).

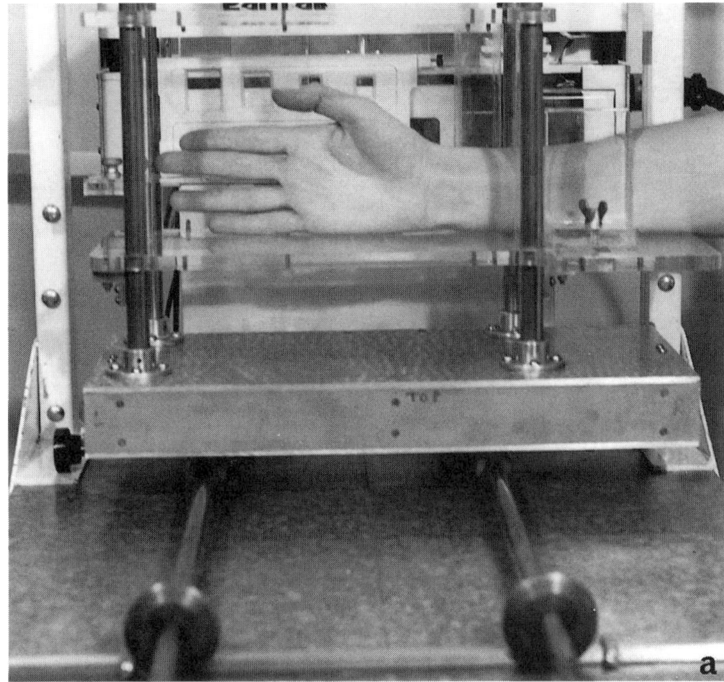

Figure 10.3 Stereotaxic device for reproducibly repositioning the hand and wrist of a patient. The platform is adjusted so as to align the part of the hand to be radiographed immediately in front of the X-ray port. Stereo-pair radiographs are obtained by displacing the metallic platform by 6 mm between successive exposures.

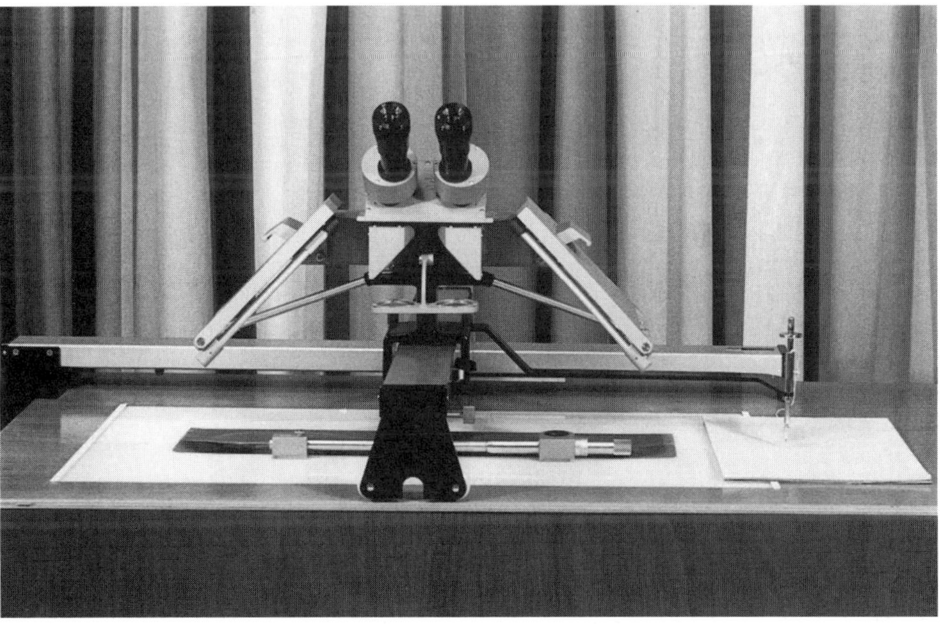

Figure 10.4 Mirror stereoscope showing movable viewing assembly situated above the light-box, with binoculars, parallax measuring bar and drawing arm in position.

Table 10.1 Suggestions for the type of X-ray film that can be used in microfocal radiography of different specimens or subjects

Specimen/subject	Film or film/screen system	Comments
Small post-mortem samples	Kodak High Resolution film SO-343	Provides micron-sized spatial resolution of object. Limitation: long exposure times.
Large post-mortem samples Small living mammals	Kodak Min-R, M, H or T film; Agfa Mamoray MR5-II	Can be used with or without cassette. Versatile film, high contrast, high resolution.
Large living mammals	SDI[a]: Cronex Hi-plus and Hi-plus cassette	
Human hand Human knee, spine	Imation[b]: Trimax XDA film/T8 screens and cassette Imation: Trimax XM film/$2 \times$ T16 screens in plastic-faced cassette	Spatial resolution limited > 0.025 mm. Spatial resolution limited > 0.050 mm. Very fast film/screen systems
Human hand to knee	SDI[a]: UVG film/Super Rapid screens in plastic-faced cassettes	Spatial resolution < 0.025 mm depending upon object size; good contrast range. Very fast film/screen systems.

[a] SDI, Sterling Diagnostic Imaging, formerly Dupont.
[b] Imation UK Ltd, formerly 3M, UK Ltd.

(b) Living tissue

Microfocal radiography of bones within animals or humans is constrained by the need to reduce the radiation dose to as low a level as possible and to obtain an exposure within as short a time as possible, minimizing any loss to image quality from subject movement. In all instances there is a need to obtain as much fine detail as is possible within the X-ray film. Most living specimens will require the use of very fast and hence coarse-grained rare earth film/screen combinations such as that used in patient examination [4] (Table 10.1). More recently, we have obtained very satisfactory results with the Sterling Diagnostic Imaging system (comprising UVG film with Super Rapid screens in plastic-faced cassettes), which was found to have an increased contrast range, higher spatial resolution and reduced exposure time compared with the systems we had tested and used previously [4]. The limitations of the large film and phosphor grain size characteristic of fast film/screen combinations upon image quality are largely overcome by the radiographic magnification, which enlarges the structural detail beyond the grain size [32].

(c) Film processing and processors

Processing methods should be standardized, with minimal variation in temperature and chemical strength to ensure reproducible conditions and hence image quality throughout the study period. Raising the processing temperature should be avoided in favour of slower processing times to retain image quality within the film. The type of processor will depend upon the type of film used and the volume of work. These can range from manual film development with dishes in a darkroom to desk-top or automatic film processors. In many instances the desk-top processors (e.g. Gevamatic 60; Agfa Gevaert Ltd.) will meet most requirements and not be too expensive.

(d) Real-time imaging

Examination of specimens using an image intensifier and camera or CCD TV camera system has the attraction of offering a direct and immediate method of examining the tissue. However, its lower image resolution and contrast range, compared with X-ray film, limit the usefulness of this method for assessing features associated with bone turnover. In practice, these systems are best used for positioning the specimen within the X-ray beam prior to recording the image on X-ray film.

10.3 SIMPLE AND LINEAR METHODS OF QUANTIFYING BONE TURNOVER: CHANGES IN THE CORTEX OR BONY CONTOUR

The use of computers and image processing offers an ever greater variety of methods for quantifying changes in the degree of mineralization within bone. Commercially available programs such as ANALYSE (Mayo Clinic, Rochester, USA) can be used or adapted to provide the appropriate method for a particular investigation. Customized image processing algorithms can be prepared, such as the textural methods described below for quantifying cancellous bone, or those for measuring changes in joint structure [27]. The methods require not only equipment for digitization of the radiographs and processing the image but also the staff or the appropriate computing skills to write the programs. The advantages of computerized image analysis of digitized radiographs are speed and increased measurement precision [27]. However, in many instances simple direct methods of measurement are best and these are described below.

10.3.1 Semi-quantitative methods

These methods consist of grading or scoring the change in the radiographic features on a defined or preset scale; for example, a four-point scale might be: absent, mild, moderate or severe. Their advantages are that they are simple, easy to use and usually require very little data manipulation. Their disadvantages are that they are insensitive and subject to inter- and intra-observer error.

10.3.2 Quantitative methods

Simple methods are described, including assessment of incidence and measurement of length and area of radiographic features.

(a) Magnifying lens
The most practical have been found to be ×5 or ×10 lenses fitted with a 10 mm graticule with 0.1 mm divisions (Agar Scientific Ltd, Stansted, Essex, UK). With a light-box to illuminate the radiograph, the lens can be placed directly on the radiograph and the width/length measured directly. This method is recommended where the distance is short or the feature is straight.

(b) Digitization tablet
A cross-wire cursor is used to trace along the length or perimeter of a feature in the macroradiograph overlying a back-illuminated digitizer tablet linked to a microprocessor (Fig. 10.5) or PC. (Translucent digitizing tablets can be obtained from Kontron Elektronik, Eching, Germany. An alternative system is the KS100 from Imaging Associates, Thame, UK.) Although the instrument can measure to an accuracy of 0.1 mm, in practice such precision is not achieved by the observer when using a handheld cursor. This system can be combined with a large format stereoscope, where the measurement requires the reliable identification of the margin of a lesion within bone (Fig. 10.6) such as erosions within the rheumatoid hand [29]. The advantage of this system is that it is easy and reliable to use when quantifying the extent of linear or surface area of an irregularly shaped radiographic features. The limitations are few.

10.3.3 Defining the margin or boundary of a radiographic feature

Using high definition macroradiography, bone turnover is detected as an alteration in the mineral content of the tissue. The main difficulty in quantifying this change arises with the observer when identifying the margin or boundary of that feature, particularly at sites of resorption or erosion. Marginal erosions such as those that occur in the rheumatoid hand may be identified by a break in the cortex, though more usually as an area of juxta-articular radiolucency (Fig. 10.6). The endosteal edge of this lesion is identified by a reduction in the mineral content and disturbance to the trabecular organization with the erosion. With disease progression there is an increase in juxta-articular osteoporosis. This loss in mineral leads to a

Figure 10.5 Large format mirror stereoscope. The right-hand viewing assembly comprises a digitizing tablet linked to a microprocessor or PC. A cross-wire cursor is used to outline the radiographic features to be measured.

loss in definition at the erosion's edge and hence a reduction in measurement precision and the ability to detect significant changes in structure from disease. With erosion progression, the lesion enlarges, forming a cavity with a total loss of cortex and disruption to the bone's shape. Measurement of the size of such a cavity requires the observer to mark on the radiograph, with a wax pencil, the former outline of the cortex, using their knowledge of the shape of the bone (either from previous radiographs or of those of other subjects). Another example of the difficulty in boundary definition is encountered with increased bony deposition such as that associated with subchondral cortical sclerosis (Fig. 10.7).

There is no simple procedure for enhancing the pattern recognition associated with defining an endosteal boundary at a site of bone resorption or formation. Three-dimensional visualization using stereoscopy can help, otherwise it must be based upon good radiographic technique to provide good image quality and very careful observation. Computer enhancement of the digitized radiographic image may prove to be an option in certain studies.

10.3.4 Reproducibility of the measurement

The validity of measurements of a radiographic feature depends not only on image quality, but also on the reproducibility of the measurement procedure. The

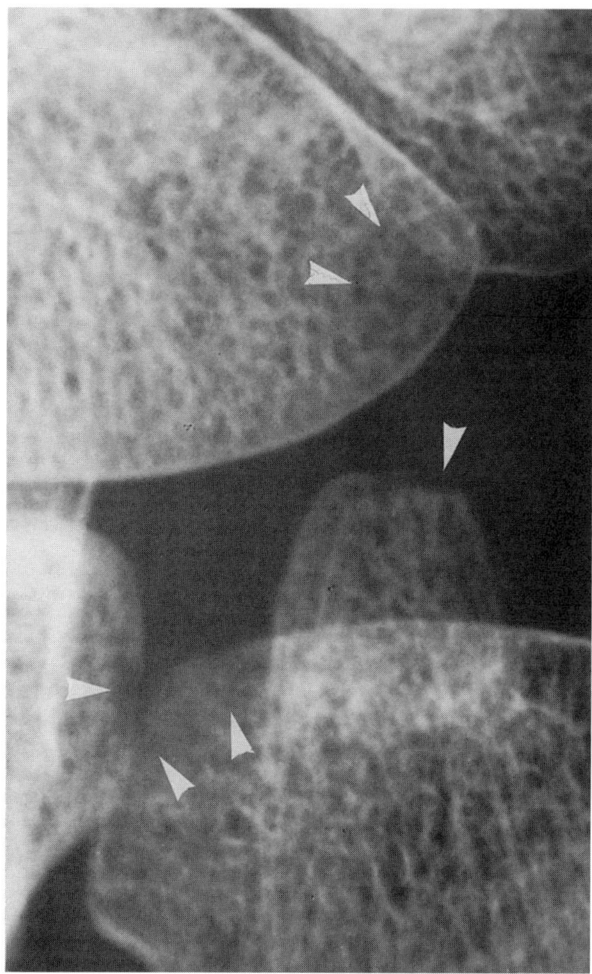

Figure 10.6 Part of a macroradiograph of the wrist of a patient with rheumatoid arthritis showing the ulna and radio-ulnar joint with the lunate above. Arrows indicate the sites of erosion. In the lunate, the endosteal margin of the lesion is identified by the change in mineral content and a disturbance in the trabecular organization within the erosion. (Original magnification ×5.)

reproducibility (or precision) for a measurement can be expressed as the standard deviation (SD) of repeat measurements of the feature within the same structure(s). To standardize the scale of precision estimates, the reproducibility of repeat measurements of the same object can frequently be expressed as a coefficient of variation (CV): the ratio of the SD to the mean of repeat measurements. Lack of attention in radiographic technique can result in a large CV – as high as 20% for repeat manual measurements [33]. In addition, it is important to distinguish between the reproducibility of the instrument used for measurement and that obtained when measuring the radiographic feature [34]. The CV for the latter will be larger and will incorporate the error associated with operating the instrument. Using the digitization tablet, the CV for length measurements has been found to range from 2.2% to 9% [25, 35] and the CV for area measurements was in the region of 5.3 to 13% [25, 34].

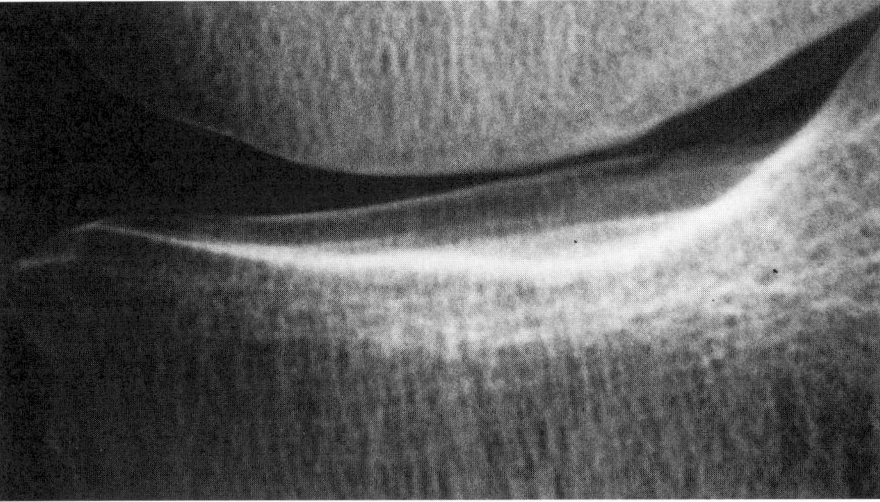

Figure 10.7 Part of a macroradiograph of a patient with knee osteoarthritis showing subchondral cortical sclerosis in the tibia. Identification of the endosteal margin of the cortex, when measuring cortical thickness, may become difficult with increasing sclerosis associated with disease progression. (Original magnification ×5, reproduced at ×4.)

10.4 APPLICATION OF LINEAR METHODS OF QUANTIFYING BONE TURNOVER

10.4.1 Bone resorption

Changes in the thickness of the metaphyseal cortex and that of trabeculae, measured with a magnifying lens fitted with a graticule, have been used to quantify the extent of osteoporosis in the limb bones of guinea pigs with an experimentally induced inflammatory arthritis [36]. Due to the greater surface area to volume ratio, trabecular thickness measurement was a far more sensitive method of quantifying bone loss than at the cortex.

The resorption of bone at the terminal phalangeal (ungual) tufts of children with renal osteodystrophy was graded on a 10- point scoring system based on the extent and depth of the lesion [37]. This method was adopted since it was not possible to determine reliably the shape of the ungual tuft prior to the loss of bone. In macroradiographs of the hands of children with renal osteodystrophy, the extent of the subperiosteal cortical resorption was measured using the digitization tablet and expressed as a percentage ratio to the total cortical length of the phalanx [37, 38]. The length of the percentage subperiosteal resorption at base line and its change during the study period correlated significantly with the level of serum parathyroid hormone levels and its change over the same period [37].

Measurement of the size of erosions in macroradiographs of the wrists and hands of patients with rheumatoid arthritis, using a digitization tablet and microprocessor, determined their distribution and incidence within the bones [39] and their rate of change during the study period [40] and showed for the first time the retardation and repair of erosions in patients treated with second-line drugs [35].

10.4.2 Bone formation

Macroradiographic examination of patients with osteoarthritis of the hands [24, 25, 41] and of the knees [26] demonstrated conclusively that the characteristic bony features of subchondral cortical sclerosis and osteophytosis appear early in the course of the disease and precede those of articular cartilage loss measured as joint space narrowing. Using the digitization tablet, subchondral cortical thickness increased in depth early in the disease and showed a variable outcome in the hand [25], whereas in the knee it showed little subsequent progression [42]. Osteophyte size increased during the course of the disease and was correlated with joint pain and increased radio-isotope uptake indicative of active bone turnover [43–45]. Non-steroidal anti-inflammatory drugs were found to halt osteophyte progression in early knee osteoarthritis [42].

10.5 DIRECT MEASUREMENT OF TRABECULAR BONE STRUCTURE *IN VITRO* USING 3D MICROCT

In three-dimensional microfocal computed tomography (3D microCT), a small tissue sample, placed close to the port of the microfocal X-ray unit, is rotated in the beam (Fig. 10.8) and its image, magnified onto a large area X-ray detector system, is digitized onto a computer, permitting the detailed organization of the internal structure of the sample to be reconstructed. The cone beam of X-rays allows a true three-dimensional CT image of the internal structure to be produced, rather than a stack of 'slices' through the object as in most CT scanners.

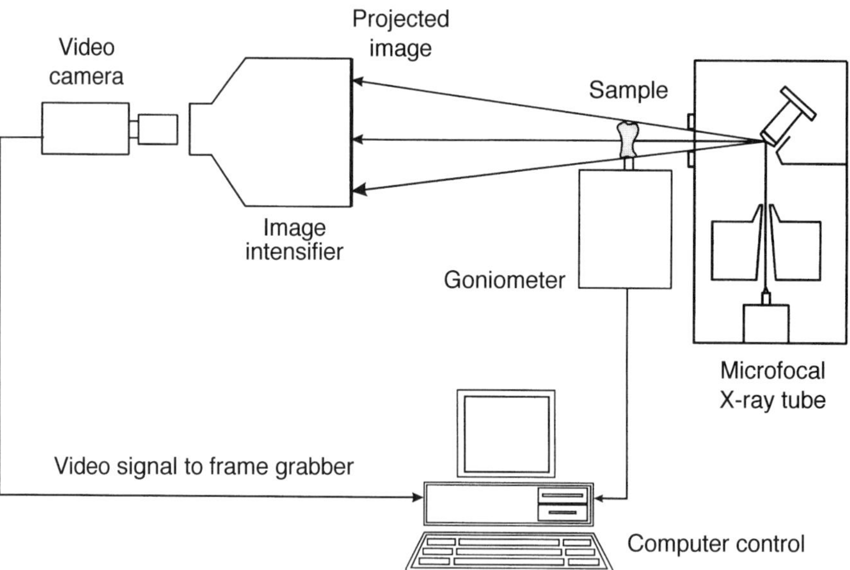

Figure 10.8   MicroCT unit.

10.5.1 3D microCT scanning

Using microfocal X-ray tubes, 3D microCT scanners have resolutions from 0.1 mm down to 10 μm [8, 9, 46, 47]. The sample, placed on a highly accurate goniometer, is rotated under computer control. A CCD video camera is focused on the back of the image intensifier and a frame grabber is used to digitize the video signal (Fig. 10.8). To improve signal-to-noise ratios, 16 video frames are averaged to produce projection images of high quality. We collect 360 images with 1° rotation of the sample between each image over a 1–2-hour period. CT image reconstruction is performed using Feldkamp's method [48] to provide images of the trabecular network at 50 μm for bone sample sizes up to about 2 cm in diameter. This technique produces accurate reconstructions for the small cone beam angles (about 2°) used in this scanner [48].

10.5.2 Analysis of 3D microCT images of trabecular bone

3D microCT images of trabecular bone samples can be converted into a binary image of the trabecular network (Fig. 10.9). Comparisons with histologically determined structure have shown that 3D microCT reconstructions of trabecular network are accurate [49]. Bone volume fraction can be calculated simply as the percentage of elements within the 3D image that are part of the trabecular network [8]. Three-dimensional connectivity can be calculated properly for a tissue sample using the Euler number [8], since the data are truly three-dimensional. Thus, there is no need to estimate indirect measures of connectivity such as in histomorphometric trabecular strut analysis [50] (Chapter 7). Trabecular orientation, or fabric, within the sample can be calculated using a 3D extension to mean intercept length calculations [51] and tensors [52].

10.5.3 Problems with 3D microCT

The sample must be very accurately positioned with reference to the X-ray beam and detectors. The axis of rotation of the sample must remain parallel to the detector face during scanning. The central ray from the X-ray tube must hit the centre of the image intensifier face perpendicularly. Procedures for this alignment are relatively simple [46] but time consuming. Image intensifier images are often distorted, and geometric techniques can be used to remedy distortion of such images before CT image reconstruction [46]. Beam hardening (the preferential absorption of low-energy X-ray photons over high-energy photons in soft tissues, when compared with mineralized tissue) can cause artefacts in microCT images, since the X-ray beam comprises a spectrum of photon energies. Various techniques can be used to correct for this, including the use of suitable filtration materials [53] placed between the source and sample. Noise and systematic errors in the acquired data can cause artefacts in reconstructed images, including ring artefacts due to detector miscalibration. These can be minimized by data correction and by improving signal-to-noise ratios in the acquired data [46].

10.5.4 Uses of 3D microCT for examining bone tissue

MicroCT of trabecular bone allows non-destructive examination of normal and disease-related changes in trabecular bone structure, such as subchondral

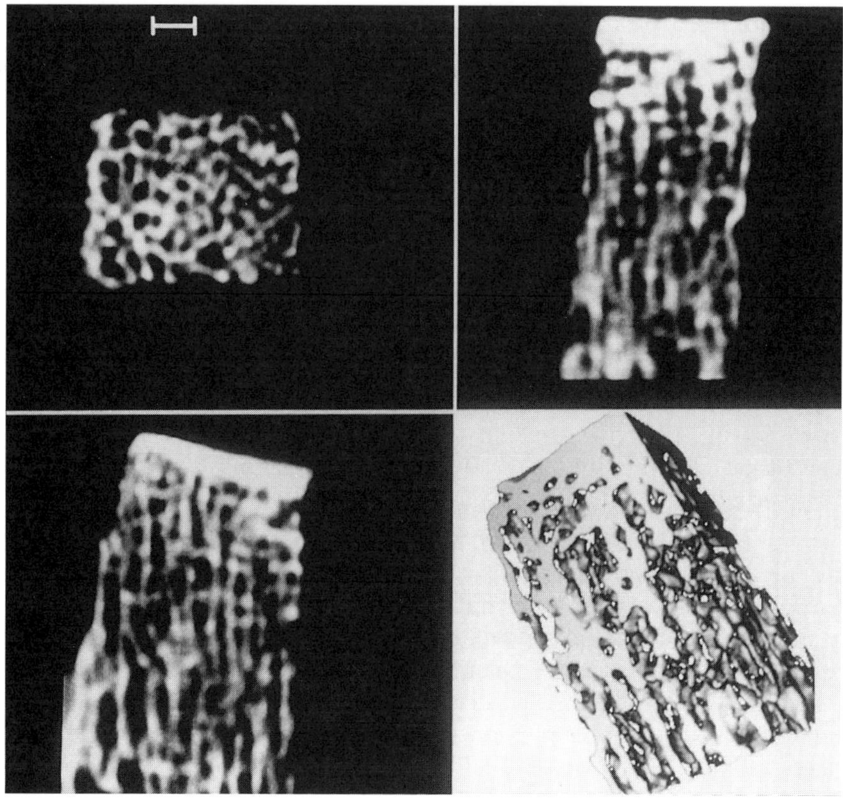

Figure 10.9   Specimen (6.5 mm × 4.5 mm × 10 mm in size) of human tibial subchondral bone, scanned using 3D microCT at 65 μm showing transverse (top left), coronal (top right) and sagittal (bottom left) slices through the three-dimensional image, together with a volume rendered view of the trabecular network (bottom right). Scale bar, 1 mm.

alterations in trabecular bone in the cruciate ligament-deficient canine model of osteoarthritis [10]. Comparisons between mechanical properties of trabecular bone and the structure determined from 3D microCT have shown that 80% of the variance in strength of cancellous bone is predicted by measurement of bone density and trabecular orientation [54]. Connectivity and trabecular plate number were related to bone density in normal samples, and not directly to bone strength, and it has been hypothesized that in osteoporotic patients this relationship may be altered [54].

10.6 IMAGE TEXTURE ANALYSIS OF TRABECULAR BONE STRUCTURE *IN VIVO*

The radiographic pattern of cancellous bone alters with bone remodelling in conditions such as osteoporosis, rheumatoid arthritis and osteoarthritis. Trabecular structure also alters as a physiological response in many metabolic diseases (e.g. renal or liver conditions). These changes can often be examined using semi-quantitative scoring systems [55–57], but these are prone to poor reproducibility and inter-observer agreement. The development of microcomputers and image analysis techniques has provided the means of analysing the subtle changes in the complex patterns on radiographs of cancellous bone, both *in vitro* and *in vivo*, using techniques of digital image texture analysis. In particular, fractal texture analysis has become widely used to examine alterations in trabecular bone [10–19, 58].

10.6.1 Fractals, objects and images

Fractals have a characteristic feature that remains constant over a range of scales. This self-similarity is a hallmark of fractals. Textured fractal images can be produced (Fig. 10.10) [59], with diffuse cloud-like structure of no particular scale. At different magnifications the appearance of the images remains fundamentally unchanged, illustrating their self-similarity. Rougher textures have high fractal dimension, and smoother images have low fractal dimension [60]. Fractal texture analysis is based on computation of the fractal dimension of an image region, calculation of whether the region is fractal, and the manner in which the region deviates from a fractal image [13, 14, 15, 60].

For image texture analysis, images are considered to be maps, where dark regions represent low ground and bright regions are high ground. The area of the ground surface is then measured at different resolutions by using rulers of different sizes. For two-dimensional fractal images, such as those in Fig. 10.10, the measured surface area A and ruler length r are related by:

$$A = A_0 r^{(2-D)} \tag{10.1}$$

where D is the fractal dimension of the object [61]. For fractal images, a log–log plot of measured surface area A against ruler length r has a constant slope

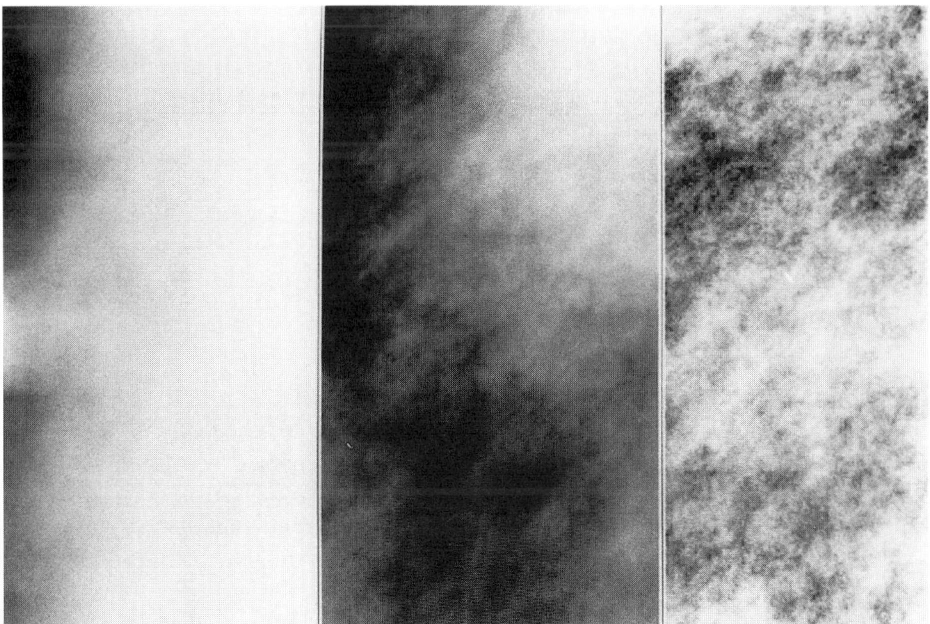

Figure 10.10   Fractal images of fractal dimension D = (a) 2.2, (b) 2.5 and (c) 2.8.

$(2 - D)$. Most real images are not completely fractal: the relationship in equation (10.1) is not obeyed, and the log–log plot is not a straight line. In fractal texture analysis, there are various options. The simplest is to fit a straight line to all the data [60]. Two straight lines can be fitted to give separate fractal dimension estimates for fine and coarse structure [13]. The third, and most complete option is to consider estimates of fractal dimension to be a function of resolution, so that, when measured with a ruler of length r, the surface area $A(r)$ is given by:

$$A(r) = A_0 r^{2-F(r)} \tag{10.2}$$

where $F(r)$ is known as the fractal signature (estimated fractal dimension as a function of resolution) [14, 15, 62].

10.6.2 Radiographic fractal analysis of bone structure

Single fractal dimension examination of bone structure from radiographs has been used *in vivo* to examine trabecular structure in the calcaneum [11], mandible [12] and hip [58]. Fine and coarse fractal dimensions have been used to examine osteoporotic changes in the spine [13], and the fractal signature has been used to examine osteoporosis of the lumbar spine [16] and osteoarthritic changes in the proximal tibia [17] of patients. Fractal signature analysis can give information about the extent and relative amounts of structure at different scales that make up an image, and quantifies features that conventional Euclidean geometry cannot.

10.6.3 Fractal signature analysis of bone structure

To measure the surface area of an image at a given resolution, the following approach is quick, easy to implement on computers, and independent of changes in X-ray exposure, film development and digitization [14, 15]. Consider $g(i,j)$ to be the intensity of the image at the pixel in column i of row j. If ϵ is the resolution at which the surface area is being measured, then an upper blanket function $u_\epsilon(i,j)$ and lower blanket function $l_\epsilon(i,j)$ of $g(i,j)$ are defined as:

$$u_\epsilon(i,j) = \max\left\{ u_{\epsilon-1}(i,j) + k, \max_{(m,n)-(i,j)\leq 1} u_{\epsilon-1}(m,n) \right\} \tag{10.3}$$

and

$$l_\epsilon(i,j) = \min\left\{ u_{\epsilon-1}(i,j) + k, \min_{(m,n)-(i,j)\leq 1} u_{\epsilon-1}(m,n) \right\} \tag{10.4}$$

where $u_0(i,j) = l_0(i,j) = g(i,j)$ where $k \geq 0$. This means that u_1 and l_1 can be calculated from g, and then u_2 and l_2 can be calculated from u_1 and l_1 and so on. It should be noted that these expressions are the mathematical morphology operations of dilation, equation (10.3), and erosion, equation (10.4), with 3×3 structuring elements [63], and have been specifically designed so that fractal signature calculations are not distorted by variations in X-ray exposure, film development and digitization [14, 15] when $k = 0$.

In particular, $u_\epsilon(i,j)$ is the maximum of the five values of $g(i-1, j)$, $g(i, j)$, $g(i+1, j)$, $g(i,j-1)$ and $g(i, j+1)$ and $l_\epsilon(i,j)$ is the minimum of the same five values of g. From

$u_\epsilon(i,j)$ and $l_\epsilon(i,j)$, the volume v_ϵ between the two blanket surfaces can be calculated from:

$$v_\epsilon(i,j) = \sum_{i,j} [u_\epsilon(i,j) - l\epsilon(i,j)] \tag{10.5}$$

and from these values, the surface area $A(\epsilon)$, measured at resolution ϵ, is given by:

$$A(\epsilon) = \frac{v_\epsilon - v_{\epsilon-1}}{2} \tag{10.6}$$

For fractal images, a plot of log $A(\epsilon)$ against log ϵ would be a straight line with a slope $2 - D$, where D is the fractal dimension. For a non-fractal image, the log–log plot will not be a straight line, but the fractal signature $F(\epsilon)$ is calculated from the slope of a regression line through the following three points:

$$\Big(\log(\epsilon{-}1),\ \log A(\epsilon{-}1)\Big) \Big(\log(\epsilon),\ \log A(\epsilon)\Big) \Big(\log(\epsilon{+}1),\ \log A(\epsilon{+}1)\Big) \tag{10.7}$$

If the slope is s_ϵ then the value of the fractal signature at resolution ϵ is:

$$F(\epsilon) = 2 - s_\epsilon \tag{10.8}$$

which is calculated for values of ϵ up to a maximum value that is as large as the largest feature of interest in the image (typically 1 mm for trabecular structures).

10.6.4 Fractal signatures for horizontal and vertical image structure

The method for fractal signature analysis given in the previous section calculates a fractal signature which examines the texture independently of direction. It can be modified to calculate the fractal signature for image texture at any particular direction [14, 15], and the method for examining horizontal and vertical image structure is now given. The calculations of the upper and lower blanket surfaces given in equations (10.3) and (10.4) are replaced by expressions:

$$u^{(H)}_\epsilon = \max\left\{ u^{(H)}_{\epsilon-1}(i,j-1),\ u^{(H)}_{\epsilon-1}(i,j),\ u^{(H)}_{\epsilon-1}(i,j+1), \right.$$
$$\left. u^{(v)}_\epsilon = \max\left\{ u^{(v)}_{\epsilon-1}(i-1,j),\ u^{(v)}_{\epsilon-1}(i,j),\ u^{(H)}_{\epsilon-1}(i+1,j) \right\} \right. \tag{10.9}$$

and:

$$l^{(H)}_\epsilon = \min\left\{ l^{(H)}_{\epsilon-1}(i,j-1),\ l^{(H)}_{\epsilon-1}(i,j),\ l^{(H)}_{\epsilon-1}(i,j+1), \right.$$
$$\left. l^{(v)}_\epsilon = \min\left\{ l^{(v)}_{\epsilon-1}(i-1,j),\ l^{(v)}_{\epsilon-1}(i,j),\ l^{(H)}_{\epsilon-1}(i+1,j) \right\} \right. \tag{10.10}$$

where $u^{(H)}_0(i,j) = u^{(V)}_0(i,j) = l^{(H)}_0(i,j) = l^{(V)}_0(i,j) = g(i,j)$.

The fractal signature $F^{(H)}(\epsilon)$ for horizontal structure is calculated by substituting $u^{(H)}_\epsilon$ and $l^{(H)}_\epsilon$ for u_ϵ and l_ϵ in equation (10.5), and calculating $A(\epsilon)$ using equation (10.6) and performing the regression line calculation as before. Similarly, the fractal signature $F^{(V)}(\epsilon)$ for vertical structure is calculated by substituting $u^{(V)}_\epsilon$ and $l^{(V)}_\epsilon$ for u_ϵ and l_ϵ in equation (10.6) and then using the same method as before. Figure 10.11 shows tibial regions digitized from ×5 magnification macroradiograph from (a) a non-arthritic and (b) an osteoarthritic knee of a patient. The fractal signatures for horizontal and vertical image structure are shown in Fig. 10.12 and illustrate the changes between horizontal and vertical trabecular structure detected in osteoarthritis of the knee [17].

Figure 10.11 Regions (20 mm × 6 mm) of subchondral cancellous bone from digitized patient macro-radiographs of a non-arthritic knee (top) an an osteoarthritic knee (bottom).

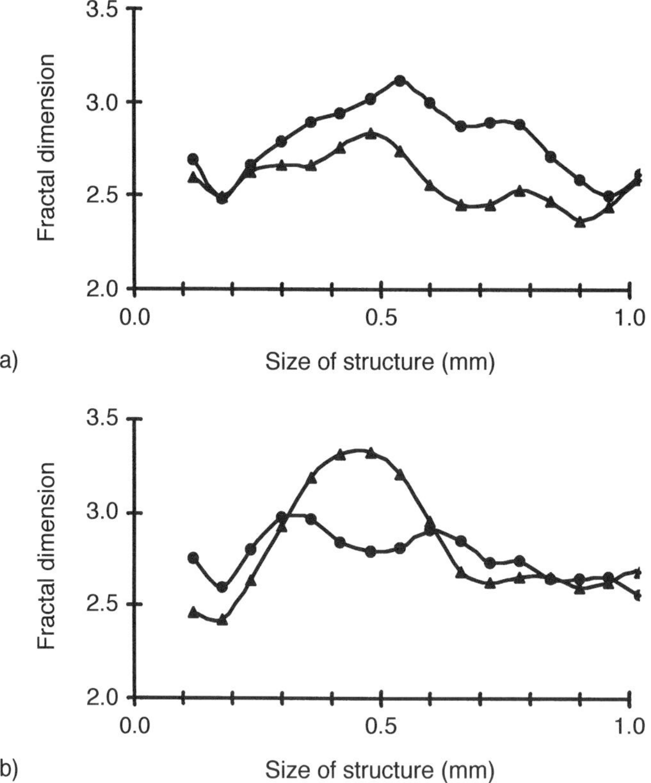

Figure 10.12 Fractal signature analysis for horizontal (filled triangles) and vertical (filled circles) structure of the patient images in Fig. 10.11 from (a) a non-arthritic knee and (b) an osteoarthritic knee.

10.6.5 Film digitization for image analysis

To digitize data from radiographic films, some means of recording the optical density of the film at numerous points is required. This is most simply done using a video camera (CCIR standard in Europe or RS-170 in USA) attached to a frame-grabber card in a PC computer. This usually digitizes a film, or region of a film, at resolutions up to 768×512 pixels with 256 different intensities. With such systems, it is important to set the gain and offset of the camera/frame-grabber system so that the useful range of optical densities within the film is recorded in the 256 different intensities. Digital video standards are being developed (RS-343 or RS-422) which will allow newer cameras to digitize at up to 1024 different intensities and resolutions of up to 1024×1024 pixels. There are also direct digital cameras, such as the Videk Megaplus (Kodak, Canandaigua, NY, USA) which can digitize X-ray films at resolutions of up to 2029×2044 with 1024 different intensities. For better quality film digitization, with up to 4096 different intensities and resolutions up to 3800×4100, laser film scanners, such as the Lumiscan (Lumisys, Sunnyvale, CA, USA) range of scanners, are required. These scanners are likely to record all the useful information on an X-ray film, and also mean that larger numbers of films can be handled with less manual intervention and in shorter periods of time. There is also the option of using direct digital radiography, but the techniques for this are expensive with new, improved methods constantly being developed.

10.6.6 Applications of fractal signature analysis

In osteoarthritis of the knee, fractal signature analysis of macroradiographs of patients has shown that alterations in bone turnover primarily affect horizontal trabeculae early in the disease process, with vertical trabeculae becoming thickened only in late disease, once most articular cartilage has been lost [17]. Post-menopausal women suffer osteoporotic loss due to alterations in bone turnover at the menopause. Fractal signature analysis of macroradiographs of the lumbar spine of early post-menopausal women has quantified age-related changes in fine horizontal trabecular structures, weight-related alterations in coarse weight-bearing vertical trabecular structures, and alterations in fine vertical trabecular structures with changes in the patient's bone mineral density. These changes in fractal signature analysis of the spine quantify trabecular anisotropy [18], which is important in the determination of the weakening of trabecular bone in osteoporosis.

REFERENCES

1. Genant, H.K and Resnick, D. (1988) Magnification radiography, in *Diagnosis of Bone and Joint Disorders*, 2nd edn, (eds D. Resnick and G. Niwayama), W.B. Saunders, Philadelphia, pp. 84–107.
2. Takahashi, S. and Sakuma, S. (1975) *Magnification Radiography*, Springer, Berlin.
3. Buckland-Wright, J.C. and Bradshaw, C.R. (1989) Clinical applications of high definition microfocal radiography. *British Journal of Radiology* **62**, 209–217.
4. Buckland-Wright, J.C. (1989) A new high definition microfocal X-ray unit. *British Journal of Radiology* **62**, 201–208.

5. Mall, J.C., Genant, H.K., Silcox, D.C. and McCarty, D.J. (1974) The efficacy of fine-detail radiography in the evaluation of patients with rheumatoid arthritis. *Radiology* **122**, 37–42.
6. Meema, H.K. and Meema, S. (1972) Comparison of microradioscopic and morphometric findings in the hand bones with densitometric findings in the proximal radius in thyrotoxicosis and in renal osteodystrophy. *Investigative Radiology* **7**, 88–96.
7. Meema, H.K., Oreopoulos, D.G. and Meema, S. (1978) A roentgenologic study of cortical bone resorption in chronic renal failure. *Radiology* **126**, 67–74.
8. Feldkamp, L.A., Goldstein, S.A., Parfitt, A.M. *et al.* (1989) The direct examination of three-dimensional bone architecture in-vitro by computed tomography. *Journal of Bone and Mineral Research* **4**, 3–11.
9. Ruegsegger, P., Koller, B. and Muller, R. (1996) A microtomographic system for the nondestructive evaluation of bone architecture. *Calcified Tissue International* **58**, 24–29.
10. Dedrick, D.K., Goldstein, S.A., Brandt, K.D. *et al.* (1993) A longitudinal study of subchondral plate and trabecular bone in cruciate-deficient dogs with osteoarthritis followed up for 54 months. *Arthritis and Rheumatism* **36**, 1460–1467.
11. Benhamou, C.L., Lespessailles, E., Jacquet, G. *et al.* (1994) Fractal organization of trabecular bone images on calcaneus radiographs. *Journal of Bone and Mineral Research* **9** 1909–1918.
12. Ruttimann, U.E., Webber, R.L. and Hazelrig, J.B. (1992) Fractal dimension from radiographs of periodontal alveolar bone – a possible diagnostic indicator of osteoporosis. *Oral Surgery Oral Medicine Oral Pathology* **74**, 98-110.
13. Caligiuri, P., Giger, M.L. and Favus, M. (1994) Multifractal radiographic analysis of osteoporosis. *Medical Physics* **21**, 503–508.
14. Lynch, J.A., Buckland-Wright, J.C. and Hawkes, D.J. (1991) Analysis of texture in macroradiographs of osteoarthritic knees using the fractal signature. *Physics in Medicine and Biology* **36**, 709–722.
15. Lynch, J.A., Buckland-Wright, J.C. and Hawkes, D.J. (1991) A robust and accurate method for calculating the fractal signature of texture in macroradiographs of osteoarthritic knees. *Medical Informatics* **16**, 241–251.
16. Buckland, J.C., Lynch, J.A., Rymer, J. and Fogelman, I. (1994) Fractal signature analysis of macroradiographs measures trabecular organization in lumbar vertebrae of postmenopausal women. *Calcified Tissue International* **54**, 106–112.
17. Buckland-Wright, J.C., Lynch, J.A. and Macfarlane, D.G. (1996) Fractal signature analysis measures cancellous bone organisation in macroradiographs of patients with knee osteoarthritis. *Annals of the Rheumatic Diseases* **55**, 749–755.
18. Lynch, J.A., Buckland-Wright, J.C., Hawkes, D.J. and Nair, S.V. (1996) Changes in anisotropy of modelled bone measured by simulated radiography and fractal signature analysis. *Transactions of the Orthopaedic Research Society* **21**, 714.
19. Wilding, R.J.C., Slabbert, J.C.G., Kathree, H. *et al.* (1995) The use of fractal analysis to reveal remodelling in human alveolar bone following the placement of dental implants. *Archives of Oral Biology* **40**, 61–72.
20. Hall, T.A., Röckert, H.O. and Saunders, R.L. de C.H. (eds) (1972) *X-Ray Microscopy in Clinical and Experimental Medicine*, C.C. Thomas, Springfield, Illinois.
21. Ely, R.V. (1980) *Microfocal Radiography*, Academic Press, London.
22. Buckland-Wright, J.C. (1980) Qualitative and quantitative assessment of tissue organisation in normal and diseased organs, in *Microfocal Radiography*, (ed. R.V. Ely), Academic Press, London, pp. 147–195.
23. Buckland-Wright, J.C. (1977) The microfocal X-ray unit: a demonstration of its potential. *Medical and Biological Illustration* **27**, 163–168.
24. Buckland-Wright, J.C., Macfarlane, D.G., Lynch, J.A. and Clark, B. (1990) Quantitative microfocal radiographic assessment of progression in osteoarthritis of the hand. *Arthritis and Rheumatism* **33**, 57–65.

25. Buckland-Wright, J.C., Macfarlane, D.G. and Lynch, J.A. (1992) Relationship between joint space width and subchondral sclerosis in the osteoarthritic hand: a quantitative microfocal study. *Journal of Rheumatology* **19**, 788–795.

26. Buckland-Wright, J.C., Macfarlane, D.G., Jasani, M.K. and Lynch, J.A. (1994) Quantitative microfocal radiographic assessment of osteoarthritis of the knee from weight bearing tunnel and semi-flexed standing views. *Journal of Rheumatology* **21**, 1734–1741.

27. Lynch, J.A., Buckland-Wright, J.C. and Macfarlane, D.G. (1993) Precision of joint space width measurement in knee osteoarthritis from digital image analysis of high definition macroradiographs. *Osteoarthritis and Cartilage* **1**, 209–218.

28. Buckland-Wright, J.C. (1983) X-ray assessment of activity in rheumatoid disease. *British Journal of Rheumatology* **22**, 3–10.

29. Buckland-Wright, J.C. (1984) Microfocal radiographic examination of erosions in the wrist and hand of patients with rheumatoid arthritis. *Annals of the Rheumatic Diseases* **43**, 160–171.

30. Bellman S. (1953) Microangiography. *Acta Radiologica* **102** (Suppl.), 1–104.

31. Hobdell, M.H. (1970) The relationship between the functional and structural organisation of bone in the jaws of mammals. PhD Thesis, London University.

32. Doi, K. and Imhof, H. (1977) Noise reduction by radiographic magnification. *Radiology* **122**, 479–487.

33. Buckland-Wright, J.C. (1994) Quantitative radiography of osteoarthritis. *Annals of the Rheumatic Diseases* **53**, 268–275.

34. Buckland-Wright, J.C., Carmichael, I. and Walker, S.R.(1986) Quantitative microfocal radiography accurately detects joint changes in rheumatoid arthritis. *Annals of the Rheumatic Diseases* **45**, 379–383.

35. Buckland-Wright, J.C., Clarke, G.S., Chikanza, I.C. and Grahame, R.(1993) Quantitative microfocal radiography detects changes in erosion area in patients with early rheumatoid arthritis treated with myocrisine. *Journal of Rheumatology* **20**, 243–247.

36. Buckland-Wright, J.C. (1981) Microfocal radiography in the quantitative assessment of experimentally induced inflammatory arthritis in guinea pigs. *Journal of Pathology* **135**, 127–145.

37. Buckland-Wright, J.C., Spring, M.W., Mak, R.H.K. *et al.* (1990) Quantitative microfocal radiography of children with renal osteodystrophy; comparison with laboratory and histological findings. *British Journal of Radiology* **63**, 609–614.

38. Wou, P.C.S., Lima, E., Turner, C. *et al.* (1993) Quantitative macroradiography with biochemical correlation of children with renal osteodystrophy: short communication. *British Journal of Radiology* **66**, 743–747.

39. Buckland-Wright, J.C. and Walker, S.R. (1987) Incidence and size of erosions in the wrist and hand of rheumatoid patients: a quantitative microfocal radiographic study. *Annals of the Rheumatic Diseases* **46**, 463–467.

40. Buckland-Wright, J.C., Clarke, G.S. and Walker, S.R. (1989) Erosion number and area progression in the wrist and hand of rheumatoid patients: a quantitative microfocal radiographic study. *Annals of the Rheumatic Diseases* **48**, 25–29.

41. Buckland-Wright, J.C., Macfarlane, D.G. and Lynch, J.A. (1995) Sensitivity of radiographic features and specificity of scintigraphic imaging in hand osteoarthritis. *Revue du Rhumatisme [English Edition]* **62** (Suppl. 1), 14S–26S.

42. Buckland-Wright, J.C., Macfarlane, D.G., Lynch, J.A. and Jasani, M.K. (1995) Quantitative microfocal radiography detects changes in OA knee joint space width in patients in placebo-controlled trial of NSAID therapy. *Journal of Rheumatology* **22**, 937–943.

43. Buckland-Wright, J.C., Macfarlane, D.G. and Lynch, J.A. (1991) Osteophytes in the arthritic hand: their incidence, size, distribution and progression. *Annals of the Rheumatic Diseases* **50**, 627–630.

44.   Buckland-Wright, J.C., Macfarlane, D.G., Fogelman, I. *et al.* (1991) Technetium 99m methylene diphosphonate bone scanning in osteoarthritic hands. *European Journal of Nuclear Medicine* **18**, 12–16.

45.   Macfarlane, D.G., Buckland-Wright, J.C., Emery, P. *et al.* (1991) Comparison of clinical, radionuclide and radiographic features osteoarthritis of the hands. *Annals of the Rheumatic Diseases* **50**, 623–626.

46.   Machin, K. and Webb, S. (1994) Cone-beam x-ray microtomography of small specimens. *Physics in Medicine and Biology*, **39**, 1639–1657.

47.   Holdsworth, D.W., Drangova, M. and Fenster, A. (1993) A highresolution XRII-based quantitative volume CT scanner. *Medical Physics* **20**, 449–462.

48.   Feldkamp, L.A., Davis, L.C. and Kress, J.W. (1984) Practical cone-beam algorithm. *Journal of the Optical Society of America A – Optics and Image Science* **1**, 612–619.

49.   Kuhn, J.L., Goldstein, S.A., Feldkamp, L.A. *et al.* (1990) Evaluation of a microcomputed tomography system to study trabecular bone-structure. *Journal of Orthopaedic Research* **8**, 833–842.

50.   Croucher, P.I., Garrahan, N.J. and Compston, J.E. (1996) Assessment of cancellous bone-structure – comparison of strut analysis, trabecular bone pattern factor, and marrow space star volume. *Journal of Bone and Mineral Research* **11**, 955–961.

51.   Turner, C.H., Cowin, S.C., Rho, J.Y. *et al.* (1990) The fabric dependence of the orthotropic elastic-constants of cancellous bone. *Journal of Biomechanics* **23**, 549–561.

52.   Harrigan, T.P. and Mann, R.W. (1984) Characterization of microstructural anisotropy in orthotropic materials using a 2nd rank tensor. *Journal of Materials Science* **19**, 761–767.

53.   Ham, Y.S., Russ, J.C., Gardner, R.P. and Verghese, K. (1993) 3-dimensional differential absorption x-ray cone-beam microtomography using balanced filters and algebraic reconstruction. *Applied Radiation and Isotopes* **44**, 1313–1320.

54.   Goldstein, S.A., Goulet, R. and McCubbrey, D. (1993) Measurement and significance of three-dimensional architecture to the mechanical integrity of trabecular bone. *Calcified Tissue International* **53** (Suppl. 1), S127–S133.

55.   Singh, M., Nagrath, A.R. and Maini, P.S. (1970) Changes in trabecular pattern of the upper end of the femur as an index of osteoporosis. *Journal of Bone and Joint Surgery*, **52A**, 457–467.

56.   Gluer, C.C., Cummings, S.R., Pressman, A. *et al.* (1994) Prediction of hip-fractures from pelvic radiographs – the study of osteoporotic fractures. *Journal of Bone and Mineral Research* **9**, 671–677.

57.   Masud, T., Jawed, S., Doyle, D.V. and Spector, T.D. (1995) A population study of the screening potential of assessment of trabecular pattern of the femoral-neck (Singh index) – the Chingford study. *British Journal of Radiology* **68**, 389–393.

58.   Caldwell, C.B., Rosson, J., Surowiak, J. and Hearn, T. (1993) Use of fractal dimension to characterize the structure of cancellous bone in radiographs of the proximal femur, in *Fractals in Biology and Medicine* (eds T.F. Nonnenmacher, G.A. Losa and E.R. Weibel), Birkhäser-Verlag, Basel, pp. 300–306.

59.   Saupe, D. (1988) Algorithms for random fractals, in *The Science of Fractal Images* (eds H.O. Peitgen and D. Saupe), Springer-Verlag, New York, pp. 71–133.

60.   Pentland, A.P. (1984) Fractal-based descriptions of natural scenes. *IEEE Transactions on Pattern Analysis and Machine Intelligence* **6**, 661–674.

61.   Voss, R.F. (1988) Random fractal forgeries, in *The Science of Fractal Images* (eds H.O. Peitgen and D. Saupe), Springer-Verlag, New York, pp. 805–836.

62.   Peleg, S., Naor, J., Hartley, R. and Avnir, D. (1984) Multiple resolution texture analysis and classification. *IEEE Transactions on Pattern Analysis and Machine Intelligence* **6**, 518–523.

63.   Sternberg, S.R. (1986) Greyscale morphology. *Computer Vision Graphics and Image Processing* **35**, 333–355.

CHAPTER ELEVEN

Bone mineral measurements by DXA in animals

Bruce H. Mitlak and Masahiko Sato

11.1 BACKGROUND

Non-invasive assessment of the small animal skeleton provides a rapid and economical means for evaluating pharmacological agents and for testing hypotheses related to bone physiology. This approach has evolved over the past two decades as investigators have begun to apply techniques, initially developed for clinical research, to estimate bone mass and density in small animals. These techniques include radiographic absorptiometry [1], single photon absorptiometry (SPA) [2], neutron activation analysis [3, 4], quantitative computerized tomography (QCT) [5–7], dual photon absorptiometry (DPA) [8], and more recently dual energy X-ray absorptiometry (DXA) [9–12]. Refinements in the ability to assess animal bones mirrors the evolution of clinical densitometry (recently reviewed by Genant *et al.* [13]). In this respect, isotope-based instruments (SPA, DPA), which provided relatively poor sensitivity and reproducibility, have been replaced by DXA instruments incorporating stable, high-flux X-ray tubes which are better suited for high resolution measurements. QCT and peripheral QCT (pQCT), which measure true bone density and can measure anatomically distinct cortical and trabecular bone compartments, have been used to a limited extent in animal studies. This use may change with the introduction of relatively low-cost scanners [5, 14] and with the demonstration that these scanners may be useful in examining mouse bones which are too small to be reliably measured by DXA [15].

Non-invasive measurements of bone mass and density by DXA provide an important 'whole animal' framework in which to evaluate our rapidly expanding understanding of bone at the cellular and molecular level. Regional and whole skeletal bone mass assessments by DXA complement the cellular and dynamic data available from histomorphometric measurements and the functional information available from biomechanical evaluation of bone. Although DXA can be used to measure bones *ex vivo*, the information can usually be obtained by much less expensive methods such as ashing or using Archimedes' principle [16]. If such analyses are not feasible (e.g. samples are needed for other testing or for specialized processing), then DXA can permit a rapid assessment of bone mass. In contrast, measurements obtained by DXA *in vivo* can significantly enhance the power and

Methods in Bone Biology. Edited by Timothy R. Arnett and Brian Henderson.
Published in 1997 by Chapman & Hall, London. ISBN 0 412 75770 2.

flexibility of studies by permitting repeated assessment of bone mass within a given animal. This approach increases the sensitivity in detecting changes in bone mass, thus permitting a smaller number of animals to be studied for a given end-point. In addition, the approach may obviate the need for sacrificing multiple groups of animals to evaluate time-dependent effects. DXA scanning before initiating experiments also permits exclusion of outliers or stratification of animals based on bone mass. This is important because matching animals for age and weight does not ensure equivalent bone mass or density between study arms prior to treatment. In this chapter, we will provide an approach we have found useful for performing non-invasive, serial measurements of bone mass by DXA. In addition, we will highlight some specific considerations for study design imposed by the rat model.

11.2 THE RAT MODEL FOR SKELETAL STUDIES

DXA studies utilizing the rat have provided extremely useful information on the effects of a wide range of physiological and pharmacological challenges to the skeleton. When the model was carefully chosen, clinically relevant insight using DXA was gained on the skeletal effects of exercise [17–19] and diabetes [20], and on interventions with PTH [14], thyroid hormone [21–24] and raloxifene [25, 26] as well as growth hormone [27], growth factors [28, 29] and cytokines and their receptor agonists [30, 31].

In particular, the ovariectomized rat has been widely used as a model of bone loss induced by oestrogen deficiency. This model has been extensively evaluated [32, 33] and reviewed [34, 35]. Using this model, investigators have demonstrated that administrations of a wide number of clinically important agents such as oestrogen [36–40], tamoxifen [37, 41], vitamin D [42], calcitonin [43], fluoride [44–46] and bisphosphonates [47–50] have yielded results that have largely paralleled those observed in the human skeleton. For this reason the ovariectomized rat model of bone loss induced by oestrogen deficiency has been identified as an essential animal model (representing a modelling species) in preclinical studies by the US Food and Drug Administration in its Guidelines for Preclinical and Clinical Evaluation of Agents Used in the Prevention or Treatment of Postmenopausal Osteoporosis [51] and by the Group for the Respect of Ethics and Excellence in Science (GREES) [52].

As we will highlight, understanding the pattern of somatic and skeletal growth and the response to gonadal steroid deficiency is central to experimental design using the rat skeleton, as well as for other non-human models. The age and sex of the experimental animals should be considered important variables in the study design. Rats reach sexual maturity at approximately 6–8 weeks of age but both male and female rats demonstrate relatively rapid somatic and skeletal growth rate until approximately 6 months of age [53, 54]. In addition, the temporal pattern of skeletal maturation differs between male and female rats. As recently reviewed [55], closure of the epiphysis is delayed in male rats, resulting in continued skeletal growth throughout life. However, in female Sprague-Dawley rats, perforation of the growth plate and foci of ossification are observed by 8–12 months of age [53, 54]. The study of aged female rats with slowed longitudinal growth is preferred because the effect of oestrogen deficiency, glucocorticoids and other compounds under study may affect bone directly, and also indirectly by affecting skeletal and/or somatic growth. The resulting pathophysiological model may, therefore, be different in a skeletally immature animal compared with an animal with a mature skeleton. Most animal suppliers are geared to providing younger animals and so the investigator must either locate cohorts of older (usually more expensive) animals or must obtain and house young animals, thus delaying the start of a study until the animals reach an appropriate age. The only immediately available older animals may be retired breeders. These animals, because of the metabolic stresses of reproduction, may have immature and relatively osteopenic skeletons [55]. Although it is difficult to match such animals (even if a detailed reproductive history is known), stratifying by bone mass determined by DXA at baseline is one possible strategy for decreasing the variability introduced by the use of these animals. It may, in addition, be necessary to follow bone mass in such animals as they adjust to a more positive calcium balance before starting a study.

11.3 TECHNICAL CONSIDERATIONS FOR DXA MEASUREMENTS

The adaptation of clinical DXA instruments for use in scanning animals has been greatly facilitated by manufacturers who have created software to permit high-resolution, small region-of-interest (ROI) scans. However, there are several unique technical issues pertaining to the measurement of bone mass in small animals by DXA which need to be appreciated by researchers. Small animal scanning imposes several basic requirements on the DXA instrument. Specifically, the DXA systems must be capable of achieving adequate spatial resolution and signal-to-noise ratio at the required scanning resolution.

The manufacturers have approached these issues in a similar fashion. Hologic addressed the problem on its 'pencil beam' instruments by employing a tightly collimated X-ray beam of approximately 1 mm in diameter and oversampling the bone region. The main disadvantage of using the high-resolution protocol is the relatively long scan time (3–15 minutes, depending on the size of the region of interest). Lunar similarly collimates the X-ray beam on its DPX systems to a width of 0.84 mm and offers a selection of pixel dimensions for its 'appendicular' fine scan mode. Hologic has also adapted its 'fan beam' multi-element detector array instruments to decrease scan times for small animal scanning. However, these 'fan beam' instruments are significantly more expensive than the pencil beam instruments. Also, a new desktop DXA instrument, the Norland pDEXA offers high-resolution scanning suitable for measurement of small animal bones.

The DXA instrument must be relatively free of non-linear beam hardening effects, a phenomenon characterized by selective removal of low-energy photons from the X-ray beam. This effect causes the X-ray beam to become more energetic (harder) and is most significant as the beam passes through the first 2.5–4 cm of soft tissue or equivalent. This issue is particularly relevant to small animal scanning, where these effects may change the measured bone density. In larger animals and humans the effect is not usually significant because sufficient soft tissue is present.

These hardening effects are difficult to 'calibrate out' of the system because they are non-linear, depend on the X-ray tube characteristics and are related to the tissues being measured. The simplest and most reliable solution is to filter the X-ray beam to simulate the attenuation provided by 3–4 cm of soft tissue. This is commonly done by scanning through a 3.8 cm acrylic block, a water bath with a depth of approximately 3 cm [9, 56] or an aluminium sheet. These materials 'pre-harden' the X-ray beam. The remaining beam hardening effects are usually within the precision of error for the DXA instrument.

11.4 SCANNING PROCEDURES

The choice of the study design and skeletal site(s) to be scanned should depend on the experimental hypothesis. A growing literature on the use of DXA in several animal models (as noted above) should provide guidance for the choice of study design and estimates of precision for power calculations. The choice of particular scanning sites may be influenced by the importance of measuring trabecular, cortical or whole skeleton mineral content or density. Clearly, because each scan requires at least several minutes of time, investigators may wish to focus on one site in 'pilot studies', especially when access to a DXA instrument is limited. The manufacturers of DXA instruments have created several scan formats in which varying resolutions and scan speeds are chosen to permit scanning of individual bones or the entire animal skeleton. While several have been optimized for the adult rat, there is some 'art' in best matching the scan protocols to other study designs.

Investigators have used DXA to examine lumbar and caudal vertebrae [9, 12, 56, 57], long bones and subregions of these bones [58] as well as the entire skeleton [11, 12, 59]. Considerations for choice of animal positioning on the DXA scanner include the bone(s) of interest, the particular skeletal geometry of the animal, and the need for reliable anatomical landmarks for identifying the scan ROI. Once the protocol has been established it is important to establish the inter- and intra-observer variability and, if possible, validate the method against a standard such as weighing of bone ash. We have found there is a significant learning curve when performing these scans. It is important for the operator to be aware of the effects of even subtle changes in animal positioning and in the choice of scan and analysis ROI to optimize study results.

Sedation is almost always required to keep the animal on the scan platform and to avoid motion artefacts during the scan. While the level of experience with anaesthetic agents should be well within that available in animal care facilities, special care should be taken to minimize the dose when repeated courses of anaesthesia are required (e.g. for serial measurements) and when anaesthesia is given to animals that are stressed (e.g. dehydrated, hypercalcaemic). Although there has been no systematic comparison of anaesthetic agents to facilitate animal scanning, we have observed that combinations of ketamine, xylazine and acepromazine appear to be well tolerated and result in minimal anaesthetic-related deaths. We have found that a combination of ketamine and xylazine results in a higher respiratory rate (resulting in potential motion artefacts) than when ketamine is combined with acepromazine. Therefore, we mix 75 mg ketamine HCl/ml with 2.6 mg acepromazine malate/ml (Aveco Co., Fort Dodge, IA) and administer the combination at a dose of 1 μl/g body weight.

11.4.1 Spine scans

Spinal vertebrae comprise a relatively high percentage of trabecular bone. Therefore, scanning the spine may increase sensitivity in detecting changes resulting from oestrogen deficiency. Spine scans can be performed with the animal sedated and positioned supine or prone [57]. We have found it easier to position the animal and ensure correct placement of the spine ROI when the animal is placed supine. Both lumbar and caudal vertebral sites have been examined. The lumbar spine was found to be a more sensitive site in detecting bone loss after the experimental animal is subjected to ovariectomy [9].

Limited information is available to help to guide the choice of which lumbar vertebrae to scan. Clearly, a desire to maximize the amount of information by scanning several vertebral bodies needs to be balanced against the resultant increases in scanning time. To address the question of which vertebrae to scan, we evaluated changes in the lumbar spine (L1–L5) bone mineral density (BMD) in 6-month-old rats, 4 weeks after undergoing ovariectomy or sham operation. As shown in Fig. 11.1, ovariectomy resulted in significantly lower (P <0.05) BMD at all vertebrae compared with sham-operated animals. The effect of treatment with 17β-oestradiol was apparent at each vertebral level but was significant only at L1–2. Therefore, there appear to be subtle differences in the response to oestrogen deficiency or treatment at different vertebral levels. However, measurements of any of these vertebrae may be informative.

One of the authors has chosen to scan L2–4 vertebral bodies using a Hologic instrument and the following procedure. The ultra-high resolution setting is selected and an acrylic platform is consistently used. The correct ROI containing the lumbar vertebrae is identified in each animal by performing a 'scout scan'. This scan can be performed at a lower resolution (which reduces scan time but will result in an unanalysable scan). After an animal is sedated, it is placed supine on the acrylic block. The 'scout scan' is performed with a scan step of 0.03–0.04, starting just below the midpoint of a imaginary line connecting the iliac crests. (The location of the start point may vary, depending on the precise alignment of the laser positioning device.) During this scout scan, the pelvis plus L6 (first vertebrae above the sacrum) through L4 are identified. Using the reposition function on the Hologic

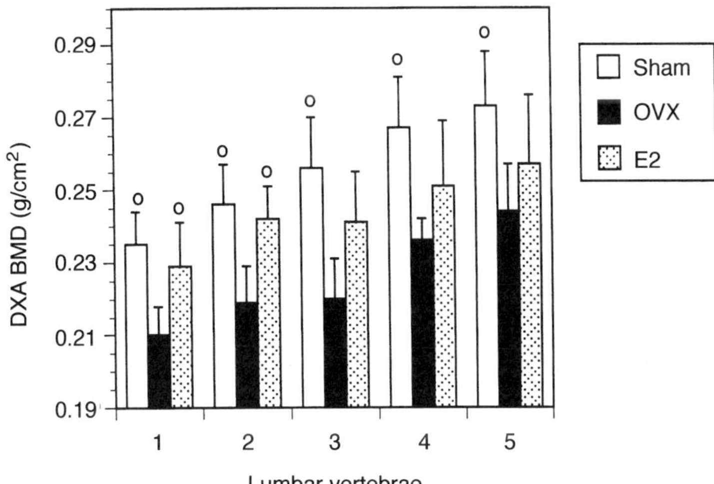

Figure 11.1 Bone mineral density (BMD) of individual lumbar vertebrae L1–5 in 6-month-old rats, 4 weeks after sham operation (Sham), ovariectomy (OVX), or ovariectomy with administered 17β-oestradiol (E2) (n = 5 groups). Groups with a mean BMD significantly different from ovariectomized rats (Fischer's PLSD, P <0.05) denoted by open circle.

instrument, the positioning arrow is placed over the middle of the L5 vertebral body. The scan is restarted at the ultra-high resolution (returning to the default scan step of 0.01). When performed in this manner, the upper portion of the L5 vertebral body and the L4–5 interspace comprises the lower portion of the scan image. When the scan is completed, the analysis ROI is set so as to include L2–L4 vertebrae and approximately the same width of soft tissue on each side of the spinal column as represented by the width of the bone (Fig. 11.2). The scan time for the final high-resolution scan including L2–4 is 3–5 minutes. Long-term precision can be enhanced by the use of the 'compare' function, which allows direct comparison to a prior (baseline) scan when performing the analysis.

11.4.2 Long bone and subregions scans

Scanning appendicular sites *in vivo* requires more attention to animal placement, including securing of the limbs in a flexed position. In addition this procedure requires more subjective operator intervention to define an analysis ROI (e.g. one that separates the proximal femur from the surrounding pelvis or the distal femur from the tibia). Also, particularly in younger animals, reliable detection of the bone edge by the software algorithms may become a problem. We have determined that a shift in the accuracy measurement of bone area (but not bone mineral) occurs when bone rods smaller than 4–5 mm are scanned *ex vivo* (Fig. 11.3). This observation suggests that DXA estimates of bone mineral content (BMC) may be more accurate than BMD in very small bones. Although this effect has not been systematically evaluated *in vivo*, we have found that the proximal tibiae and distal femora of 270 g rats are approximately 5–7 mm thick and therefore should be above the threshold for this effect.

Several investigators have evaluated the usefulness of scanning tibiae [9, 57] or femora [10, 21, 58]. Further, these investigators have demonstrated an enhanced

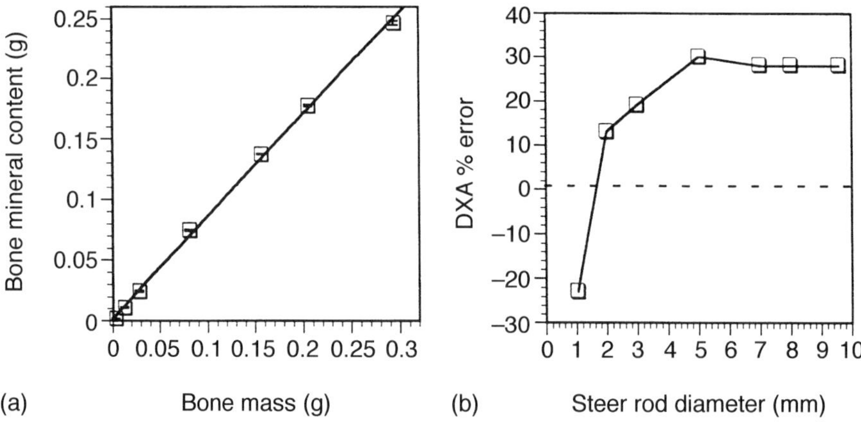

Figure 11.2   (a) Correlation between bone mineral content, determined by DXA, of rods milled from the cortex of a steer bone and bone mass determined by Archimedes' principle. Bone mass can be converted to calcium content by multiplying by 0.23, as determined by atomic absorption spectrophotometry. (b) Percentage error (deviation from expected) as a function of bone diameter for steer bone rods 1.04–09.6 mm in diameter. Values calculated from projected area determined by DXA and compared with diameter measured directly by calipers. (Reproduced with permission from *Bone*, 1995, **17**, 1575–1625.)

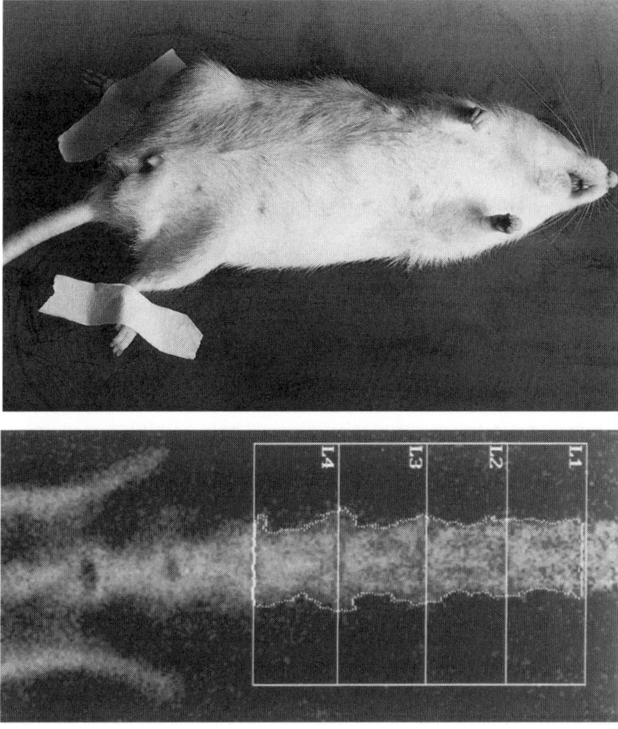

Figure 11.3   (a) Animal positioned prone for acquisition of a spine scan and (b) resultant scan image with analysis region of interest (ROI) set to include spinal vertebrae L2–4. Scan ROI enlarged to demonstrate anatomical landmarks.

sensitivity in detecting changes induced by ovariectomy or PTH treatment by the analysis of a subregion of the distal femur including the metaphysis [10, 21, 58], the location of a high proportion of trabecular bone. One of these groups [10] demonstrated the precision of repeated measurement of the proximal tibiae was lower than that for other sites. However, they did not evaluate whether this was balanced by greater sensitivity to detect changes. Ammann *et al.* [9] did find that measurements of the proximal tibiae site changed more quickly than the spine after oestrogen deficiency *in vivo*. Long bone subregions were defined by a specific length from the end of a bone, fractions of the bone length or by anatomic landmarks. Although subregion analysis may enhance sensitivity in detecting changes in bone mass, investigators need to consider that femoral or tibial dimensions may change during a study and indeed may be influenced by the intervention under study. Because of this, subregions defined by length may not include precisely the same bone region over time.

11.4.3 Whole skeleton scans

The whole skeleton can be scanned to evaluate skeletal mass and to obtain an assessment of cortical bone. This approach for determining cortical bone mass and density is a compromise between very simple, reliable animal placement and the additional sensitivity that can be gained by measurement sites entirely composed of cortical bone (i.e. mid one-third of the femur). For this whole-body measurement, animals are placed prone on top of the acrylic platform on the scanner, with the limbs extended so as to avoid superimposition of bones. The tail is included in the scan but is looped up next to the animal without superimposition of the limbs or body (Fig. 11.4). Although animals can also be scanned supine, we have found that this position causes them more distress and therefore lightly sedated animals are more likely to move during the scan. If the animals are expected to grow during the course of the study, we choose a single scan window size large enough to accommodate all subsequent scans. The resultant scan time for rats up to 500 g is approximately 12–15 minutes. With the newer DXA instruments the option to obtain a body composition assessment during the whole-animal scan may be available. While DXA does not measure actual fat or muscle mass (soft tissue is partitioned into fat and lean 'compartments' based on average X-ray attenuation) it may nonetheless be a useful measurement. However, careful validation in small animals has not yet been reported.

11.4.4 Accuracy and precision of the technique

For DXA scanning to provide useful and reliable information, attention needs to be focused on developing a consistent method for animal positioning plus selection of the region of interest for scanning and analysis. We have evaluated the

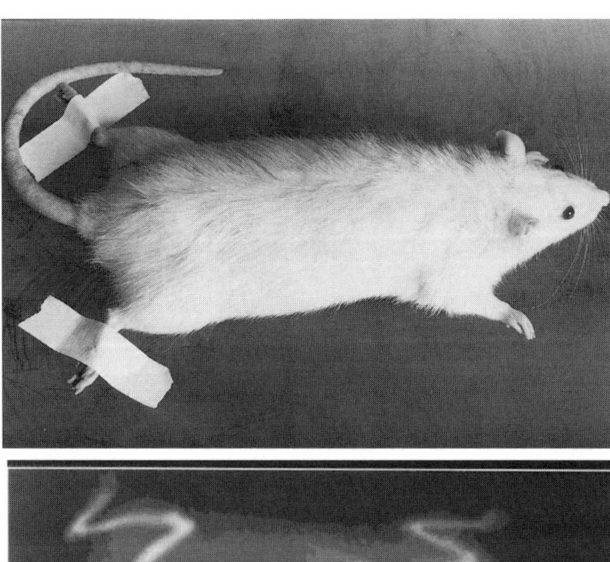

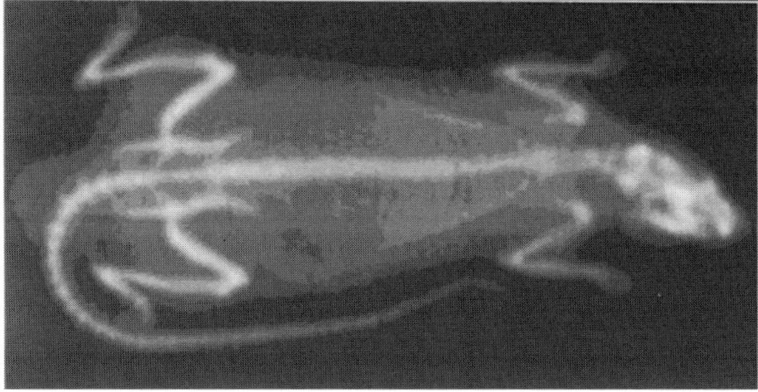

Figure 11.4   (a) Animal positioned supine for acquisition of whole skeleton scan and (b) resultant scan image.

precision (reproducibility) and accuracy (ability to predict ash weight) of high-resolution DXA for measurement of rats in the following way [12].

To assess precision of the technique, the short-term variability was determined by measuring spine and whole skeleton BMC in 38 rats twice in one day, with repositioning between the scans, using a Hologic QDR-1000. The male and female animals were selected to include a range of ages and weights encompassing a significant portion of the animals' adult life range (1–10 months of age). The spine scan was measured using the ultra-high resolution software. Whole-animal scans were then performed using the rat whole-body software before the animal awoke from sedation as described above. To evaluate long-term variability a frozen rat was measured once daily for 27 days. Use of a frozen animal minimized variability from changes that might occur within the animal itself over the period of study. The coefficients of variation were estimated by analysis of variance.

Table 11.1 demonstrates the range of values and the estimated CV for the measured BMC and BMD. The reproducibility of paired measurements was impressive and was consistent with that observed by others who performed similar analyses [10, 60–62]. Correcting the measured BMC by projected area (i.e. estimated BMD) slightly but significantly improved the reproducibility of the spine measurements, but not measurement of the whole skeleton.

At the end of the study, the animals were sacrificed. Lumbar vertebrae L2–4 were identified by dissection, and were removed and ashed (480°C for 36–48 hours in a muffle furnace). The remainder of the skeleton was ashed separately and the relationship, measured by DXA, between BMC and ash weight was determined. Consistent with the findings of others, regression analysis demonstrated that ash weight and lumbar spine BMC, determined by DXA, were closely and linearly related in animals weighing between 175 and 850 g (Fig. 11.5). A similar relationship between ash weight and the whole skeleton BMC, measured by DXA, existed for the same animals (Fig. 11.6). The intercept terms were not significantly different from zero. Therefore, the data were reanalysed assuming an intercept of zero, resulting in the following relationships:

- Spine BMC (g) = (0.91 ± 0.02) × spine ash weight (g).
- Whole skeleton BMC (g) = (0.86 ± 0.01) × whole skeleton ash weight (g).

The slope of the regression lines was not affected by the sex, weight or age of the rats, by analysis of covariance, but could be affected by changes in the scanning platform, thus demonstrating the importance of uniform scanning conditions. These relationships are similar to those found by others [5, 10, 11, 56, 57, 62].

Table 11.1   Coefficient of variation estimated from paired measurements

	Range	*Short-term[a]* *CV (%)*	*Long-term[b]* *CV (%)*
Whole-skeleton BMC	3.23–18.77 g	0.69	1.03
Whole-skeleton BMD	0.087–0.179 g/cm^2	0.66	1.15
Lumbar spine BMC	0.083–0.745 g	1.26	1.28
Lumbar spine BMD	0.097–0.289 g/cm^2	0.57	0.88

[a] Estimated from duplicate same-day measurements of 38 rats aged 1–10 months.
[b] Estimated from 27 daily measurements of a frozen rat.
BMC, bone mineral content; BMD, bone mineral density.

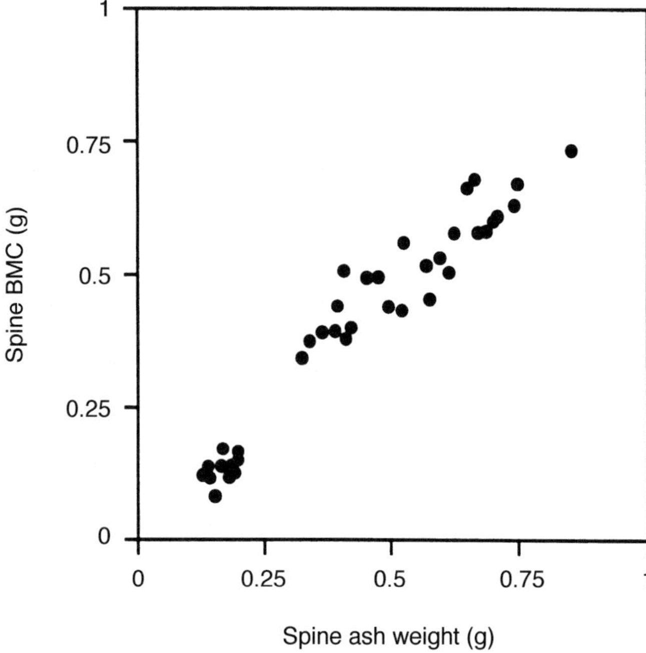

Figure 11.5 Correlation between bone mineral content (mean of two measurements) by DXA and the subsequently measured ash weight of spinal vertebrae L2–4 in 38 male and female rats. (Reproduced with permission from *Journal of Bone and Mineral Research*, 1994 [12].)

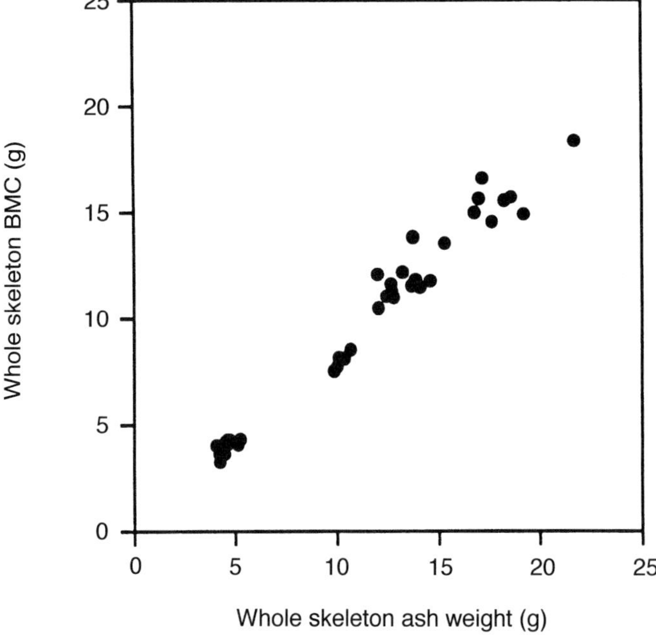

Figure 11.6 Correlation between bone mineral content (mean of two measurements) by DXA and the subsequently measured ash weight of the whole skeleton in 38 male and female rats. (Reproduced with permission from *Journal of Bone and Mineral Research*, 1994 [12].)

Because assessment of long-term variability may not be adequately assessed by measurement of a frozen animal (e.g. it will not reflect true physiological and positioning variability), an additional analysis was performed. In this analysis, assuming a linear rate of change and a different slope and intercept for each animal, we fitted a model for BMC and BMD to measurements obtained from seven female rats subjected to sham operation or seven vehicle-treated male animals. The residual variation around these lines (root mean squared error, RMSE) was used to estimate the variability of the technique. We estimated that, determined in this manner, the average *in vivo* CVs for BMC were 3% for the whole skeleton and 5% for the spine. For BMD, the corresponding figures were 2.8% and 3.2%. This long-term *in vivo* within-animal variation significantly exceeds the short-term (same day paired measurements) and long-term (frozen rat measurements) variation of the technique. Although the source of this variability was unclear, it should be considered when designing studies requiring repeated measurements over time.

Whether to report BMC, which is directly measured by DXA, or BMD is sometimes unclear. As noted, BMD is not volumetric density but BMC normalized by projected bone area. The use of the calculated BMD is particularly useful when comparing bones of different sizes but is more sensitive than BMC to the positioning of the animal during the scan. In particular, BMD may be affected when asymmetrical bones are rotated or when there is overlap of bones in the scan. In our studies (Table 11.1), we found spinal BMD to be more reproducible than BMC. In contrast, the whole-skeleton measurements of BMC and BMD had equivalent reproducibility. BMC usually mirrors bone tissue mass but BMC and tissue mass can diverge if there are major changes in the completeness of mineralization (e.g. osteomalacia). Smaller divergences in the ratio of BMC to bone tissue mass occur when the rate of bone remodelling changes. Because secondary mineralization is less complete in recently formed bone, BMC may underestimate bone tissue when the proportion of young osteons increases in high turnover states [63].

11.5 CHARACTERIZING BMD IN THE RAT DURING GROWTH AND THE EFFECT OF OVARIECTOMY

As noted, less well appreciated aspects of the rat model are the delay of epiphyseal fusion and the persistent skeletal growth after the animal reaches sexual maturity. Despite the continued skeletal growth, the rate of change decreases relative to somatic growth. The following data (Fig. 11.7) [56] demonstrate the dissociation between continued somatic growth over a 56-week period and spinal BMC and BMD as determined by DXA. Spinal BMC and BMD which peak at approximately 23 weeks remain relatively stable through the 56-week period. Therefore when studies using DXA are performed on rats younger than 4 months of age, investigators need to be aware that an interaction between rapid skeletal growth and the treatment or intervention is likely.

A corollary to the observed differences in the sexual and skeletal developmental profiles in the rat is that the effect of ovariectomy on BMD may be different when young, skeletally immature animals are compared with older, similarly treated animals. As seen in Fig. 11.8, both the 8- and 23-week-old ovariectomized animals have a significantly lower lumbar spinal BMD (L1–6) and BMC (not shown) than the sham-operated animals. However, the 8-week-old ovariectomized rats continue

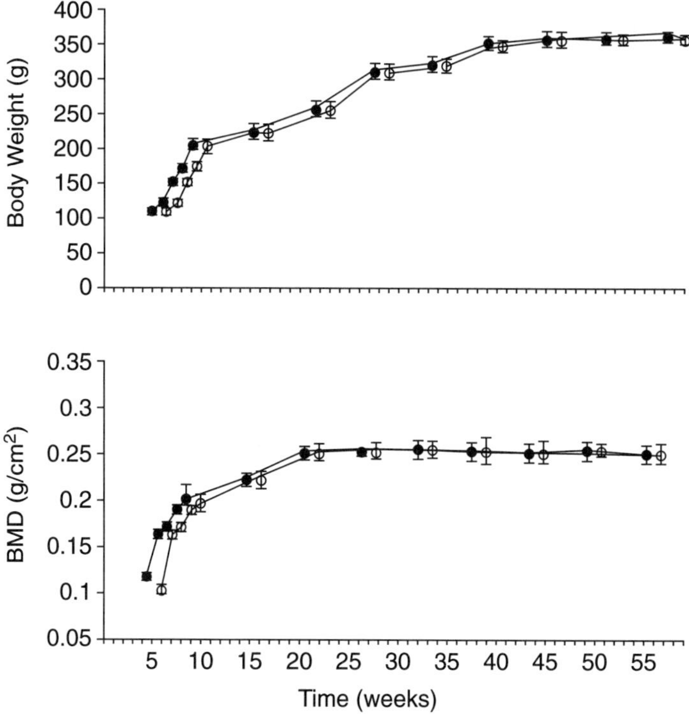

Figure 11.7   Changes in (a) body weight and (b) bone mineral density by DXA of spinal vertebrae L1–6 measured over time *in vivo* (filled circles) (*n* = 5) and from comparable rats *in vitro* (open circles). (Reproduced with permission from *Journal of Bone and Mineral Research*, 1995 [56].)

to increase spinal BMD while the 23-week-old animals actually lose BMD and BMC compared with their pre-ovariectomy value. Further, performing ovariectomy on 11-week-old animals did not result in loss of spinal BMD compared with sham-operated animals. Therefore, these data demonstrate that ovariectomy will result in actual bone loss (rather than failure to gain bone) only if sufficiently aged animals are studied. Of course, factors such as nutritional status and developmental and reproductive history may also affect the response to ovariectomy. These factors should be considered in the planning of experiments.

11.6 THE USE OF DXA IN OTHER SPECIES

Skeletal studies have been performed utilizing sheep, dogs, monkeys, ferrets and mini-pigs. These models have recently been reviewed [55, 64]. In general, the substantially greater expense and the difficulty with animal handling result in these species being less frequently studied than the rat. However, they can provide access to a species where bone remodelling occurs and therefore may be of particular interest to skeletal researchers. The general approach toward use of DXA in other species remains the same as outlined for the rat. The consideration of model physiology and, in particular, skeletal geometry of the animal will impact on the choice of skeletal site(s) to be examined. The choice of scanning protocol will depend on

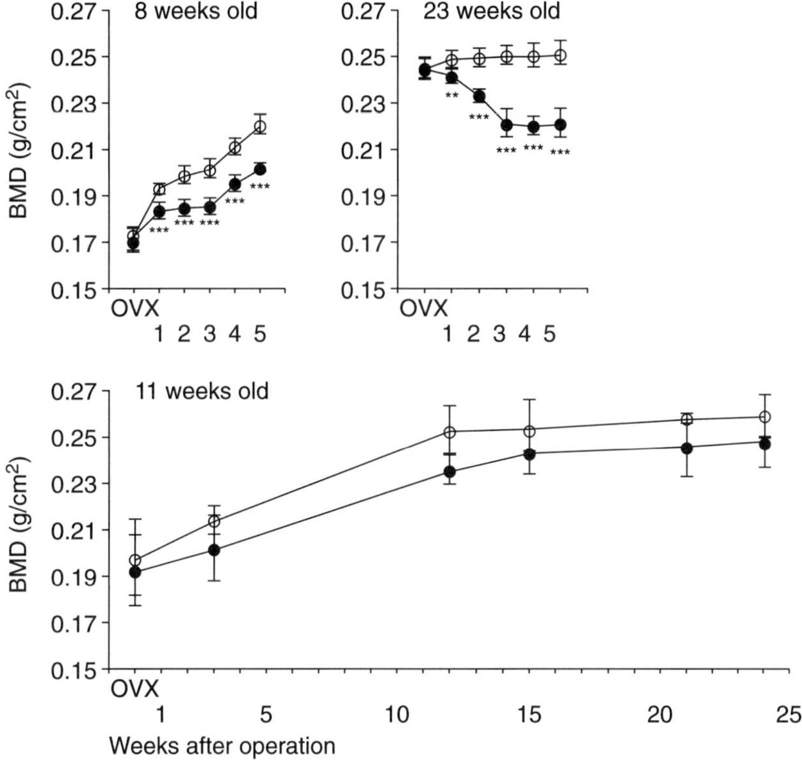

Figure 11.8 Changes in bone mineral density of spinal vertebrae L1–6, determined by serial DXA measurements in rats after ovariectomy (filled circles) or sham operation (open circles) performed at 8, 11 or 23 weeks of age. (Reproduced with permission from *Journal of Bone and Mineral Research*, 1995 [56].)

the size of the ROI. Indeed, the use of clinical scan protocols (e.g. infant, forearm, spine or hip protocols) may be most useful for scanning. Once a study paradigm is established with one of these species, the precision and accuracy of the method should be evaluated.

11.7 QUALITY ASSURANCE (QA) ISSUES IN LONGITUDINAL ANIMAL STUDIES

The use of DXA in clinical studies, particularly in multi-site pharmaceutical trials, has resulted in careful consideration of quality assurance strategies [65, 66]. Long-term study of animal models is recommended by the US Guidelines for Preclinical and Clinical Evaluation of Agents in the Prevention and Treatment of Osteoporosis. However, little attention has been focused on the need for QA procedures for longitudinal studies in animals using DXA. Arguably, relatively small changes in machine performance may have an even greater impact on high-resolution serial scanning than routine clinical scanning. This hypothesis has not been formally evaluated. Usual precautions, such as avoiding changes in instruments, software during a long-term study, or hardware (x-ray tube, detectors, collimators) are essential as is the daily scanning of a human anthropomorphic spine phantom to verify machine performance. The development of a phantom comprising animal bones of comparable size to that of the experimental animals may further ensure the stability of instrument performance in the high-resolution mode over time, but such phantoms are not commercially available and there are no data to establish their value as yet.

REFERENCES

1. Cosman, R., Herrington, B. and Himmelstein, S. (1996) Radiographic absorptometry: a simple method for determining bone mass. *Osteoporosis International* **2**, 34–38.
2. Nordsletten, L., Kaastad, T.S., Skjeldal, S. *et al.* (1994) Fracture strength prediction in rat femoral shaft and neck by single photon absorptiometry of the femoral shaft. *Bone and Mineral* **25**(1), 39–46.
3. Hefti, E., Trechsel, U., Bonjour, J.P. *et al.* (1982) Increase of whole-body calcium and skeletal mass in normal and osteoporotic rats treated with parathyroid hormone. *Clinical Science* **62**, 389–396.
4. Evans, H.J., LeBlanc, A.D. and Johnson, P.C. (1976) Feasibility study; in vivo neutron activation analysis for regional measurement of calcium using californium 252. *Medical Physics* **3**, 148–152.
5. Sato, M., Kim, J., Short, L.L. *et al.* (1995) Longitudinal and cross-sectional analysis of raloxifene effects on tibiae from ovariectomized aged rats. *Journal of Pharmacology and Experimental Therapeutics* **272**(3), 1252–1259.
6. Kinney, J.H., Lane, N.E. and Haupt, D.L. (1995) In vivo, three-dimensional microscopy of trabecular bone. *Journal of Bone and Mineral Research* **10**(2), 264–270.
7. Muller, R., Hahn, M., Vogel, M. *et al.* (1996) Morphometric analysis of noninvasively assessed bone biopsies: comparison of high resolution computed tomagraphy and histologic sections. *Bone* **18**, 215–220.
8. Kimmel, D.B. and Wronski, T.J. (1990) Nondestructive measurement of bone mineral in femurs from ovariectomized rats. *Calcified Tissue International* **46**(2), 101–110.
9. Ammann, P., Rizzoli, R., Slosman, D. and Bonjour, J.P. (1992) Sequential and precise in vivo measurement of bone mineral density in rats using dual-energy x-ray absorptiometry. *Journal of Bone and Mineral Research* **7**(3), 311–316.
10. Griffin, M.G., Kimble, R., Hopfer, W. *et al.* (1993) Dual-energy x-ray absorptiometry of the rat: accuracy, precision, and measurement of bone loss. *Journal of Bone and Mineral Research* **8**(7), 795–800.
11. Casez, J.P., Muehlbauer, R.C., Lippuner, K. *et al.* (1994) Dual-energy X-ray absorptiometry for measuring total bone mineral content in the rat: study of accuracy and precision. *Bone and Mineral* **26**(1), 61–68.
12. Mitlak, B.H., Schoenfeld, D. and Neer, R.M. (1994) Accuracy, precision, and utility of spine and whole-skeleton mineral measurements by DXA in rats. *Journal of Bone and Mineral Research* **9**(1), 119–126.
13. Genant, H.K., Engelke, K., Fuerst, T. *et al.* (1996) Noninvasive assesment of bone mineral and structure: state of the art. *Journal of Bone and Mineral Research* **11**(6), 707–730.
14. Mitlak, B.H., Burdette-Miller, P., Schoenfeld, D. and Neer, R.M. (1996) Sequential effect of chronic human PTH (1-84) treatment of estrogen-deficiency osteopenia in the rat. *Journal of Bone and Mineral Research* **11**(4), 430–439.
15. Beamer, W.G., Donahue, L.R., Rosen, C.J. and Baylink, D.J. (1996) Genetic variability in adult bone density among inbred strains of mice. *Bone* **18**(5), 397–403.
16. Saville, P.D. and Smith, R. (1966) Bone density, breaking force and leg muscle mass as functions of weight in bipedal rats. *American Journal of Physical Anthropology* **25**, 35–40.

17. Yeh, J.K., Aloia, J.F., Tierney, J.M. and Sprintz, S. (1993) Effect of treadmill exercise on vertebral and tibial bone mineral content and bone mineral density in the aged adult rat: determined by dual energy X-ray absorptiometry. *Calcified Tissue International* **52**(3), 234–238.

18. Yeh, J.K., Aloia, J.F., Chen, M.M. *et al.* (1993) Influence of exercise on cancellous bone of the aged female rat. *Journal of Bone and Mineral Research* **8**(9), 1117–1125.

19. Chen, M.M., Yeh, J.K., Aloia, J.F. *et al.* (1994) Effect of treadmill exercise on tibial cortical bone in aged female rats: a histomorphometry and dual energy x-ray absorptiometry study. *Bone* **15**(3), 313–319.

20. Mori, Y., Yokoyama, J., Nemoto, M. *et al.* (1993) Studies of WBN/Kob rat, 18th report: effect of 1 alpha-OH-D3 on diabetic osteopenia. *Nippon Naibunpi Gakkai Zasshi* **69**(9), 989–996. (In Japanese.)

21. Rosen, H.N., Middlebrooks, V.L., Sullivan, E.K. *et al.* (1994) Subregion analysis of the rat femur: a sensitive indicator of changes in bone density following treatment with thyroid hormone or bisphosphonates. *Calcified Tissue International* **55**(3), 173–175.

22. Ongphiphadhanakul, B., Alex, S., Braverman, L.E. and Baran, D.T. (1992) Excessive L-thyroxine therapy decreases femoral bone mineral densities in the male rat: effect of hypogonadism and calcitonin. *Journal of Bone and Mineral Research* **7**(10), 1227–1231.

23. Kung, A.W. and Ng, F. (1994) A rat model of thyroid hormone-induced bone loss: effect of antiresorptive agents on regional bone density and osteocalcin gene expression. *Thyroid* **4**(1), 93–98.

24. Rosen, H.N., Sullivan, E.K., Middlebrooks, V.L. *et al.* (1993) Parenteral pamidronate prevents thyroid hormone-induced bone loss in rats. *Journal of Bone and Mineral Research* **8**(10), 1255–1261.

25. Turner, C.H., Sato, M. and Bryant, H.U. (1994) Raloxifene preserves bone strength and bone mass in ovariectomized rats. *Endocrinology* **135**(5), 2001–2005.

26. Black, L.J., Sato, M., Rowley, E.R. *et al.* (1994) Raloxifene (LY139481 HCI) prevents bone loss and reduces serum cholesterol without causing uterine hypertrophy in ovariectomized rats. *Journal of Clinical Investigation* **93**(1), 63–69.

27. Wright, N.M., Renault, J., Hollis, B. *et al.* (1995) Effect of growth hormone on bone: bone mineral density, trabecular bone volume, and alkaline phosphatase improve or are restored in the dwarf rat treated with growth hormone. *Journal of Bone and Mineral Research* **10**(1), 127–131.

28. Benedict, M.R., Adiyaman, S., Ayers, D.C. *et al.* (1994) Dissociation of bone mineral density from age-related decreases in insulin-like growth factor-I and its binding proteins in the male rat. *Journal of Gerontology* **49**(5), B224–230.

29. Mitlak, B.H., Finkelman, R.D., Hill, E.L. *et al.* (1996) The effect of systemically administered PDGF-BB on the rodent skeleton. *Journal of Bone and Mineral Research* **11**(2), 238–247.

30. Kimble, R.B., Matayoshi, A.B., Vannice, J.L. *et al.* (1995) Simultaneous block of interleukin-1 and tumor necrosis factor is required to completely prevent bone loss in the early postovariectomy period. *Endocrinology* **136**(7), 3054–3061.

31. Kimble, R.B., Vannice, J.L., Bloedow, D.C. *et al.* (1994) Interleukin-1 receptor antagonist decreases bone loss and bone resorption in ovariectomized rats. *Journal of Clinical Investigation* **93**(5), 1959–1967.

32. Wronski, T.J., Dann, L.M. and Horner, S.L. (1989) Time course of vertebral osteopenia in ovariectomized rats. *Bone* **10**(4), 295–301.

33. Wronski, T.J., Walsh, C.C. and Ignaszewski, L.A. (1986) Histologic evidence for osteopenia and increased bone turnover in ovariectomized rats. *Bone* **7**(2), 119–123.

34. Frost, H.M. and Jee, W.S. (1992) On the rat model of human osteopenias and osteoporoses. *Bone and Mineral* **18**(3), 227–236.

35. Kalu, D.N. (1991) The ovariectomized rat model of postmenopausal bone loss. *Bone and Mineral* **15**, 175–192.

36.   Vanin, C.M., MacLusky, N.J., Grynpas, M.D. and Casper, R.F. (1995) The effect of three hormone replacement regimens on bone density in the aged ovariectomized rat. *Fertility and Sterility* **63**(3), 643–651.

37.   Kalu, D.N., Salerno, E., Liu, C.C. *et al.* (1991) A comparative study of the actions of tamoxifen, estrogen and progesterone in the ovariectomized rat. *Bone and Mineral* **15**(2), 109–123.

38.   Durbridge, T.C., Morris, H.A., Parsons, A.M. *et al.* 1990) Progressive cancellous bone loss in rats after adrenalectomy and oophorectomy. *Calcified Tissue International* **47**(6), 383–387.

39.   Wronski, T.J., Cintron, M., Doherty, A.L. and Dann, L.M. (1988) Estrogen treatment prevents osteopenia and depresses bone turnover in ovariectomized rats. *Endocrinology* **123**(2), 681–686.

40.   Turner, R.T., Vandersteenhoven, J.J. and Bell, N.H. (1987) The effects of ovariectomy and 17 beta-estradiol on cortical bone histomorphometry in growing rats. *Journal of Bone and Mineral Research* **2**(2), 115–122.

41.   Turner, R.T., Wakley, G.K., Hannon, K.S. and Bell, N.H. (1987) Tamoxifen prevents the skeletal effects of ovarian hormone deficiency in rats. *Journal of Bone and Mineral Research* **2**(5), 449–456.

42.   Wronski, T.J., Halloran, B.P., Bikle, D.D. *et al.* (1986) Chronic administration of 1,25-dihydroxyvitamin D3: increased bone but impaired mineralization. *Endocrinology* **119**(6), 2580–2585.

43.   Wronski, T.J., Yen, C.F., Burton, K.W. *et al.* (1991) Skeletal effects of calcitonin in ovariectomized rats. *Endocrinology* **129**(4), 2246–2250.

44.   Sogaard, C.H., Mosekilde, L., Schwartz, W. *et al.* (1995) Effects of fluoride on rat vertebral body biomechanical competence and bone mass. *Bone* **16**(1), 163–169.

45.   Cheng, P.T. and Bader, S.M. (1990) Effects of fluoride on rat cancellous bone. *Bone and Mineral* **11**(2), 153–161.

46.   Turner, R.T., Francis, R., Brown, D. *et al.* (1989) The effects of fluoride on bone and implant histomorphometry in growing rats. *Journal of Bone and Mineral Research* **4**(4), 477–484.

47.   Wronski, T.J., Dann, L.M., Scott, K.S. and Crooke, L.R. (1989) Endocrine and pharmacological suppressors of bone turnover protect against osteopenia in ovariectomized rats. *Endocrinology* **125**(2), 810–816.

48.   Toolan, B.C., Shea, M., Myers, E.R. *et al.* (1992) Effects of 4-amino-1-hydroxybutylidene bisphosphonate on bone biomechanics in rats. *Journal of Bone and Mineral Research* **7**(12), 1399–1406.

49.   Lauritzen, D.B., Balena, R., Shea, M. *et al.* (1993) Effects of combined prostaglandin and alendronate treatment on the histomorphometry and biomechanical properties of bone in ovariectomized rats. *Journal of Bone and Mineral Research* **8**(7), 871–879.

50.   Seedor, J.G., Quartuccio, H.A. and Thompson, D.D. (1991) The bisphosphonate alendronate (MK-217) inhibits bone loss due to ovariectomy in rats. *Journal of Bone and Mineral Research* **6**(4), 339–346.

51.   Thompson, D.D., Simmons, H.A., Pirie, C.M. and Ke, H.Z. (1995) FDA Guidlines and animal models for osteoporosis. *Bone* **17**(4), 125s–133s.

52.   Reginster, J.Y., Compston, J.E., Jones, E.A. *et al.* (1995) Recommendations for the registration of new chemical entities used in the prevention and treatment of osteoporosis. *Calcified Tissue International* **57**, 247–250.

53.   Turner, R.T., Hannon, K.S., Demers, L.M. *et al.* (1989) Differential effects of gonadal function on bone histomorphometry in male and female rats. *Journal of Bone and Mineral Research* **4**(4), 557–563.

54.   Kimmel, D.B. (1991) Quantitative histologic changes in the proximal tibial growth cartilage of aged female rats. *Cells and Materials* (Suppl. 1), 11–18.

55. Kimmel, D.B. (1996) Animal models for in vivo experimentation in osteoporosis research, in *Osteoporosis* (eds R. Marcus, D. Feldman and J. Kelsey), Academic Press, San Diego, pp. 671–690.

56. Yamauchi, H., Kushida, K., Yamazaki, K. and Inoue, T. (1995) Assessment of spine bone mineral density in ovariectomized rats using DXA. *Journal of Bone and Mineral Research* **10**(7), 1033–1039.

57. Rozenberg, S., Vandromme, J., Neve, J. *et al.* (1995) Precision and accuracy of in vivo bone mineral measurement in rats using dual-energy X-ray absorptiometry. *Osteoporosis International* **5**(1), 47–53.

58. Pastoureau, P., Chomel, A. and Bonnet, J. (1995) Specific evaluation of localized bone mass and bone loss in the rat using dual-energy X-ray absorptiometry subregional analysis. *Osteoporosis International* **5**(3), 143–149.

59. Hagiwara, S., Lane, N., Engelke, K. *et al.* (1993) Precision and accuracy for rat whole body and femur bone mineral determination with dual X-ray absorptiometry. *Bone and Mineral* **22**(1), 57–68.

60. Sato, M., McClintock, C., Kim, J. *et al.* (1994) Dual-energy x-ray absorptiometry of raloxifene effects on the lumbar vertebrae and femora of ovariectomized rats. *Journal of Bone and Mineral Research* **9**(5), 715–724.

61. Sievanen, H., Kannus, P. and Jarvinen, M. (1994) Precision of measurement by dual-energy X-ray absorptiometry of bone mineral density and content in rat hindlimb in vitro. *Journal of Bone and Mineral Research* **9**(4), 473–478.

62. Lu, P.W., Briody, J.N., Howman-Giles, R. *et al.* (1994) DXA for bone density measurement in small rats weighing 150–250 grams. *Bone* **15**(2), 199–202.

63. Russell, J.E. and Avioli, L.V. (1972) Effect of experimental chronic renal insufficiency on bone mineral and collagen maturation. *Journal of Clinical Investigation* **51**, 3072–3079.

64. Miller, S.C., Bowman, B.M. and Jee, W.S. (1995) Available animal models of osteopenia – small and large. *Bone* **17**(4, Suppl.), 117s–123s.

65. Faulkner, K.G. and McClung, M.R. (1995) Quality control of DXA instruments in multicenter trials. *Osteoporosis International* **5**(4), 218–227.

66. Lu, Y., Mathur, A.K., Blunt, B.A. *et al.* (1996) Dual energy x-ray absorptimetry quality control: comparison of visual examination and process-control charts. *Journal of Bone and Mineral Research* **11**(5), 626–637.

CHAPTER TWELVE

Animal models for the investigation of the action of factors on bone metabolism

Colin R. Dunstan and Brendan F. Boyce

12.1 INTRODUCTION

There is increasing interest from the world's pharmaceutical industry in the problems of bone pathology, and many of the large companies and many smaller biotechnology companies have active bone research programmes. This chapter has been structured to encompass the requirements for both investigational (i.e. research) and preclinical assessment of factors with putative action (therapeutic or otherwise) on bone resorption or formation associated with bone modelling and remodelling. Models of fracture and repair of segmental defects are beyond the scope of this chapter. Investigational requirements centre around identification of an activity or validation of an activity identified in *in vitro* systems. Preclinical assessment centres around application of an activity to human disease. Under normal circumstances there is a progression from the investigational studies into preclinical assessment.

The animal models reviewed in this chapter have been conducted under current animal-handling legislation in the United States. Investigators in other countries must seek guidance from their own authorities before carrying out the procedures described.

12.2 INVESTIGATIONAL MODELS

Rats or mice are typically used in investigational models because they are inexpensive and easy to handle, and large numbers of them can be included in experiments and housed in small spaces in most animal facilities. Factors can be administered either locally or systemically, with local administration providing more economical use of factors and a faster screening method to determine effects on bone cells. In the investigational stages of the study of factors (e.g. recombinant proteins), only small quantities may be available for testing or the cost of obtaining sufficiently large amounts for *in vivo* studies may be prohibitive. In these

Methods in Bone Biology. Edited by Timothy R. Arnett and Brian Henderson.
Published in 1997 by Chapman & Hall, London. ISBN 0 412 75770 2.

circumstances, local administration can be used to produce *in vivo* effects at much lower doses than those required for systemic effects, even if the eventual application of the factor is expected to be systemic. Furthermore, it may be difficult to administer some factors systemically due to their poor uptake from soft tissues, their rapid clearance from the circulation, or undesired effects on other organs. The choice of model will also depend on whether effects are expected at the growth plate or on periosteal or endosteal surfaces.

12.1.1 *In vivo* models using local administration of factors

(a) Supra-calvarial administration in rodents
The calvarial bones are formed by intramembranous bone formation and consist predominantly of the left and right parietal bones, which are joined along the midline by the sagittal suture. These bones are separated from the smaller frontal bones by the frontal suture and from the occipital bone posteriorly by the lambdoid suture (Chapter 4 gives a detailed description of the calvaria). The sutures can be identified readily, both macroscopically and microscopically, and thus provide useful landmarks to standardize sites for injection of factors and for subsequent histomorphometry. The outer periosteal surface of the parietal bones is covered by loosely attached skin with no intervening muscle and thus is readily accessible for subcutaneous injection. Most of the central part of the parietal bones consists of solid lamellar bone matrix laid down by periosteal osteoblasts and the bone marrow spaces are largely confined to the edges of the bones adjacent to the sutures (Fig. 12.1).

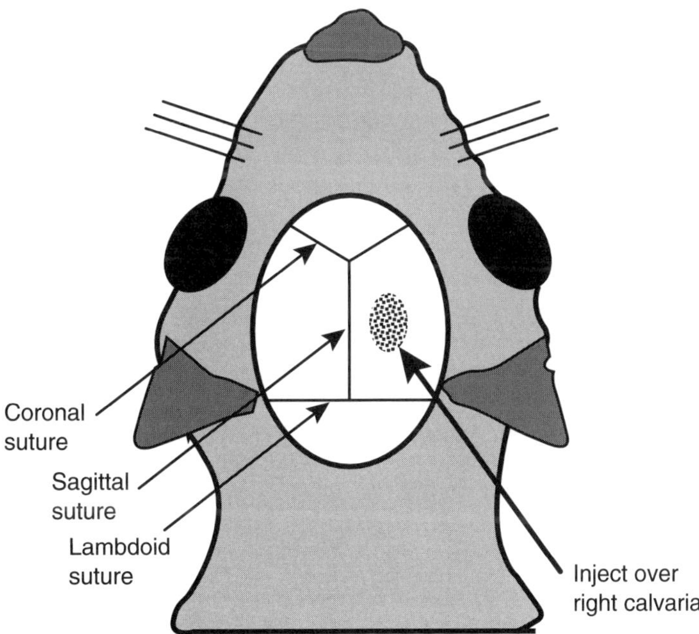

Figure 12.1   Diagram of exposed calvaria of a mouse showing sutures that are useful landmarks and site for injection of bone active factors.

The subcutaneous surface of the mouse calvaria is a convenient site for the introduction of factors with either resorptive or anabolic effects. Small volumes (10–50 μl) of factors can be injected over the right or left parietal bone using a Hamilton syringe (Reno, Nevada). Hamilton syringes are available in a number of sizes, and those holding 10 μl or 20 μl have a long, thin, pliable needle that can be passed easily through the skin to allow accurate delivery of agents to one side of the calvaria without scratching or damaging the underlying periosteal surface. A mouse can be picked up and held in one hand and head movement can be minimized by placing an index finger at the side of the jaw. The injection can be given without local or systemic anaesthetic, using the other hand, and sudden head movement at the time of injection can be restricted by the correct hold or minimized by distracting the mouse's attention – for example, by blowing exhaled air over its head. These simple, practical distraction techniques to ensure that an agent is delivered to the same site on the calvarial surface can help to determine whether a factor's activity is restricted to the injection site or not. Note that in certain countries (the UK, for example) mice would have to be anaesthetized before an injection was given.

The endosteal surfaces of the marrow spaces near the calvarial sutures are the sites where bone resorption typically is first seen following either local or systemic administration of bone resorbing agents such as interleukin 1 (IL-1) [1], parathyroid hormone (PTH) or PTH-related peptide (PTHrP) [2]. Osteoclasts increase the size of these spaces and extend them centripetally within each parietal bone by resorbing channels through the solid matrix in the centres of the bones. Thus, the relative activity of factors could be assessed by determining the extent of progression of these bone marrow channels from the lambdoid suture anteriorly. When bone resorbing factors are administered at high concentrations, or are given for prolonged periods, bone resorption also occurs along the outer periosteal surface, a feature that is seen most characteristically following local injection of PTHrP [2].

The cells on the periosteal surface of the mouse and rat are sensitive to a number of anabolic factors, including transforming growth factor beta [3, 4] and acidic and basic fibroblast growth factor (FGF) [5]. The initial response of these cells to the local injection of such factors for 3 or 4 days is to proliferate and increase the thickness of the periosteum. Following factor withdrawal, the increased numbers of osteoblasts produce a substantial band of new bone over the susequent 2 to 3 weeks and this can be distinguished easily from the original bone by its predominantly woven structure (Fig. 12.2). Factors can also be delivered to this site in nude mice using cell lines, such as Chinese hamster ovarian (CHO) cells, transfected with genes coding for putative osteoblast stimulating factors. CHO cells transfected with the empty vector for the transfected genes have little stimulatory effect on bone resorption or bone formation when grown against bone, but when producing growth factors such as bone morphogenetic protein (BMP) they produced prolific new periosteal bone growth (Dunstan, in preparation).

The advantages of the supra-calvarial model are:

- the negligible effects of the injection procedure on the underlying bone;
- the low background level of bone remodelling;
- the ability to deliver a factor to the same site many times;
- the ease of processing, aligning and sectioning the calvarial bone after sacrifice.

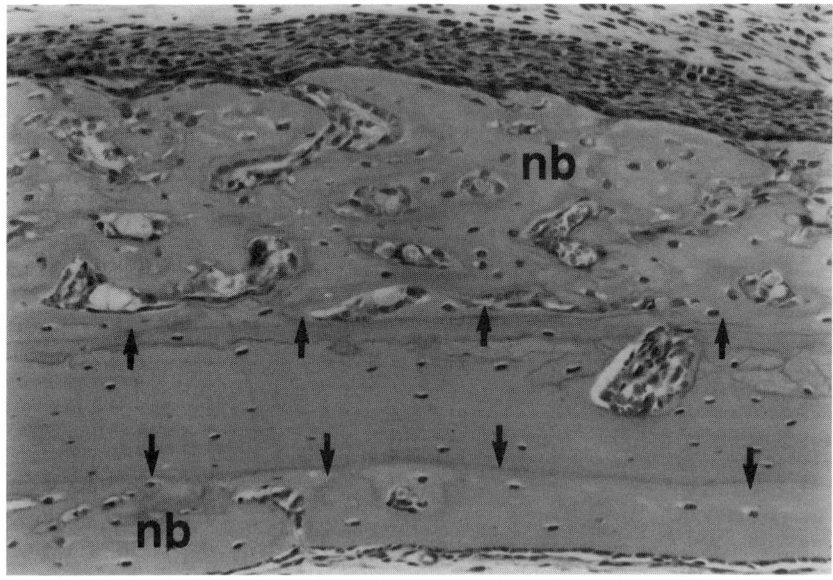

Figure 12.2 Photomicrograph showing effect of supra-calvarial injection of FGF2 over calvaria of mice. New bone formation (nb) can be seen on upper and lower surface of original calvarial bone and is separated from new bone by a cement line (arrows).

In some circumstances the non-injected side of the calvaria can be used as a control in addition to the calvariae of vehicle-treated animals, but the validity of this depends on the use of a factor that can be dissolved in small volumes and that does not diffuse readily over the entire calvarial surface. The calvaria can also be used to examine the effects of anti-resorptive agents, but only if it is given in the presence of a resorption stimulating factor. This is because the background level of resorption in the unstimulated calvaria is so low that, realistically, it is not possible to demonstrate inhibition. This model has not been used to examine the effects of sex hormone withdrawal or of immobility on osteoclast activity and recruitment, but could be useful, given the low basal levels of resorption.

(b) Intra-osseous administration into the rat
An alternative to the administration of small amounts of factors locally over the calvaria is injection or infusion of them into the bone marrow cavity of the long bones of the rat. Infusion can be done by cannulation of the appropriate artery [6] or by using osmotic mini-pumps and fine-calibre tubing passed through a small hole drilled with a fine drill through the femoral cortex proximally. This approach allows examination of the effects of factors on the cancellous bone at the lower end of the femur [7] or on the endosteal bone lining the inner surface of the femoral shaft. Factors may also be introduced into the metaphyses of long bones by direct injection [8] or by means of a titanium or other inert implant [9]. These bone compartments are different from the periosteal surface of the calvaria, and the basal level of resorption in the cancellous bone of the growth plate of the rat is much higher than that of the endosteal surface of the bone marrow spaces inside the calvaria. The advantage of these bone sites is that bone factors are introduced directly into an endosteal region where therapeutics for diseases such as

osteoporosis will need to have their effect. The growth plate region is obviously active during growth and such activity does not completely cease with ageing. Depending on the age of the rat, the factor can be introduced to an environment of rapid bone formation and resorption occurring in the primary and secondary spongiosae, where anti-resorptive factors in particular could be expected to have a profound effect. The disadvantages of these models are that disruption of the cortex and the marrow causes significant local effects on bone formation and the procedures themselves are quite complex. If the growth plate is the target, as it may be to assess the activity of an anti-resorptive agent, then a young, rapidly growing rat would be used to produce a maximum effect behind the advancing growth plate.

12.2.2 Systemic administration of factors in the investigational phase

(a) The mouse
The mouse has the advantages of being small, inexpensive and easy to handle and only small amounts of material are required to see a biological effect. Intraperitoneal and subcutaneous administration of factors under test is simple. In contrast, intravenous administration, particularly in young growing animals, or repeated injections are more difficult. An alternative method of factor delivery is the use of tumours formed from injected cells transfected to produce the factor of interest. CHO cells form tumours in nude mice when injected intramuscularly and have little effect on bone metabolism. For example, CHO tumours transfected to produce parathyroid hormone-related protein [10], interleukin 6 (IL-6) [11] or tumour necrosis factor alpha (TNFα) [12] have been used successfully to evaluate the effects of these proteins on bone. The advantages of CHO cells are that they produce the protein of interest (and thus production and purification of proteins is avoided) and ensure its continuous delivery. The two main disadvantages are that it is difficult to control for interactions between the expressed protein and other tumour products or host responses to the tumour, and that the expression levels vary within and between animals and are not easy to quantify.

The mouse is also invaluable because it can easily be manipulated genetically – for example, to produce gene knockout transgenic animals. The *fos* [13] and *src* [14] knockout mice highlighted the previously unsuspected role for these proteins in osteoclast function and generation. A transgenic mouse over-expressing interleukin 4 has been found to have an osteoporotic phenotype due to suppressed bone formation [15]. A potential difficulty with knockout transgenic mice is that the effects of a particular factor may be different in development from those seen in mature animals and the impact on development may be pre- or neonatally lethal.

Young growing mice have a rapid growth rate over the first 6 weeks of life. Elongation of the vertebral and long bones is by endochondral bone formation, which involves rapid bone formation for the primary and secondary spongiosae, and removal of mineralized cartilage and later the primary and secondary spongiosae to produce an open marrow cavity. At this time there is also rapid modelling of the bone to shape it. Inhibition of bone resorption is able to produce profound changes in density of the bone in the spongiosae and in the shape of the bone, and to produce easily measurable effects on bone volume (Fig. 12.3). The presence of the spongiosae and a considerable osteoblast population make this a potential site for rapid response to the anabolic agents. As mice age, growth decreases and the

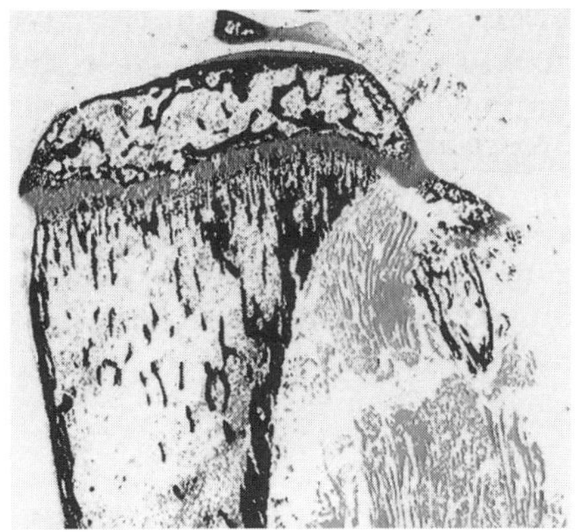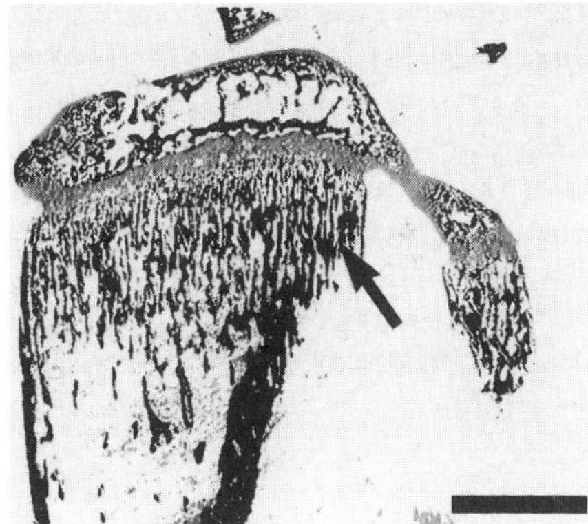

Figure 12.3   Photomicrograph showing increase in bone density produced at proximal tibial metaphysis in a young growing mouse treated with systemic pamidronate (b) compared with control (a). Note
also change in shape of bone below growth plate (arrowed), an indication that the pamidronate was
also blocking periosteal resorption associated with modelling of the bone (bar = 1 mm).

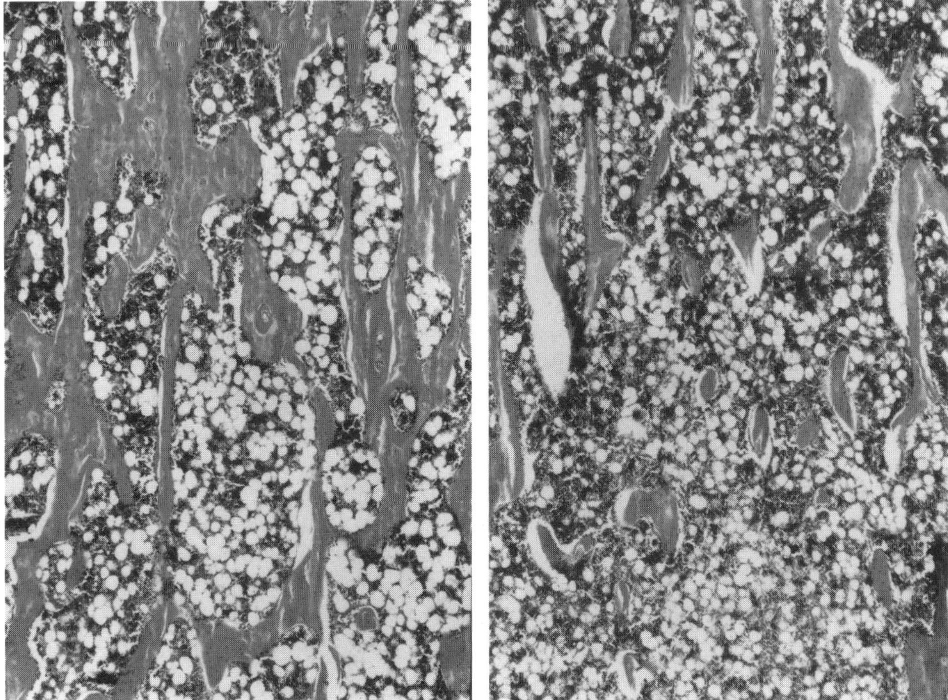

Figure 12.4   Photomicrograph showing (a) normal bone density in proximal tibial metaphysis of sham-
operated rat and (b) decreased bone density in metaphysis of littermate 3 months after ovariectomy.

animals become less responsive to factors affecting bone resorption, as indicated by the recovery from some osteopetrotic phenotypes. In addition, the amount of spongiosa in the long bone metaphyses becomes greatly diminished and in consequence little bone surface is available for resorption or for the action of anabolic factors. In older mice the vertebral bodies, the long bone epiphyses and the calvaria become better sites to look at for the effects of bone resorption and formation.

Dual X-ray absorptiometry techniques (DXA Hologic, USA) are not available to determine changes in bone density in the spine and hind quarters in the mouse *in vivo*. Peripheral QCT (Stratec, Germany) can be used [16] to assess bone density accurately in the mouse but these machines are also not widely available (see Chapter 11 for discussion of the use of DXA in small animals).

(b) The rat
Systemic administration in the rat can be used if the greater dosing requirements can be accommodated. Similar site and age constraints apply as in the mouse (above) but bone density can be easily followed in these animals using DXA measurements. The sites most easily measured and with the most rapid changes are the distal femoral and proximal tibial metaphyses.

Larger animal models are not usually used in the investigational phase due to the cost of animals and of the large doses of compounds required to achieve biological effects. Alternative small animal models have roles in more specific areas of research, such as in vitamin C research in the guinea pig and the study of the effects of glucocorticoids in the rabbit

12.3 PRECLINICAL MODELS

The progression from investigational to preclinical models requires the introduction of some challenge to reproduce a human pathological condition. The shift from investigational to preclinical can be blurred, particularly where the investigation of the action of a bone-modulating factor requires the use of a challenge to unmask its activity. There are several sorts of challenge that can be used, depending on the human pathology being modelled. The most significant human bone disease is osteoporosis, resulting from ageing and postmenopausal bone loss, and the most established bone models relate to it. This chapter will concentrate on these models. Two other well studied areas of bone loss are those associated with immobilization or malignancy (e.g. hypercalcaemia of malignancy and bone metastases). These will be dealt with only briefly.

12.3.1 Postmenopausal bone loss

In women, maintenance of bone mass depends on an adequate circulating level of oestrogen. Bone loss commences when oestrogen levels fall, whether due to natural or surgical menopause, or with over-rigorous exercise or inadequate nutrition. Restoration of premenopausal oestrogen levels is able to arrest bone loss but not reverse it. Bone mass is lost most briskly from the cancellous bone, with changes in the vertebral bodies occurring most rapidly. Cortical bone is also lost, though more slowly by cortical thinning and by increasing porosity through the activity of Haversian systems. An effective therapeutic should be able to inhibit the bone loss occurring with oestrogen deprivation. The questions being addressed in this situation will determine the best model to use.

The ideal animal model would have the following properties:

- Rapid bone loss following ovariectomy that is prevented but not reversed by oestrogen.
- Loss of bone that is not complicated by concurrent skeletal growth at the site of interest.
- Loss of bone occurring in true cancellous bone as opposed to the temporary spongiosa adjacent to a growth plate.
- Haversian remodelling so that the changes in bone distribution from cortical to cancellous bone through increases in cortical porosity can be determined.
- Reasonable cost in animal maintenance and factor dosing requirements.

There are no animal models at the moment that satisfy all of these requirements. The appropriate model is of necessity a compromise.

(a) The mouse

The first requirement for an adequate animal model is rapid bone loss following ovariectomy that is blocked but not reversed by oestrogen. The mouse loses cancellous bone following ovariectomy but its usefulness is limited for two reasons:

- Oestrogen is strongly anabolic in the mouse and exogenous administration produces a profound osteosclerosis via reduction in bone resorption and stimulation of bone formation [17].
- Cancellous bone loss in the mouse occurs largely in the distal femoral and proximal tibial metaphyses following cessation of bone growth after 6 weeks of age. However, by 6 months of age, there is very little cancellous bone remaining at the sites and thus older mice generally are poor models for study of post-menopausal bone loss.

For these two reasons the mouse is not a good model for postmenopausal bone loss.

(b) The rat

The rat undergoes rapid bone loss after ovariectomy, particularly in the proximal tibial and distal femoral metaphyses [18]. Bone loss continues at these sites over a 3-month period, with significant bone loss that can be measured by DXA or by histomorphometry after 1 month. Oestrogen inhibits the bone loss but is not anabolic, as in the mouse, and administration is not able to produce a recovery from the ovariectomy-induced relative osteopenia [19]. The rat, compared with larger animals, is substantially more economical and has been designated by the US Federal Drug Authority as the appropriate small animal model of osteoporosis and postmenopausal bone loss [20].

The rat lacks Haversian remodelling and so it cannot be used to evaluate the effects of a therapy on this aspect of human bone loss. The growth plates do not fuse in rats, as occurs in the higher animals, and so it can be difficult to separate the effects of a factor on oestrogen deficiency-induced bone loss from effects on growth rate. For example, treatments that increase growth rate will increase the extent of the primary and secondary spongiosae even if the rates of remodelling and removal remain unaltered. This complication can be minimized by using rats that are at least 3 months of age, when growth rate has greatly diminished. For these reasons the preclinical assessment of new treatments for postmenopausal bone loss must also be assessed in a larger animal. What is required in a larger animal model is power to give a definitive answer as to biological efficacy and results need to have regulatory acceptability.

(c) Primate models

The model with the greatest validity is the primate model, as adult primates show significant bone loss after ovariectomy and their Haversian bone structure is similar to that of humans. Ovariectomy in young monkeys tends to stabilize bone mass instead of reducing it while bone mass in the sham animals continues to increase. Thus it is better to use mature animals. Diet and nutritional history of the animals can have profound effects and need to be carefully controlled. Cynomolgus monkeys from the wild, when placed on a diet containing 0.6% calcium, gained bone density whereas domesticated monkeys showed an actual rather than a relative bone loss [21]. Baboons [22] and cynomolgus monkeys [23] show bone loss and increased bone turnover following ovariectomy. Rhesus monkeys also show bone loss [24] though the evaluation of markers is less complete. The choice of which primate to use is often determined by price, availability and dosing constraints.

Primate studies are very expensive and primate experimentation raises ethical questions. Therefore a variety of alternative large animal models have been evaluated to some extent. The larger non-primate animals most used are the dog, the mini-pig and the sheep. There is considerable uncertainty whether any of these have predictably adequate bone loss with ovariectomy to permit study of drug efficacy, whether the bone structure is similar enough to that of humans and whether regulatory bodies will be satisfied with any of these models as an alternative to primate studies.

(d) The dog

The traditional dog breed used in the study of bone metabolism has been the beagle. While some studies in beagles have shown an ovariectomy-related bone loss [25], this decrease has not been reproducible. Frequently the effects of ovariectomy on bone turnover have been transient and changes in bone density not significant [26, 27]. In addition the dog is not considered satisfactory as a large animal model by the regulatory authorities in the United States (FDA guidelines) [20].

e) The mini-pig

The pig has an oestrous cycle of 21 days, which is close to that of humans. Skeletal maturity is reached at a reasonably young age but the costs of upkeep are moderately expensive. The pig has bone loss in vertebrae over the 6 months following ovariectomy. The extent of loss depends on maintenance of a carefully defined calcium intake. Ovariectomy itself does not appear to produce a sustained increase in bone turnover, with no significant effects 6 months after ovariectomy [28]. In the study of Mosekilde *et al.* [28], it was necessary to use one tailed statistical analysis to obtain significance for the bone loss, possibly indicating that the response was not robust, and the power of the model to identify therapeutic action of a factor being tested may not be strong. It would be encouraging if further studies were able to confirm a sustained loss of bone over a period longer than 6 months.

(f) The sheep
The sheep is reasonably cheap to maintain but has not been extensively studied. Sheep have considerable seasonal variation in their bone turnover and their oestrous cycling. However, in some breeds the oestrous cycle is not seasonal and these would appear to be the breeds to work with [29]. Bone loss following ovariectomy is not substantial but does appear to be inhibited by oestrogen [30]. The greatest advantage of the sheep is the ready availability and lower price of aged females and the ease of handling and maintaining the animals.

Both the pig and the sheep have the potential problem that the cortical long bones have a plexiform rather than Haversian structure, making direct comparison of cortical bone loss difficult, though there is some development of Haversian remodelling with ageing. Evidence demonstrating increased cortical porosity, similar to that seen in humans with ageing, is required to validate the sheep and the pig more fully as models able to illuminate the whole spectrum of human skeletal changes with oestrogen loss, ageing and intervention or preventative therapy.

(g) The ferret
A recently suggested alternative animal is the ferret [31]. It is an attractive possibility as this animal is small and its skeleton does exhibit Haversian remodelling, and the seasonal nature of oestrogen cycling can be controlled by controlling light exposure. The ferret appears to undergo bone loss following ovariectomy though not enough has been done to establish the validity of this model.

12.3.2 Models of established osteoporosis

Osteoporosis in humans is predominantly a disease of the aged. Bone mass is diminished in the cancellous and cortical bone and the cortical bone has increased porosity. Cancellous bone loss is marked by the loss of trabecular elements. For this reason the ideal animal model would be an aged osteopenic animal with low circulating oestrogen levels that has Haversian remodelling of cortical bone.

(a) The rat
The rat becomes significantly osteopenic with ageing and this process can be accelerated with ovariectomy. The 9-month-old rat ovariectomized at 3 months shows osteopenia in the vertebrae and almost no bone in the tibial and femoral metaphyses [32]. Strongly anabolic factors such as FGF1 [5] or FGF2 [33] are able in the rat to induce the filling of the marrow cavities with new trabecular-like bone structure. These factors are able to overcome the difficulty in increasing bone mass when no structure remains on which to build bone. Whether this generation of new trabecular bone can occur in higher animals or with weaker anabolic agents has not been reported. Treatment with anti-resorptives would not be expected to increase the density of trabecular bone in these animals, though effects may be observed on the cortical and vertebral bone. As it takes a long time to produce these rats, an alternative is to use retired breeding stock. However, although such rats do have diminished bone density there is substantial individual variation, resulting in a requirement for larger group sizes. Thus aged rats do provide a good small animal model for established osteoporosis.

Table 12.1 Summary of comparison of different animal models of postmenopausal osteoporosis with that in humans

Model	Ovariectomy effect of BMD	Bone turnover after ovariectomy	Oestrogen response	Growth arrest at skeletal maturity	Haversian bone remodelling	Seasonal variation a problem	Pathological fractures
Human	Rapid loss	Increased	Blocks bone loss	Yes	Yes	No (if vitamin D replete)	Yes
Mouse	Moderate loss	Increased	Osteosclerosis	No	No	No	No
Rat	Rapid loss	Increased	Blocks bone loss	No	No	No	No
Ferret	Strong over 3 months	Not studied	Blocks bone loss	Yes	Yes	Yes (can be controlled by light)	No
Dog	Weak	Transient increase	Blocks bone loss	Yes	Yes	No	No
Mini-pig	Moderate	Transient or no increase	Blocks bone loss	Yes	Yes	No	No
Sheep	Small	Can be masked by seasonal variation	Blocks bone loss	Yes	No (some in old sheep)	Yes	No
Old world primates	Strong	Increased	Blocks bone loss	Yes	Yes	No	No

(b) Large animal models of established osteoporosis
The same problems apply to models of established osteoporosis as to post-menopausal bone loss. The difficulty and expense involved in obtaining significantly osteopenic primates make a primate model even more prohibitively expensive and time consuming than for postmenopausal bone loss. Thus, if a primate has to be kept to maturity, ovariectomized and then kept 6–9 months before treatment can start, it is clear that the expense in time and money will be substantial.

It is perhaps in this situation that the sheep becomes a viable alternative. Aged sheep are readily available are relatively inexpensive and have reduced bone density. Ovariectomy can be carried out to reduce the seasonal cycle effects on bone in these animals. Breeding can be carefully controlled to minimize genetic variation, and handling and maintenance are not difficult. It is possible that the effects of therapies on Haversian systems could be studied, because in aged sheep there is evidence of Haversian remodelling [29]. The problems with seasonal variation will require carefully-controlled studies.

The aged dog is also a possible model, though genetic control is more difficult and maintenance costs are higher than for the sheep.

12.3.3 Immobilization

Immobilization of individual limbs or of the whole body can be achieved by a number of procedures that all produce significant bone loss and that model different causes of pathology in humans. Methods used include de-enervation of a limb [34], tail suspension [35], immobilization by mechanical restraint [36] or cast [37] and use of a zero-gravity environment [38] (an expensive animal model system). Therapies can be introduced in these systems to test the ability to oppose the increased bone resorption and/or decreased bone formation that occur in these situations. In humans a significant proportion of the bone loss is related to Haversian systems in cortical bone producing increased cortical porosity. While bone loss occurs in rodents, the optimum animal models should be able to reproduce this type of bone loss. Good results have been obtained in the dog [39] and the primate [36].

12.3.4 Hypercalcaemia of malignancy

Hypercalcaemia of malignancy can arise from humoral production of a bone active factor, most commonly PTHrP, or from the action of lytic bone metastases. The mechanism of the hypercalcaemia can be related to PTHrP or cytokine production or be of undetermined cause. To model the humoral production of resorptive factors by tumours, one can grow one of the many cell lines that have been reported to produce hypercalcaemia in syngeneic or in immune-compromised animals and that have been reported to produce PTHrP [40, 41] or cytokines [42]. To model the metastatic process, tumour-derived cells such as the breast cancer cell line MDA231 can be introduced into the vascular system to approximate the development of bone metastases [43]. Two potential problems with these models are that the available cell lines available do not always continue to reproduce the originally reported effects, and that more than one active factor may be produced by the tumour, or secondarily by the host. Thus the reasons for success or failure of a single factor to impact on the bone resorption or hypercalcaemia may not be easy to define. An alternative model is to administer specific factors such as PTHrP systemically and evaluate the ability of agents under investigation to counteract their effects [10].

12.4 CONCLUSIONS

It is possible to use *in vivo* models to identify and define the actions of natural and synthetic factors on bone metabolism in health and disease. Each model tends to have characteristic advantages and limitations. The final confirmation of the action of a factor in humans is when it is tested in that setting. However, animal models are very valuable in establishing the safety and efficacy of a factor as long as their limitations are taken into account.

REFERENCES

1. Boyce, B.F., Aufdemorte, T.B., Garrett, I.R. *et al.* (1989) Effects of interleukin-1 on bone turnover in normal mice. *Endocrinology* **125**, 1142–1150.
2. Yates, A.J., Gutierrez, G.E., Smolens, P. *et al.* (1988) Effects of a synthetic peptide of a parathyroid hormone-related protein on calcium homeostasis, renal tubular calcium reabsorption, and bone metabolism in vivo and in vitro in rodents. *Journal of Clinical Investigation* **81**, 932–938.
3. Noda, M. and Camilliere, J.J. (1989) In vivo stimulation of bone formation by transforming growth factor-beta. *Endocrinology* **124**, 2991–2994.
4. Marcelli, C., Yates, A.J. and Mundy, G.R. (1990) In vivo effects of human recombinant transforming growth factor beta on bone turnover in normal mice. *Journal of Bone and Mineral Research* **10**, 1087–1096.
5. Dunstan, C.R., Garrett, I.R., Adams, R. *et al.* (1995) Systemic fibroblast growth factor (FGF-1) prevents bone loss, increases new bone formation, and restores trabecular architecture in ovariectomized rats. *Journal of Bone and Mineral Research* **10** (Suppl. 1), s198.
6. Schlechter, N.L., Russell, S.M., Greenberg, S. *et al.* (1986) A direct growth effect of growth hormone in rat hindlimb shown by arterial infusion. *American Journal of Physiology* **250**, E231–235.
7. Takano-Yamamoto, T. and Rodan, G.A. (1990) Direct effects of 17 beta-estradiol on trabecular bone in ovariectomized rats. *Proceedings of the National Academy of Sciences USA* **87**, 2172–2176.
8. Russell, S.M. and Spencer, E.M. (1985) Local injections of human or rat growth hormone or on human somatomedin-C stimulate unilateral tibial epiphyseal growth in hypophysectomized rats. *Endocrinology* **116**, 2563–2567.
9. Wang, J.S. and Aspenberg, P. (1994) Basic fibroblast growth factor increases allograft incorporation. Bone chamber study in rats. *Acta Orthopaedica Scandinavica* **65**, 27–31.
10. Guise, T.A., Chirgwin, J.M., Favarato, G. *et al.* (1992) Chinese hamster ovarian cells transfected with human parathyroid hormone-related protein cDNA cause hypercalcemia in mice. *Laboratory Investigation* **67**, 477–485.
11. De la Mata, J., Uy, H.L., Guise, T.A. *et al.* (1995) Interleukin-6 enhances hypercalcemia and bone resorption mediated by parathyroid-hormone related protein in vivo. *Journal of Clinical Investigation* **95**, 2846–2852.

12. Yates, A.J., Boyce, B.F., Favaroto, G. *et al.* (1992) Expression of human transforming growth factor alpha by Chinese hamster ovarian tumors in nude mice causes hypercalcemia and increased osteoclastic bone resorption. *Journal of Bone and Mineral Research* **7**, 847–853.

13. Wang, Z.Q., Ovitt, C., Grigoriadis, A.E. *et al.* (1992) Bone and haematopoietic defects in mice lacking c-fos. *Nature* **360**, 741–745.

14. Soriano, P., Montgomery, C., Geske, R. and Bradley, A. (1991) Targeted disruption of the c-src proto-oncogene leads to osteopetrosis in mice. *Cell* **64**, 693–702.

15. Lewis, D.B., Liggitt, H.D., Effman, E.L. *et al.* (1993) Osteoporosis induced in mice by overproduction of interleukin 4. *Proceedings of the National Academy of Sciences USA* **90**, 11618–11612.

16. Beamer, W.G., Donahue, L.R., Rosen, C.J. and Baylink, D.J. (1996) Genetic variability in adult bone density among inbred strains of mice. *Bone* **18**, 397–403.

17. Liu, C.C. and Howard, G.A. (1991) Bone-cell changes in estrogen-induced bone-mass increase in mice: dissociation of osteoclasts from bone surfaces. *Anatomical Record* **229**, 240–250.

18. Wronski, T.J., Lowry, P.L., Walsh, C.C. and Ignaszewski, L.A. (1985) Skeletal alterations in ovariectomized rats. *Calcified Tissue International* **37**, 324–328.

19. Kalu, D.N. (1991) The ovariectomized rat model of postmenopausal bone loss. *Bone and Mineral* **15**, 175–191.

20. Food and Drug Administration (1994) Guidelines for preclinical and clinical evaluation of agents used in the prevention of post menopausal osteoporosis. Division of Metabolism and Endocrine Drug Products. Draft, April.

21. Jerome, C.P., Lees, C.J. and Weaver, D.S. (1995) Development of osteopenia in ovariectomized cynomolgus monkeys (*Macaca fascicularis*). *Bone* **17** (Suppl.), 402S–408S.

22. Balena, R., Toolan, B.C., Shea, M. *et al.* (1993) The effects of 2-year treatment with the aminobisphosphonate alendronate on bone metabolism, bone histomorphometry, and bone strength in ovariectomized nonhuman primates. *Journal of Clinical Investigation* **92**, 2577–2586.

23. Jerome, C.P., Carlson, C.S., Register, T.C. *et al.* (1994) Bone functional changes in intact, ovariectomized, and ovariectomized, hormone-supplemented adult cynomolgus monkeys (*Macaca fascicularis*) evaluated by serum markers and dynamic histomorphometry. *Journal of Bone and Mineral Research* **9**, 527–540.

24. Longcope, C., Hoberg, L., Steuterman, S. and Baran, D. (1989) The effect of ovariectomy on spine bone mineral density in rhesus monkeys. *Bone* **10**, 341–344.

25. Malluche, H.H.,Faugere, M.C., Friedler, R.M. and Fanti, P. (1988) 1,25-dihydroxyvitamin D_3 corrects bone loss but suppresses bone remodeling in ovariohysterectomized beagle dogs. *Endocrinology* **122**, 1998–2006.

26. Boyce, R.W., Franks, A.F., Jankowsky, M.L. *et al.* (1990) Sequential histomorphometric changes in cancellous bone from ovariohysterectomized dogs. *Journal of Bone and Mineral Research* **5**, 947–953.

27. Shen, V., Dempster, D.W., Birchman, R. *et al.* (1992) Lack of changes in histomorphometric, bone mass, and biochemical parameters in ovariectomized dogs. *Bone* **13**, 311–316.

28. Mosekilde, L., Weisbrode, S.E., Safron, J.A. *et al.* (1993) Evaluation of the skeletal effects of combined mild dietary calcium restriction and ovariectomy in Sinclair S-1 minipigs: a pilot study. *Journal of Bone and Mineral Research* **8**, 1311–1321.

29. Hornby, S.B., Ford, S.L., Mase, C.A. and Evans, G.P. (1995) Skeletal changes in the ovariectomized ewe and subsequent response to treatment with 17 estradiol. *Bone* **17** (Suppl.), 389s–394s.

30. Turner, A.S., Mallinckrodt, C.H., Alvis, M.R. and Bryant, H.U. (1995) Dose response effects of estradiol implants on bone mineral density in ovariectomized ewes. *Bone* **17** (Suppl.), 421S–427S.

31.   Mackey, M.S., Stevens, M.L., Ebert, D.C. *et al.* (1995) The ferret as a small animal model with BMU-based remodeling for skeletal research. *Bone* **17** (Suppl.), 191S–196S.

32.   Kalu, D.N., Liu, C.C., Hardin, R.R. and Hollis, B.W. (1989) The aged rat model of ovarian deficiency bone loss. *Endocrinology* **124**, 7–16.

33.   Mayahara, H., Ito, T., Nagai, H. *et al.* (1993) In vivo stimulation of endosteal bone formation by basic fibroblast growth factor in rats. *Growth Factors* **9**, 73–80.

34.   Weinreb, M., Rodan, G.A. and Thompson, D.D. (1989) Osteopenia in the immobilized rat limb is associated with increased bone resorption and decreased bone formation. *Bone* **10**, 187–194.

35.   LeBlanc, A., Marsh, C., Evans, H. *et al.* (1985) Bone and muscle atrophy with suspension in the rat. *Journal of Applied Physiology* **58**, 1669–1675.

36.   Young, D.R., Niklowitz, W.J., Brown, R.J. and Jee, W.S.S. (1986) Immobilisation-associated osteoporosis in primates. *Bone* **7**, 109–117.

37.   Waters, D.J., Caywood, D.D. and Turner, R.T. (1991) Effect of tamoxifen citrate on canine immobilization (disuse) osteoporosis. *Veterinary Surg* **20**, 392–396.

38.   Morey-Holton, E.R. and Arnaud, S.B. (1991) Skeletal responses to spaceflight. *Adv Space Biol Med* **1**, 37–69.

39.   Sietsema, W.K. (1995) Animal models of cortical porosity. *Bone* **17** (Suppl.), 297S–305S.

40.   Insogna, K.L., Stewart, A.F., Vignery, A.M. *et al.* (1984) Biochemical and histomorphometric characterization of a rat model for humoral hypercalcemia of malignancy. *Endocrinology* **114**, 888–896.

41.   Hilgard, P. Schmitt, W., Minne, H. and Ziegler, R. (1970) Acute hypercalcemia due to Walker carcinosarcoma 256 in the rat. *Horm Metab Res* **2**, 255–256.

42.   Strassman, G., Jacob, C.O., Fong, M. and Bertolini, D.R. (1993) Mechanism of paraneoplastic syndromes of colon-26: involvement of interleukin 6 in hypercalcemia. *Cytokine* **5**, 463–468.

43.   Sasaki, A., Boyce, B.F., Story, B. *et al.* (1995) Bisphosphonate risedronate reduces metastatic human breast cancer burden in bone in nude mice. *Cancer Research* **55**, 3551–3557.

Index

Page numbers in **bold** refer to illustrations; those in *italic* refer to tables.